Practical Applications of Machine Learning and AI:

Medicine, Environmental Science, Transportation, and Education

Toufik Mzili
Chouaib Doukkali University, Morocco

Adarsh Kumar Arya
Harcourt Butler Technical University, India

IGI Global
Scientific Publishing
Publishing Tomorrow's Research Today

Vice President of Editorial Melissa Wagner
Managing Editor of Acquisitions Mikaela Felty
Managing Editor of Book Development Jocelynn Hessler
Production Manager Mike Brehm
Cover Design Phillip Shickler

Published in the United States of America by
IGI Global Scientific Publishing
701 East Chocolate Avenue
Hershey, PA, 17033, USA
Tel: 717-533-8845
Fax: 717-533-8661
Website: https://www.igi-global.com E-mail: cust@igi-global.com

Library of Congress Cataloging-in-Publication Data

Names: Mzili, Toufik, 1995- editor. | Arya, Adarsh, 1979- editor.
Title: Practical applications of machine learning and AI : medicine, environmental science, transportation, and education / edited by Toufik Mzili, Adarsh Kumar Arya.
Description: Hershey, PA : IGI Global Scientific Publishing, [2025] | Includes bibliographical references and index. | Summary: "The book will explore various aspects of optimization, machine learning, and artificial intelligence, covering topics such as algorithmic advancements, application domains, theoretical frameworks, and practical implementations"-- Provided by publisher.
Identifiers: LCCN 2024059911 (print) | LCCN 2024059912 (ebook) | ISBN 9798337313993 (hardcover) | ISBN 9798337314006 (paperback) | ISBN 9798337314013 (ebook)
Subjects: LCSH: Artificial intelligence--Industrial applications. | Medical care--Data processing. | Environmental engineering--Data processing. | Transportation engieering--Data processing. | Education--Data processing.
Classification: LCC TA347.A78 P73 2025 (print) | LCC TA347.A78 (ebook) | DDC 006.3/1--dc23/eng/20250220
LC record available at https://lccn.loc.gov/2024059911
LC ebook record available at https://lccn.loc.gov/2024059912

British Cataloguing in Publication Data
A Cataloguing in Publication record for this book is available from the British Library.

Table of Contents

Detailed Table of Contents

Devanshi Srivastava, Department of Chemical Engineering, Harcourt Butler Technical University, Kanpur, India
Adarsh Kumar Arya, Department of Chemical Engineering, Harcourt Butler Technical University, Kanpur, India
Toufik Mzili, Chouaib Doukkali University, Morocco
Murali Pujari, Department of Chemical Engineering, University of Petroleum & Energy Studies, Bidholi, India

The extensive reliance on conventional fossil fuels escalates greenhouse gas emissions and exacerbates environmental pollution, necessitating an urgent energy shift. Consequently, the advancement and use of renewable energy are of paramount significance. Hydrogen is expected to progress in the following years as a replacement for fossil fuels among various renewable sources. However, the high cost of the hydrogen supply network (HSN) has provoked the researchers to use optimization techniques to optimize the (HSN). The present study analyzes modern optimization methods for (HSN) and thoroughly examines the 'Green hydrogen supply chain' (GHSC), including transportation, production, storage and consumption, emphasizing metaheuristic optimization (MO) applications. The difficulties inherent in each phase are highlighted, and the capacity of MO techniques to mitigate these difficulties is examined. The study also examines multi-objective methodologies for optimizing issues within this area.

Recent advancements in deep learning (DL) and machine learning (ML) have opened doors for revolutionary applications in healthcare, with better patient care, diagnosis, and treatment. Bio-inspired algorithms, drawing inspiration from natural processes, have gained attention for their potential to enhance ML and DL models in this field. This paper explores current research directions and challenges in utilizing bio-inspired algorithms for advancing healthcare ML and DL models. We investigate their applications and facilitating feature selection, while acknowledging limitations such as scalability, interpretability, and robustness to noisy healthcare data. Ethical considerations surrounding their use in sensitive healthcare contexts are discussed. Through interdisciplinary collaboration and innovative algorithmic approaches, we strive to overcome these challenges and fully unlock the potential of bio-inspired algorithms in healthcare, ultimately aiming to revolutionize healthcare delivery for improved patient outcomes, personalized treatment strategies, and more accurate diagnoses.

Neurodegenerative diseases are one of the most debilitating diseases associated with progressive neuronal dysfunction and protein accumulation. The increasing prevalence of Alzheimer's disease (AD), Parkinson's disease (PD), and amyotrophic lateral sclerosis (ALS) in population is leading to the development of neuroinnovative techniques for early recognition, classification and treatment. Various computer-aided software programs based on machine learning (ML) studies have been developed in recent decades in medical scope. ML is a branch of artificial intelligence that recognizes and models different data sets and algorithms. Different disciplines such as neural networks, support vector machines, Decision Trees, Random Forests, logistic regression, k-Nearest Neighbor techniques such as multi-layer perceptron, Neural Networks, Gaussian Mixture Models, and Boosting methods have been introduced in neurology. This section will discuss neuropathological and neuroradiological

In the realm of wastewater treatment, Wastewater Stabilization Ponds (WSPs) are highly regarded for their efficiency, particularly when incorporating baffle walls (BWs) into their design. These BWs play a crucial role in minimizing space requirements and improving contaminant removal, albeit at the expense of increased construction material usage. To find the best configuration, a thorough analysis was first carried out utilizing typical methodology (TM) design worksheets for WSPs. This phase validated the effectiveness of BWs by showing a considerable reduction in both WSPs area and hydraulic retention time (RT). Within the MM framework, additional reductions in WSPs area and RT were attained, coupled with a 5% drop in necessary concrete volume compared to TM, by employing the interior-point approach in MATLAB and the generalized reduced gradient (GRG) algorithm in MS Excel Solver. While these findings highlight the effectiveness of MM in optimizing WSPs, there is potential for exploring alternative algorithms to achieve more favorable results.

The paper examines the latest optimization models for the hydrogen transportation and production. Optimization plays a crucial role in systematic methodologies that aid in hydrogen expansion. Energy for most systems comes from non-renewable sources like fossil fuels, which take a long time to recharge and never quite reach their full potential. Energy problems, climate change, and environmental degradation have all resulted from the widespread usage of fossil fuels. One of the many benefits of hydrogen is its abundance of clean, efficient sources. In order to back up hydrogen expansion, systematic techniques must include optimization. Some of the recent developments in the transportation, storage, and generation of hydrogen are covered, including the use of clean processes such as photo electrolysis, water electrolysis, pyrolysis, and partial oxidation. Researchers interested in applying optimization approaches to hydrogen supply chain networks may find this study to be a useful blueprint.

Deforestation poses a significant threat to global biodiversity and climate stability, necessitating effective monitoring and management strategies. It is highly necessary for an effective monitoring strategy to mitigate deforestation as it possesses a potential threat to climate stability and global biodiversity. A novel deep learning technique with Convolutional Neural Networks (CNNs) and Recurrent Neural Networks (RNNs) is proposed to identify the deforestation and monitoring the forest. CNN is deployed to identify the deforested areas by extracting spatial features and RNN are used to capture the patterns of forest dynamics processing the time series satellite data. This is a novel mechanism where the spatial and temporal analysis is done for the prediction.

The urban population's rapid growth has heightened traffic congestion challenges, necessitating innovative solutions. This paper explores the potential of Graph-Based Machine Learning (GBML) in transforming traffic management. It focuses on enhancing efficiency to reduce travel times and congestion, and improving mobility for a better user experience and urban accessibility. We assess the effectiveness of logistic regression and random forest for classifying traffic situations. While logistic regression offers interpretability, random forest outperforms it in accuracy. Random forest excels in capturing complex traffic data relationships, aiding tasks like traffic flow prediction and anomaly detection. Its adaptability to diverse datasets ensures

Preface

INTRODUCTION

In an era defined by rapid technological advancement, machine learning and artificial intelligence (AI) have emerged as transformative forces across diverse fields. From revolutionizing medicine with AI-driven diagnostics and personalized treatments to enhancing environmental science through the analysis of complex datasets, these technologies are reshaping the way we approach critical challenges. Their applications in transportation—such as optimizing logistics and enabling autonomous vehicles—and education—through tools that personalize learning experiences—highlight their versatility and far-reaching impact. This book explores these intersections of innovation, presenting both theoretical foundations and practical implementations.

The chapters in this volume provide a comprehensive examination of how optimization, machine learning, and AI are driving smarter decision-making and fostering efficiency across disciplines. With a focus on real-world applications, we aim to bridge the gap between theory and practice, showcasing how these technologies contribute to solving pressing global issues. From climate predictions enabled by machine learning to the integration of AI in enhancing civic education, this book underscores the potential of these tools to create meaningful change.

As editors, our goal has been to curate a collection that not only informs but also inspires. We believe the case studies and research presented here will serve as a valuable resource for a diverse audience, including computer scientists, researchers, educators, and policymakers. By illuminating the transformative possibilities of AI and machine learning, we hope to stimulate further exploration and innovation in these critical areas.

CHAPTER OVERVIEW

This book is structured to provide a nuanced exploration of the applications of machine learning and AI across four primary domains: medicine, environmental science, transportation, and education. Each chapter delves into cutting-edge research and practical implementations:

1. **Hydrogen Supply Networks**: This chapter examines optimization techniques for designing green hydrogen supply chains. It highlights key challenges in transportation and storage, focusing on strategies to enhance efficiency and sustainability in renewable energy systems.
2. **Healthcare 6.0**: By exploring bio-inspired algorithms, this chapter discusses their integration into machine learning models to improve patient care. Case studies showcase how these algorithms aid in early diagnosis and personalized treatment plans.
3. **Neurodegenerative Disease Diagnosis**: A deep dive into the use of machine learning algorithms for early detection of neurodegenerative diseases such as Alzheimer's and Parkinson's. The chapter outlines approaches for classification and potential impacts on healthcare.
4. **Polyp Segmentation**: This chapter reviews advancements in medical imaging, particularly the use of U-Net architecture in detecting colorectal cancer. It demonstrates the role of machine learning in enhancing diagnostic accuracy.
5. **Wastewater Treatment**: Focusing on industrial processes, this chapter explores how AI models are optimizing wastewater treatment operations, reducing costs, and improving environmental outcomes.
6. **Stabilization Ponds**: This chapter investigates innovative machine learning approaches to design and improve the performance of wastewater stabilization ponds, offering solutions for sustainable water management.
7. **Hydrogen Energy**: It highlights computational chemistry advancements that drive progress in hydrogen production and storage, discussing their implications for the future of clean energy technologies.
8. **Deforestation Monitoring**: With the application of CNN and RNN models, this chapter addresses global deforestation challenges, emphasizing the importance of machine learning in environmental conservation efforts.
9. **Traffic Management**: This chapter applies graph-based machine learning techniques to optimize urban traffic systems, offering insights into reducing congestion and improving city infrastructure.
10. **Pothole Detection**: By leveraging YOLO models, this chapter discusses real-time detection systems designed to enhance road safety and maintenance efficiency.

11. **Anomaly Detection in the Metaverse**: This chapter delves into Python-based models for ensuring data integrity and reliability within virtual environments, highlighting their importance in the expanding metaverse.
12. **AIoT in Education**: Exploring the integration of AI with IoT, this chapter examines applications aimed at fostering critical thinking and communication skills among students in diverse educational settings.
13. **AI in Civic Education**: This chapter demonstrates the use of ChatGPT to enhance civic education, showcasing how AI can promote interaction and engagement in learning processes.
14. **Student Performance Prediction**: It investigates the potential of graph machine learning to predict and improve educational outcomes, providing insights into tailored learning strategies.

CONCLUSION

The chapters compiled in this book collectively illustrate the profound impact of machine learning and AI in addressing global challenges. By providing insights into diverse applications, from optimizing renewable energy systems to revolutionizing education, this work underscores the potential of these technologies to reshape industries and improve lives. The research presented here not only demonstrates current capabilities but also points toward future possibilities, encouraging inter-disciplinary collaboration and innovation.

As we conclude, we wish to express our gratitude to the contributors whose expertise and dedication have made this book possible. Their work reflects the dynamic and rapidly evolving nature of these fields. We also thank the readers, whose engagement with this material will undoubtedly drive new ideas and applications.

We hope this book serves as a foundation for ongoing exploration and inspires further advancements in machine learning and AI. By embracing the opportunities presented by these technologies, we can collectively work toward a smarter, more sustainable, and equitable future.

Toufik Mzili
Chouaib Doukkali University, Morocco

Adarsh Kumar Arya
Harcourt Butler Technical University, India

Acknowledgement

I would like to express my sincere gratitude to my university for fostering an environment conducive to academic excellence and innovation. My most sincere thanks go to the Faculty of Science, which has been the cornerstone of my educational journey, and to the Department of Computer Science for its unfailing support and guidance throughout my studies and research. I am deeply grateful to my dear friends, Professors Ilyass Mzili, Hamid Nebdi, Hassan Silakn, Mohammed Essaid Riffi, Jalal Eddine Bahbouhi and Saif Islam Bouderba, whose encouragement, collaboration and camaraderie have been invaluable. Their insights and camaraderie have enriched my personal and academic development, and I thank my PhD students Dr. Fatima Ezzahra Zamzam, Dr. Akdim Doha, Dr.Otmane Houdaif and Dr. Safae Mougare for their help. This acknowledgement is a tribute to all those who have contributed to the success of my work. Thank you for your inspiration and support.

Chapter 1
Recent Trends in Optimizing Hydrogen Supply Networks

Devanshi Srivastava
Department of Chemical Engineering, Harcourt Butler Technical University, Kanpur, India

Adarsh Kumar Arya
https://orcid.org/0000-0002-2378-3832
Department of Chemical Engineering, Harcourt Butler Technical University, Kanpur, India

Toufik Mzili
https://orcid.org/0000-0002-5733-3119
Chouaib Doukkali University, Morocco

Murali Pujari
Department of Chemical Engineering, University of Petroleum & Energy Studies, Bidholi, India

ABSTRACT

The extensive reliance on conventional fossil fuels escalates greenhouse gas emissions and exacerbates environmental pollution, necessitating an urgent energy shift. Consequently, the advancement and use of renewable energy are of paramount significance. Hydrogen is expected to progress in the following years as a replacement for fossil fuels among various renewable sources. However, the high cost of the hydrogen supply network (HSN) has provoked the researchers to use optimization techniques to optimize the (HSN). The present study analyzes modern optimization methods for (HSN) and thoroughly examines the 'Green hydrogen sup-

DOI: 10.4018/979-8-3373-1399-3.ch001

ply chain'(GHSC), including transportation, production, storage and consumption, emphasizing metaheuristic optimization (MO) applications. The difficulties inherent in each phase are highlighted, and the capacity of MO techniques to mitigate these difficulties is examined. The study also examines multi-objective methodologies for optimizing issues within this area.

1. INTRODUCTION

In light of the official initiation of the Paris Climate Change Agreement, which took place on November 4, 2016, there has been a notable increase in the expectations surrounding hydrogen fuel, as it is being recognized as an essential element in the broader strategy for achieving environmental sustainability. The ambitious 2°C global warming scenario is designed to create a carbon release trajectory that effectively limits the accumulation of greenhouse gases within our atmosphere, thereby compelling various governments and advocates for climate action, especially in developing nations, to diligently pursue the critical goal of significantly reducing emissions while simultaneously ensuring that economic growth remains unhindered and robust. Hydrogen is increasingly viewed as a vital component that enhances the sustainability and resilience of our power systems and plays a crucial role in providing the necessary flexibility that modern energy infrastructures require to adapt to fluctuating demands and supply conditions(Arya et al., 2022). Consequently, the ongoing exploration and investment in hydrogen fuel technologies present a promising avenue for addressing the dual challenges of climate change and economic development in an era that increasingly prioritizes environmental stewardship alongside economic viability. The International Energy Agency (IEA), in its publication titled "Hydrogen and Fuel Cells," released in the year 2015, made a noteworthy observation that the utilization of hydrogen has the potential to seamlessly integrate a multitude of energy sectors alongside the intricate logistics and distribution (T&D) networks, thereby significantly enhancing the operational flexibility and adaptability of the low-carbon energy systems that are anticipated to emerge shortly. The current energy system predominantly relies on fossil fuels; besides co-generation, there are few interconnections among the various transmission and distribution systems. In a forthcoming system, Hydrogen may play a crucial role in connecting different tiers of infrastructures within a low-carbon energy framework. Hydrogen products and technology have advanced considerably over several decades and are now being integrated into the market. The infrastructure's inadequacy is a significant obstacle to reaching the hydrogen economy. Consequently, extensive infrastructure investment initiatives grounded in formulating novel techniques should be implemented. Furthermore, the business has sought to enhance the expansion of hydrogen utilization. In January 2017, 13 prominent energy, transport, and industrial firms initiated a global endeavour—the "Hydrogen Council"—to articulate a cohesive vision and long-term objective for utilizing Hydrogen to facilitate energy transition ("Hydrogen Council Launched in Davos as 13 Major Companies Join to Promote Hydrogen," 2017). In its updated roadmap released in November 2017 (Http://Ljournal.Ru/Wp-Content/Uploads/2017/03/a-2017-023.Pdf, 2017)(Lin et al., 2008), the Hydrogen Council projected that by the year 2050, Hydrogen could fulfil 18% of global energy requirements and decrease CO2 emissions across many sectors, including residential, transportation, and industry, by 40 to 60 per cent. Furthermore, governments and scholars persist in examining the feasibility and challenges of transitioning from an unsustainable carbon-derived economy to a green hydrogen-based economy. A primary priority is developing a strategy to actualize this vision; the responses to the "4W questions" encapsulate this strategy. The temporal and spatial determinations about establishing hydrogen infrastructures—specifically, the timing, location, scale, and technologies employed 4. The supply chain network's (SCND) design, or strategic supply chain planning, is a critical issue in managing supply chains (SCM) (Govindan et al., 2017). The comprehensive examination and analysis of the implementation and establishment of hydrogen infrastructures are currently being meticulously explored and researched through the innovative framework and methodology known as the hydrogen supply chain network design (HSCND) concept, which seeks to optimize and facilitate the effective distribution and utilization of hydrogen as a sustainable energy source. The initiation of this intricate process is marked by the utilization of appropriate feedstock, culminating in the eventual sale of Hydrogen at various filling stations strategically located for consumer accessibility. Within each distinct area of this comprehensive framework, a plethora of options is available for consideration and selection. For instance, water, when combined with electricity, biomass derived from organic materials, and coal, can all serve as a viable feedstock for the diverse production methodologies associated with electrolysis, biomass gasification (commonly referred to as BG), and coal gasification (designated as CG), respectively. Following this, the Hydrogen generated through these advanced processes can be efficiently transported to designated terminals utilizing various conveyance methods, including but not limited to trucks, trains, or pipelines, each offering unique logistical advantages. Furthermore, within these terminals, different storage methodologies can be selected based on the specific physical state of hydrogen, thus ensuring optimal storage conditions tailored to its characteristics. The intrinsic nature of a supply chain ensures that each component is interrelated rather than separate. Establishing a competitive market in all forms of energy from renewable sources necessitates intricate design, planning, and optimization approaches. These have been seen in many optimization approaches used in the domain of sustainable and environmentally friendly energy. The elaborate nature of the issue in the HSCND (Baños et al., 2011) is contingent upon modelling the interactions among the various components of the HSCN. The production processes rely on the regionally distinct properties of feedstocks; it is essential to note that transportation and storage methods are closely linked to the physical form of Hydrogen. Furthermore, the (HSN) configuration substantially influences the location and the refuelling station technology. The optimization-based methodology has significantly enhanced our understanding of these technological and spatial linkages. Following two decades of development, a growing body of research on the HSCND is now accessible to elucidate the future of HSCN in the transportation sector. This study intends to evaluate the literature related to HSCND models published in scholarly journals. Analysis of the supply chain for hydrogen manufacturing is represented in Figure 1. The extant gaps in the current literature are explicitly delineated to provide research possibilities and future initiatives. A classification of optimization-based studies on HSCND is conducted, followed by selecting articles for further investigation. The concluding part proposes a list of prospective obstacles for future study avenues.

Figure 1. Analysis of the hydrogen energy generation supply chain from (Doniavi et al., 2023) under Creative Commons CC.BY license.

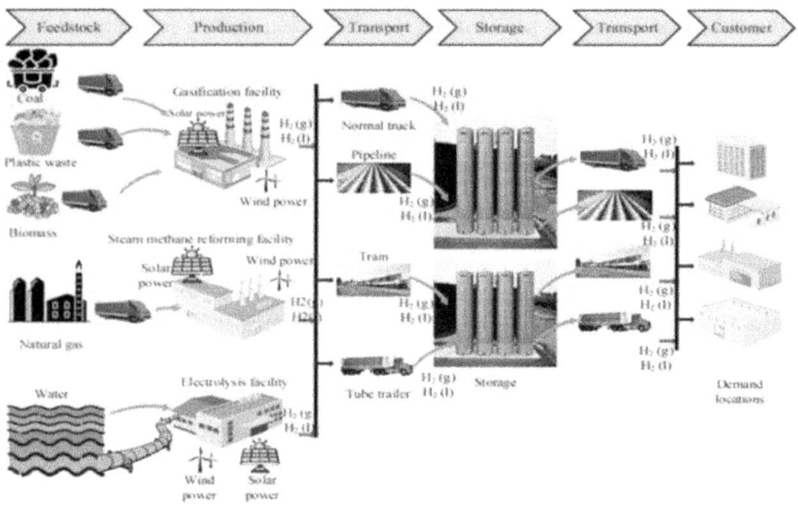

2. OPTIMIZATION TECHNIQUES

Optimization techniques are fundamental tools used across various disciplines to find the best possible solution to a problem within a given set of constraints(Arya & Honwad, 2018). The diverse methodologies employed in the resolution of the issues can be systematically and thoroughly categorized into discrete classifications that account for several pivotal elements, such as the intrinsic attributes of the issue at hand, the specific types of variables pertinent to the analysis, and the particular techniques utilized to identify and determine the most effective and optimal solutions to the challenges encountered (Arya & Honwad, 2016). The present section summarizes the optimization techniques and their primary applications.

2.1 Types of Optimization Problems

i. **Linear vs. Non-linear Optimization**

   Linear Optimization: Involves objective functions and linear constraints. This type is widely applied and fundamental in optimization theory.

   Non-linear Optimization: Deals with problems where the objective function or constraints are non-linear, making them more complex to solve

ii. Discrete vs. Continuous Optimization

   Discrete Optimization: Variables can only take on discrete values, often integers. This includes combinatorial optimization problems

   Continuous Optimization: Variables can take any value within a range, making the solution space continuous

iii. Convex vs. Non-convex Optimization

   Convex Optimization: Both the objective function and the feasible region are convex, ensuring that any local minimum is a global minimum

   Non-convex Optimization: Involves non-convex functions or constraints, which can have multiple local minima, making the problem more challenging

iv. Deterministic vs. Stochastic Optimization

   Deterministic Optimization: All parameters are known and fixed, leading to a predictable outcome.

   Stochastic Optimization: Involves random variables and uncertainty, requiring methods that can handle variability in the data(Sarkar & Arya, 2022)

v. Single-objective vs. Multi-objective Optimization

   Single-objective Optimization: Focuses on optimizing one objective function

Multi-objective Optimization: Involves multiple conflicting objectives, requiring a tradeoff to find an optimal solution(Arya et al., 2023).

2.2 Common Optimization Techniques

i. Gradient Descent

A local optimization method that iteratively adjusts parameters to minimize a cost function. It is widely used in machine learning for training models (Arya, 2021)

ii. Simulated Annealing:

A global optimization technique inspired by the annealing process in metallurgy. It allows for occasional uphill moves to escape local minima, making it practical for complex landscapes

iii. Simplex Algorithm

Designed for linear programming, this algorithm iteratively moves towards the optimal solution by traversing the edges of the feasible region

iv. Genetic Algorithms

Evolutionary algorithms that mimic natural selection processes to find optimal solutions. They are beneficial for complex, multi-modal problems (Arya, 2022a)

v. Particle Swarm Optimization

A global optimization technique that simulates the social behaviour of birds flocking or fish schooling to find optimal solutions

vi. Lagrange Multipliers

Used for problems with equality constraints, transforming the constrained problem into an unconstrained one to find the optimum

2.2 Applications of Optimization Techniques

Optimization techniques are applied in numerous fields to solve practical problems:

i. **Engineering:** Used for system design, resource allocation, and performance optimization
ii. **Economics:** Helps in maximizing profits or minimizing costs under given constraints
iii. **Healthcare**: Optimizes wait times, resource allocation, and treatment plans
iv. **Supply Chain Management:** Enhances logistics, inventory management, and distribution networks
v. **Energy Systems:** Aims to reduce energy consumption and operational costs

Figure 2 summarizes optimization techniques reported in the last few years.

Figure 2. Various Techniques for optimization from (Arya, 2022b) under Creative Commons CC.BY license.

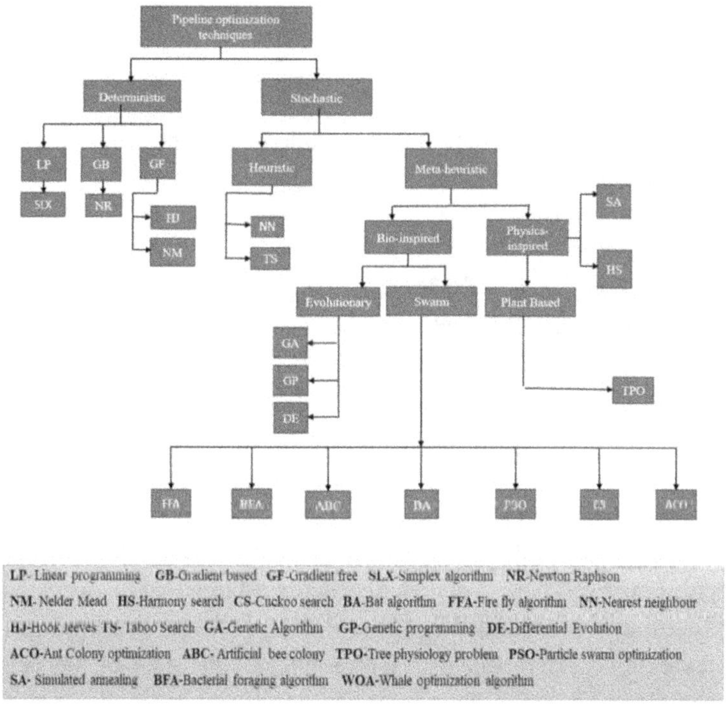

Since metaheuristics initially emerged, several advancements have been made to enhance overall performance. The creation of new algorithms or the upgrading of existing ones are examples of these advancements. The graphic depicting the evolution of metaheuristic algorithms across time is shown in Figure 3.

Figure 3. A chronology outlining the evolution of optimization techniques from (Nassef et al., 2023) under Creative Commons CC.BY license.

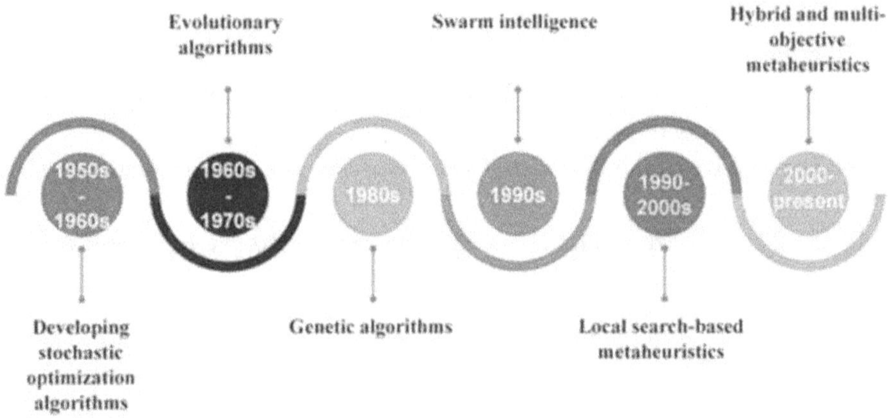

In the next section, we present the methodologies to optimize the various parts of the hydrogen supply chain.

3. APPLICATION OF OPTIMIZATION TO HYDROGEN PRODUCTION

MO algorithms have recently attracted much interest because of their ability to tackle large-scale, complex, and severely limited problems. The sustainable generation of hydrogen systems, especially hydrogen microgrids, presents several challenges that need effective, scalable, and flexible optimization methods. These systems' degree of detail and non-linearity require advanced optimization procedures to achieve optimum efficiency. A metaheuristic problem that involves the intricate process of optimization is fundamentally composed of three essential components, which include an objective function that delineates the goal of the optimization endeavour, various obstacles that may impede the progress, and a comprehensive search space that outlines the parameters within which potential solutions can be explored. The objective function serves as the cornerstone of the optimization process, representing the specific target that the optimization seeks to achieve, which frequently entails minimizing or maximizing a particular value that interests the researcher or practitioner. Constraints are defined as the limitations or prerequisites that any viable solution must satisfy to be deemed acceptable. At the same time, the search space encompasses all potential solutions that could theoretically resolve the problem at hand. A multitude of widely recognized metaheuristic techniques, such as 'Particle Swarm Optimization' (PSO), 'Simulated Annealing' (SA), 'Ant Colony Optimization' (ACO), 'Genetic Algorithms' (GA), and Differential Evolution, have been rigorously applied and extensively utilized to effectively tackle a diverse array of optimization challenges that arise in various fields of study and practical applications. These methods possess a common framework yet diverge in their particular variation operator and selection mechanisms.

In contrast to other optimization techniques, Particle Swarm Optimization (PSO) methodically adjusts the locations of particles by taking into account not only their individual speeds but also the most frequently occurring positions within the swarm, subsequently evaluates the overall effectiveness of these positions, and enhances the collective performance of the entire population by systematically replacing the previous positions with newly calculated ones that exhibit improved potential. Furthermore, PSO algorithms utilize both their own best-known positions and the globally recognized best positions to intelligently guide the search process, which effectively directs the movements of the particles toward the more advantageous regions of the search space that are likely to yield optimal solutions. Consequently, the various components of the 'objective function', along with any applicable constraints and additional parameters embedded within the 'pseudo-code', can be meticulously tailored and customized to effectively address the unique challenges associated with the production of green hydrogen, ensuring that the optimization process is both comprehensive and precisely aligned with the desired outcomes. Consequently, metaheuristic approaches can be utilized in multiple facets of the production of green Hydrogen.

i. Design of system and dimensioning

The goal is to reduce the entire system cost, encompassing capital, operating, and repair expenses while adhering to energy generation, storage capacity, and system dependability restrictions.

ii. The instantaneous reaction of the electrolyzer

The goal is to reduce the discrepancy between the electrolyzer's intended and tangible outcome, constrained by the electrolyzer's input power and operational limits.

iii. Allocation of energy resources

The goal function may be formulated to optimize the production of green Hydrogen or to reduce energy expenditures while conforming to restrictions on renewable energy' accessibility, storage capacity, and demand for consumption.

iv. Restoration of waste heat

The objective function is to maximize waste heat recovery efficiency, subject to the constraints of heat exchanger limitations and the available waste heat.

v. Optimization of water use

The goal function is to minimize water usage in producing green hydrogen procedures, considering constraints related to water accessibility, quality standards, and electrolyzer limitations.

4. APPLICATION OF OPTIMIZATION TO HYDROGEN STORAGE

This section examines ' MO ' approaches in storing Hydrogen, particularly addressing the mentioned problems. We revisit the generic pseudo-code and ' MO ' strategies outlined in the preceding segment, focusing on each challenge's objective functions and restrictions.

i. Optimization of storage facility layout

The target function is to reduce capital and operating expenditures, optimize space efficiency, and reduce energy utilization during storage and retrieval while complying with limits on storage capacity, safety rules, and facility dimensions.

ii. Optimization of liquefaction and compression of Hydrogen

The goal function is to minimize energy consumption, enhance output, and improve operational parameters while complying with limitations concerning pressure and temperature thresholds, instrument specifications, and safety regulations.

ii. Optimization of pressure and temperature control

The goal function can be formulated to guarantee the reliability of Hydrogen storage, reduce energy dissipation from escape or evaporation, and reduce energy consumption for 'temperature and pressure' regulation while complying with constraints on temperature limits, pressure and specifications for equipment and safety regulations.

iii. Selection of optimal storage materials

The objective function may prioritize maximizing storage capacity, lowering costs, reducing weight, and assuring safety while accounting for constraints related to compatibility with Hydrogen, material availability, and efficiency requirements. In summary, employing ' MO ' methods for storing Hydrogen can enhance system efficiency, economic viability, and sustainability. The topological structure of the hydrogen energy storage system comprises an atmospheric power production system, a solar power development system, a battery stored system, a hydrogen storage system, and the hydrogen demand end. It is connected with the wind and solar power production systems. Figure 3 illustrates a topological configuration of wind and solar energy production integrated with a hydrogen storage system.

Figure 4. The topological configuration of wind and solar energy production is integrated with a hydrogen storage system (Chen et al., 2022) under Creative Commons CC, BY license.

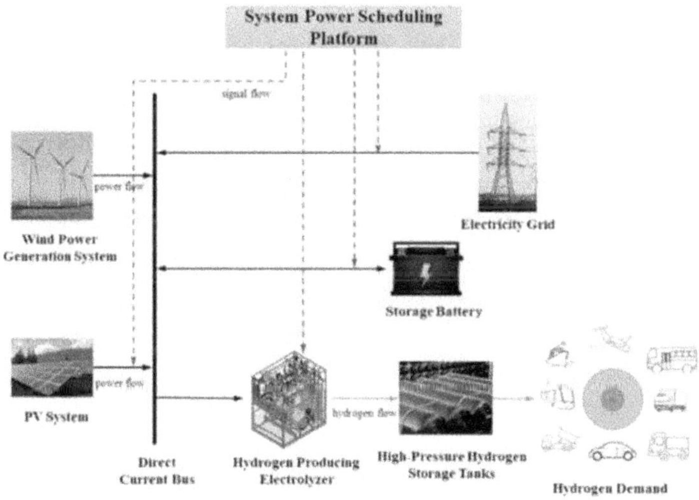

With the increasing need for green Hydrogen, researchers and practitioners can investigate innovative approaches to strengthen the development of advanced and efficient hydrogen storage technologies. Future research topics may encompass examining evolving 'metaheuristic algorithms' and formulating hybrid methodologies that leverage the advantages of several methods, during which machine learning integration can augment optimization efficacy.

5. APPLICATION OF OPTIMIZATION TO HYDROGEN TRANSPORTATION

MO approaches have been effectively utilized in multiple facets of hydrogen transportation, assisting in overcoming the constraints outlined in the preceding subsection. This segment addresses the utilization of MO approaches in several aspects of hydrogen conveyance.

(i) Optimization of different forms of transportation

Optimizing multi-modal transportation systems aims to minimize the overall transportation expenses for Hydrogen, encompassing costs related to various infrastructure and forms of transportation. Limitations may encompass distances, transportation capabilities, and the accessibility of different transportation options. Utilizing MO approaches enables the identification of an ideal combination of transportation systems that satisfies the requisite demand while reducing expenses.

(ii) Scheduling of Supply of Hydrogen

The goal of planning for the shipment of Hydrogen is to reduce 'total operational costs, encompassing production transportation, storage, and expenses while satisfying the demand for Hydrogen among consumers. In this context, constraints may pertain to transportation and storage capacities and temporal fluctuations in demand and supply. MO methods can be employed to identify effective scheduling of supplies that adhere to restrictions and reduce operating expenses.

(iii) Design of a hydrogen pipeline network

The target function of the design is to minimize the total network cost, encompassing operational, capital, and repair expenses. Constraints may encompass pipeline path limitations, safety rules, and material specifications. ' MO ' methods can ascertain an ideal network of pipeline configurations that adhere to limitations and reduce total costs.

(iv) Planning for the location of refuelling stations

The target operating mechanism for filling station placement planning aims to reduce the costs associated with setting up and sustaining filling stations while making sure enough coverage and accessibility are available. Constraints may encompass the stations' quantity, capacity, and geographical limitations. ' MO ' approaches can identify the best sites for refuelling stations that adhere to limitations and save expenses. Global transportation of green Hydrogen, emphasizing the critical areas where MO approaches can be utilized to tackle the issues. In conclusion, MO methods can significantly enhance the performance and affordability of hydrogen transportation networks. By tackling diverse problems in transportation strategy, planning, and creation, these methods facilitate establishing resilient and dependable networks for transporting Hydrogen.

6. APPLICATION OF OPTIMIZATION TO CONSUMPTION OF HYDROGEN

This segment examines the application part of ' MO ' approaches in the consumption of Hydrogen. Fuel cell performance enhancement, design of H_2-based heating systems, selection of energy conversion technologies (including optimization of combustion processes), and H_2 integration within multi-energy systems (encompassing optimization of chemical processes).

(i) Efficiency enhancement for fuel cells

The aim is to enhance the efficacy and longevity of fuel cells while reducing their cost. ' MO ' strategies may enhance parameters, including pressure, operating temperature, flow rates, and the design of specific elements such as membranes, catalysts, and electrodes. Constraints encompass material characteristics, longevity, and safety standards.

(ii) Design of hydrogen-based heating systems

The objective is to develop effective, economical, sustainable hydrogen-powered heating systems. Optimal dimensions, designs, and control techniques for these systems may be found by optimization approaches that consider heat demand, energy pricing, and greenhouse gas emissions, among other things. 'Constraints' may encompass spatial limitations, financial restrictions, and regulatory obligations.

(iii) Selection of energy conversion technology

Choosing the optimal conversion of energy technology for a particular application is essential for enhancing efficiency while reducing costs and emissions. ' MO ' techniques can evaluate and identify the optimal technology based on efficiency, capital expenditure, operational costs, and environmental consequences. Constraints encompass technological feasibility, existing infrastructure, and regulatory limitations.

(iv) Integration of H_2 in multi-energy systems, encompassing optimization of chemical processes

This particular domain is dedicated to the strategic and effective integration of Hydrogen across a diverse array of energy applications, which encompasses not only the development and deployment of micro-grids but also the intricate manufacturing processes, all with the primary objective of significantly enhancing both system effectiveness and adaptability while concurrently striving to minimize operational expenses and mitigate any adverse ecological repercussions that may arise from such activities. The implementation of 'MO' methods holds the potential to meticulously ascertain the precise control, optimal scheduling, and thoughtful design of various components related to hydrogen storage, production, and consumption, as well as ensuring the most efficient utilization of Hydrogen in critical chemical processes, including but not limited to the production of ammonia and methanol, which are vital for numerous industrial applications. The constraints that may come into play in this context are likely to encompass a variety of factors, such as the availability of resources, fluctuations in energy demand, and the rigorous framework of environmental regulations that govern these operations. In conclusion, the application of MO methods appears to be exceedingly promising for enhancing the performance, efficiency, and overall ecological footprint of devices designed to consume hydrogen, highlighting their importance in the transition towards sustainable energy solutions. As we witness continued advancements in these technologies, it will become increasingly essential to develop optimizing methodologies to unlock Hydrogen's full potential as a renewable and environmentally friendly energy carrier, thereby contributing to a more sustainable future.

6.1. Multi-Objective Problems in HSC

Multi-objective problems typically necessitate identifying a representative collection of 'Pareto optimal solutions' (Marler & Arora, 2004). The predominant method, the ε-constraint approach, is used in the literature; it entails lowering one goal function and adding the other objective functions to the model as constraints, with a level ε limiting their incorporation. While making a choice, if it possesses a predetermined preference, a designated ε (Marler & Arora, 2004), value is assigned, and a singular solution is identified. Nevertheless, in most instances, the model is executed for several ε values. (Guillén-Gosálbez et al., 2009) employ the ε-constraint technique to identify tradeoff solutions that balance the minimization of total system costs with the maximization of 'Life Cycle Inventory (LCI)' for the HSN. To significantly accelerate the computational efficiency associated with deriving 'Pareto solutions', the researchers undertook the strategic implementation of a Multi-level breakdown framework, which effectively partitions their complex problem into two distinct components: a master problem and a subproblem. Within the context of the master problem, the integer variables that represent the various manufacturing plants and storage facilities are systematically eliminated. In contrast, the variables that pertain to the transmission aspects are judiciously relaxed to facilitate the analysis. The master problem serves to pinpoint a specific group of technologies related to manufacturing, storage, and transportation that are pertinent to the Hydrogen Supply Chain (HSC). It is noteworthy that the master problem possesses a greater significance compared to the original formulation, as it contains a reduced number of integer variables and strategically selects a defined group of processes for optimization. In the subsequent phase, referred to as the secondary issue or the planning problem, the focus shifts to identifying the chosen array of technologies that collectively establish an upper limit for the total system cost—this was achieved through the development of a sophisticated multi-objective model aimed at designing a comprehensive hydrogen supply chain specifically tailored for Gyeongsang-do Province, South Korea, utilizing the steam reforming process of biomethane amid the challenges posed by uncertain hydrogen demand.They solve their model for each ε using the commercial solver ("ODH|CPLEX Now Included in GAMS," 2018) (Sabio et al., 2010) also examine a stochastic multi-objective model for developing an HSN in Spain. They use the ε-constraint framework to divide the issue into an enslaver and enslaved person problem, expediting a resolution of each 'Pareto solution'. The weighted constraint method consolidates all goal functions into a singular composite function using weighted coefficients, each summing to one.

The weighting coefficient significantly influences the resolution of the multi-objective issue.(Dagdougui et al., 2012) employ the weighted sum method to formulate a singular objective function to maximize hydrogen demand satisfaction from refuelling stations and the industrial sector, optimize energy sales to the power grid, and maximize hydrogen storage in a tank. Lexicographic methods entail prioritizing objective functions and sequentially addressing single-objective issues, each constrained by an upper limit derived from the optimal value of the preceding optimization challenge (Marler & Arora, 2004). (De-León Almaraz et al., 2013) employ the lexicographic approach succeeded by the ε-constraint method. Nevertheless, it is important to note that, to date, no comprehensive studies have exclusively utilised lexicographic approaches in isolation to effectively address a multi-objective problem, which remains a significant gap in the existing literature. Genetic algorithms, often recognized as complex metaheuristic strategies, leverage diverse cohorts of candidate solutions to systematically refine multifaceted challenges while gradually moving towards increasingly optimal solutions over time. It is crucial to understand that a 'Genetic algorithm (GA)' does not inherently guarantee optimality in its outcomes and, as a result, may require multiple iterations and refinements to achieve the best possible answer in any given scenario ultimately (Konak et al., 2006). Two articles (Robles et al., 2020) (Reyes-Barquet et al., 2022) employ a (GA) to construct Pareto optimal solutions, notably a non-dominating genetic algorithm II variant.

7. A SUSTAINABLE HYDROGEN SYSTEM VIA MULTI-OBJECTIVE METAHEURISTIC OPTIMIZATION

'Multi-objective optimization' is essential for addressing difficult situations with numerous competing goals. Dominant or Pareto-optimal solutions signify the best tradeoffs among these goals. Multi-objective optimization may enhance efficiency, cost-effectiveness, performance, and safety (Alirahmi et al., 2021) (Sharafi & ELMe-kkawy, 2014) (Xu et al., 2020) within green hydrogen supply chains. The following segment offers two comprehensive examples of optimization with multi-objective challenges, concentrating on different phases of the supply chain and highlighting an algorithm-agnostic methodology.

8.1. Transportation and Production

Using a multi-objective methodology, the emphasis is on sustainable hydrogen optimization generation and transportation phases. The goals for the optimization issue are:

(i) Reducing the overall 'cost of production of hydrogen (C1)';
(ii) Minimizing the overall 'cost of transportation of hydrogen (C2)';
(iii) Optimizing the 'efficiency of production of hydrogen (E1)' and
(iv) Maximizing the 'safety of transportation of hydrogen (E2)'.

The optimization issue may be articulated as follows:

Minimize: $C_1 + C_2$; Optimize: E_1; Optimize: E_2

Considering the limitations associated with green hydrogen production, including renewable energy availability, 'electrolyzer capacity', and water consumption, as well as transportation factors like refuelling station location, pipeline capacity and 'transportation mode availability', a method for optimizing many objectives using the Pareto principle is applicable for addressing this issue, as outlined in algorithm 2. This example illustrates how optimization of the dominant solution seeks to determine the optimal balance between competing objectives. Identifying 'Pareto-optimal solutions' enables decision-makers to allocate resources effectively and make informed choices regarding the transportation and production of green Hydrogen, all while considering environmental, economic, and safety factors.

8.2. Consumption and Storage

This emphasizes optimizing eco-friendly processes for hydrogen storage and use through a multi-objective methodology. The Objectives of this optimization issue are:

(i) Reducing the overall expense associated with hydrogen storage (C_3);
(ii) Reducing the overall expense related to hydrogen consumption (C_4);
(iii) Enhancing the efficiency of the 'storage system' (E_3); and
(iv) Improving the efficiency of conversion of energy (E_4).

The optimization issues may be expressed in the following manner:

Minimize: $C_3 + C_4$; Optimize: E_3; Optimize: E_4

Constraints associated with the storage of green Hydrogen, including 'storage capacity, pressure and temperature regulation, and storage materials', as well as consumption factors like 'fuel cell performance, H2-based heating system design, and energy conversion technology selection', can be addressed using a 'Pareto-based multi-objective optimization technique', as outlined in algorithm 2. This example illustrates that optimization of the most popular solutions seeks to discover Pareto-optimal solutions, which embody the most effective tradeoffs between cost minimization C3 and C4 and efficiency maximization E3 and E4. Recognizing these solutions enables decision-makers to make well-informed selections regarding the storage of Hydrogen and its consumption strategies, all while enhancing efficient and cost-effective factors.

9. CHALLENGES AND FUTURE RECOMMENDATION

This article comprehensively reviews evolving developments in the hydrogen supply chain, highlighting essential aspects of 'production, storage, transportation, and consumption.

9.1 Production

Technological advancements in water electrolysis, including 'waste heat recovery' and solid-state electrolyzers, enhance efficiency and reduce costs in green hydrogen generation. Furthermore, the design and optimization of electrolyzer systems include advancing more efficient and resilient catalysts, membranes, and the system's thermal control. Incorporating these manufacturing techniques with Direct Current (DC) microgrids may improve efficiency since batteries, Photovoltaic (PV) panels, and electrolyzers function as DC sources and electrical loads. Employing more efficient materials like gallium nitride (GaN) and superconductors may enhance power conversion processes' efficiency, reducing energy losses and optimizing overall system performance. The following elements highlight these trends:

(i) Enhanced emphasis on green Hydrogen
(ii) Sophisticated water electrolysis technology (Saufi Sulaiman et al., 2019)
(iii) Renewable energy sources (Patel et al., 2021)
(iv) Direct current microgrid (Moradi et al., 2022)
(v) Advanced materials for power conversion (Shahbazi et al., 2021).

9.2 Storage

Research on hydrogen storage concentrates on advancing sophisticated storage materials, including solid-state materials, organic-metal frameworks, and nanostructured materials, which promise enhanced safety, storage capacities, and efficiency. Subzero temperatures and intense pressure container improvements will boost safety and storage capabilities. Furthermore, as these sophisticated technologies and materials increase, economies of scale and technological progress will result in reduced prices for hydrogen storage. The following is a summary of these trends:

(i) Advancement of sophisticated storage materials (Srivastava et al., 2022).
(ii) Enhanced safety and storage capacity and
(iii) Reduced expenses

9.3 Transportation

The enhancement of infrastructure for Hydrogen, including ships, pipelines, and trucks, is anticipated to optimize hydrogen delivery. The quantity of hydrogen refuelling stations will rise as separate layouts, and expedited technology for refuelling will enhance the availability of hydrogen fuel to consumers. Advanced transportation technologies, including autonomous cars and transportation management systems online, will improve the transportation procedure, as detailed below:

(i) Development of infrastructure for Hydrogen
(ii) Enhancement of refuelling stations for Hydrogen and
(iii) Sophisticated transportation systems

9.4 Consumption

Regarding consumption, integrating into current energy systems, including integrated networks and intelligent control of energy consumption, will be essential for optimizing energy use. Progress in fuel cell technology includes durability, efficiency, and cost reduction improvements. Developing a more effective and extensible power source based on hydrogen-generating technologies, including sophisticated engines that run on Hydrogen and internal combustion, is crucial for Hydrogen's widespread use and integration across diverse transportation sectors and manufacturing. The following trends are summarized:

(i) The incorporation of Hydrogen into existing energy infrastructure
(ii) Progress in fuel cell technology (Suresh et al., 2022)
(iii) Increasing need for Hydrogen in industrial applications

10. CONCLUSION

This study thoroughly reviews the hydrogen distribution network, emphasizing the problems inherent in each phase and exploring the potential for MO methods. The principal approach for green hydrogen generation, namely electrolysis of water with sustainable power, was delineated, along with the need to enhance these techniques to enhance efficiency while reducing costs and environmental repercussions. The discussion addressed key storage options; optimization is necessary for safety, capacity, and performance in liquid Hydrogen, compressed gas, and material-based storage. Various choices for the transportation of Hydrogen were examined, including pipelines, vehicles, and ships, along with the variables influencing the selection of transportation techniques in diverse places. Furthermore, many hydrogen consumption techniques were examined, along with their corresponding issues, including fuel cell performance enhancement, design of heating systems based on H_2, and selection of transformation of energy technologies. The report emphasized the potential of MO approaches for tackling these difficulties and enhancing the effectiveness and long-term viability of energy systems powered by Hydrogen.

REFERENCES

http://ljournal.ru/wp-content/uploads/2017/03/a-2017-023.pdf. (2017). LJournal. DOI: 10.18411/a-2017-023

Alirahmi, S. M., Razmi, A. R., & Arabkoohsar, A. (2021). Comprehensive assessment and multi-objective optimization of a green concept based on a combination of hydrogen and compressed air energy storage (CAES) systems. *Renewable & Sustainable Energy Reviews*, 142, 110850. DOI: 10.1016/j.rser.2021.110850

Arya, A. K. (2021). Optimal operation of a multi-distribution natural gas pipeline grid: An ant colony approach. *Journal of Petroleum Exploration and Production Technology*, 11(10), 3859–3878. Advance online publication. DOI: 10.1007/s13202-021-01266-3

Arya, A. K. (2022a). A comparison of the MOGA and NSGA-II optimization techniques to reduce the cost of a biomass supply network. *Materials Today: Proceedings*, 57, 1553–1558. Advance online publication. DOI: 10.1016/j.matpr.2021.12.161

Arya, A. K. (2022b). A critical review on optimization parameters and techniques for gas pipeline operation profitability. *Journal of Petroleum Exploration and Production Technology*, 12(11), 3033–3057. DOI: 10.1007/s13202-022-01490-5

Arya, A. K., Gautam, S., & Yadav, S. (2022). Impact of Hydrogen Embrittlement in Pipeline Structures—A Critical Review. In *Springer Proceedings in Materials* (Vol. 15). DOI: 10.1007/978-981-19-2572-6_31

Arya, A. K., & Honwad, S. (2016). Modeling, simulation, and optimization of a high-pressure cross-country natural gas pipeline: Application of an ant colony optimization technique. *Journal of Pipeline Systems Engineering and Practice*, 7(1), 04015008. Advance online publication. DOI: 10.1061/(ASCE)PS.1949-1204.0000206

Arya, A. K., & Honwad, S. (2018). Optimal Operation of a Multi Source Multi Delivery Natural Gas Transmission Pipeline Network. *Chemical Product and Process Modeling*, 13(3). Advance online publication. DOI: 10.1515/cppm-2017-0046

Arya, A. K., Kumar, A., Pujari, M., & Pacheco, D. A. J. (2023). Improving natural gas supply chain profitability: A multi-methods optimization study. *Energy*, 282, 128659. DOI: 10.1016/j.energy.2023.128659

Baños, R., Manzano-Agugliaro, F., Montoya, F. G., Gil, C., Alcayde, A., & Gómez, J. (2011). Optimization methods applied to renewable and sustainable energy: A review. *Renewable & Sustainable Energy Reviews*, 15(4), 1753–1766. DOI: 10.1016/j.rser.2010.12.008

Chen, K., Peng, H., Gao, Z., Zhang, J., Chen, P., Ruan, J., Li, B., & Wang, Y. (2022). Day-Ahead Operation Analysis of Wind and Solar Power Generation Coupled with Hydrogen Energy Storage System Based on Adaptive Simulated Annealing Particle Swarm Algorithm. *Energies*, 15(24), 9581. DOI: 10.3390/en15249581

Dagdougui, H., Ouammi, A., & Sacile, R. (2012). Modelling and control of hydrogen and energy flows in a network of green hydrogen refuelling stations powered by mixed renewable energy systems. *International Journal of Hydrogen Energy*, 37(6), 5360–5371. DOI: 10.1016/j.ijhydene.2011.07.096

De-León Almaraz, S., Azzaro-Pantel, C., Montastruc, L., Pibouleau, L., & Senties, O. B. (2013). Assessment of mono and multi-objective optimization to design a hydrogen supply chain. *International Journal of Hydrogen Energy*, 38(33), 14121–14145. DOI: 10.1016/j.ijhydene.2013.07.059

Doniavi, E., Babazadeh, R., & Hasanzadeh, R. (2023). Optimization of Renewable Energy Supply Chain for Sustainable Hydrogen Energy Production from Plastic Waste. *Sustainability (Basel)*, 15(24), 16813. DOI: 10.3390/su152416813

Govindan, K., Fattahi, M., & Keyvanshokooh, E. (2017). Supply chain network design under uncertainty: A comprehensive review and future research directions. *European Journal of Operational Research*, 263(1), 108–141. DOI: 10.1016/j.ejor.2017.04.009

Guillén-Gosálbez, G., Mele, F. D., & Grossmann, I. E. (2009). A bi-criterion optimization approach for the design and planning of hydrogen supply chains for vehicle use. *AIChE Journal. American Institute of Chemical Engineers*, 56(3), 650–667. DOI: 10.1002/aic.12024

Hydrogen and Fuel Cells. (2015). In *IEA Technology Roadmaps*. OECD. DOI: 10.1787/9789264239760-en

(2017). Hydrogen Council launched in Davos as 13 major companies join to promote hydrogen. (2017). *Fuel Cells Bulletin*, 1(1), 1. DOI: 10.1016/S1464-2859(17)30001-9

Konak, A., Coit, D. W., & Smith, A. E. (2006). Multi-objective optimization using genetic algorithms: A tutorial. *Reliability Engineering &amp. Reliability Engineering & System Safety*, 91(9), 992–1007. DOI: 10.1016/j.ress.2005.11.018

Lin, Z., Chen, C.-W., Ogden, J., & Fan, Y. (2008). The least-cost hydrogen for Southern California. *International Journal of Hydrogen Energy*, 33(12), 3009–3014. DOI: 10.1016/j.ijhydene.2008.01.039

Marler, R. T., & Arora, J. S. (2004). Survey of multi-objective optimization methods for engineering. *Structural and Multidisciplinary Optimization*, 26(6), 369–395. DOI: 10.1007/s00158-003-0368-6

Moradi, A., Gorji, S. A., Hakemi, A., Ganjavi, A., & Sera, D. (2022). Study of a DC Micro-Gird Configuration to Produce Hydrogen (DCMG-H2). In *2022 IEEE 7th Southern Power Electronics Conference (SPEC)* (Vol. 24, pp. 1–5). IEEE. DOI: 10.1109/SPEC55080.2022.10058435

Nassef, A. M., Abdelkareem, M. A., Maghrabie, H. M., & Baroutaji, A. (2023). Metaheuristic-Based Algorithms for Optimizing Fractional-Order Controllers—A Recent, Systematic, and Comprehensive Review. *Fractal and Fractional*, 7(7), 553. DOI: 10.3390/fractalfract7070553

Patel, H. V., Gorji, S. A., Shahi, S. S. M., & Love, J. G. (2021). Implementation of a Lab-Scale Green Hydrogen Production System with Solar PV Emulator and Energy Storage System. In *2021 11th International Conference on Power and Energy Systems (ICPES)* (Vol. 1, pp. 201–208). IEEE. DOI: 10.1109/ICPES53652.2021.9683797

Reyes-Barquet, L. M., Rico-Contreras, J. O., Azzaro-Pantel, C., Moras-Sánchez, C. G., González-Huerta, M. A., Villanueva-Vásquez, D., & Aguilar-Lasserre, A. A. (2022). Multi-Objective Optimal Design of a Hydrogen Supply Chain Powered with Agro-Industrial Wastes from the Sugarcane Industry: A Mexican Case Study. *Mathematics*, 10(3), 437. DOI: 10.3390/math10030437

Robles, J. O., Azzaro-Pantel, C., Garcia, G. M., & Lasserre, A. A. (2020). Social cost-benefit assessment as a post-optimal analysis for hydrogen supply chain design and deployment: Application to Occitania (France). *Sustainable Production and Consumption*, 24, 105–120. DOI: 10.1016/j.spc.2020.06.010

Sabio, N., Gadalla, M., Guillén-Gosálbez, G., & Jiménez, L. (2010). Strategic planning with risk control of hydrogen supply chains for vehicle use under uncertainty in operating costs: A case study of Spain. *International Journal of Hydrogen Energy*, 35(13), 6836–6852. DOI: 10.1016/j.ijhydene.2010.04.010

Sarkar, A., & Arya, A. K. (2022). A Survey on Optimization Parameters and Techniques for Crude Oil Pipeline Transportation. In *Smart Innovation* (Vol. 292). Systems and Technologies., DOI: 10.1007/978-981-19-0836-1_43

Saufi Sulaiman, M., Singh, B., & Mohamed, W. A. N. W. (2019). Experimental and theoretical study of thermoelectric generator waste heat recovery model for an ultra-low temperature PEM fuel cell powered vehicle. *Energy*, 179, 628–646. DOI: 10.1016/j.energy.2019.05.022

Shahbazi, M., Dehghan-Manshadi, A., Hossain, M. D., Ravishankar, R., & Mackinnon, I. (2021). Microstructure and Superconducting Properties of Copper Sheathed MgB_2 Wires With Ti Barrier Fabricated by in Situ Powder-in-Tube Method. *IEEE Transactions on Applied Superconductivity*, 31(5), 1–5. DOI: 10.1109/TASC.2021.3065634

Sharafi, M., & ELMekkawy, T. Y. (2014). Multi-objective optimal design of hybrid renewable energy systems using PSO-simulation based approach. *Renewable Energy*, 68, 67–79. DOI: 10.1016/j.renene.2014.01.011

Srivastava, S., Shet, S. P., Shanmuga Priya, S., Sudhakar, K., & Tahir, M. (2022). Molecular simulation of copper based metal-organic framework (Cu-MOF) for hydrogen adsorption. *International Journal of Hydrogen Energy*, 47(35), 15820–15831. DOI: 10.1016/j.ijhydene.2022.03.089

Suresh, R., Rajendran, S., Kumar, P. S., Dutta, K., & Vo, D.-V. N. (2022). Current advances in microbial fuel cell technology toward removal of organic contaminants – A review. *Chemosphere*, 287, 132186. DOI: 10.1016/j.chemosphere.2021.132186 PMID: 34509759

Xu, C., Ke, Y., Li, Y., Chu, H., & Wu, Y. (2020). Data-driven configuration optimization of an off-grid wind/PV/hydrogen system based on modified NSGA-II and CRITIC-TOPSIS. *Energy Conversion and Management*, 215, 112892. DOI: 10.1016/j.enconman.2020.112892

Chapter 2
Bio-Inspired Algorithms-Based Machine Learning and Deep Learning Models in Healthcare 6.0

Shugufta Fatima
https://orcid.org/0009-0005-4396-4947
Stanley College of Engineering and Technology for Women, India

C. Kishor Kumar Reddy
Stanley College of Engineering and Technology for Women, India

Marlia Mohd Hanafiah
Universiti Kebangsaan Malaysia, Malaysia

ABSTRACT

Recent advancements in deep learning (DL) and machine learning (ML) have opened doors for revolutionary applications in healthcare, with better patient care, diagnosis, and treatment. Bio-inspired algorithms, drawing inspiration from natural processes, have gained attention for their potential to enhance ML and DL models in this field. This paper explores current research directions and challenges in utilizing bio-inspired algorithms for advancing healthcare ML and DL models. We investigate their applications and facilitating feature selection, while acknowledging limitations such as scalability, interpretability, and robustness to noisy healthcare data. Ethical considerations surrounding their use in sensitive healthcare contexts are discussed. Through interdisciplinary collaboration and innovative algorithmic approaches, we strive to overcome these challenges and fully unlock the potential of bio-inspired algorithms in healthcare, ultimately aiming to revolutionize healthcare

DOI: 10.4018/979-8-3373-1399-3.ch002

delivery for improved patient outcomes, personalized treatment strategies, and more accurate diagnoses.

1. INTRODUCTION

Computer techniques that take cues from natural phenomena, procedures or actions noticed in biological systems are referred to as bio-inspired or nature-inspired algorithms these algorithms leverage the principles governing the functioning of various natural systems to resolve complex refinement and search problems. They are widely useful in a wide range of fields such as optimization, machine learning, robotics, and computational biology. Some of the key bio-inspired algorithms like Genetic Algorithms, Particle Swarm Optimization, Ant Colony Optimization, Artificial Immune Systems, and Swarm Intelligence **(Kar, 2016)** are discussed below.

Genetic Algorithms (GA)

- Driven by natural selection and genetics, GA simulates the evolutionary basics of survival of the fittest.
- Solutions are represented as chromosomes (often binary strings), Along with a population of solutions that evolves over generations through processes like selection, crossover, and mutation.
- GA is used in optimization problems where the search space is vast and complex, such as in engineering design, scheduling, and financial modeling.

Particle Swarm Optimization (PSO).

- PSO uses a population of particles to search across a search space in quest of the best answer. It is inspired by the social behavior of fish schools and bird flocking.
- Based on its own experiences and those provided by its neighbors, each particle modifies its position.
- PSO is used in optimization problems, particularly in continuous and multidimensional search spaces, such as in engineering design, neural network training, and robotics.

Ant Colony Optimization (ACO).

- ACO mimics the collective behavior of ants in determining the shortest path between their nest and a food source, which is inspired by the foraging activity of ants.
- ACO algorithms use Pheromone pathways deposited via ants to mark good paths, guiding other ants to explore promising routes.

- ACO is commonly applied to optimization problems related to routing, scheduling, and network design.

Artificial Immune Systems (AIS).

- Motivated by the immunological system of humans, AIS algorithms model the processes of recognition, response, and memory formation to detect and eliminate foreign entities (e.g., viruses, bacteria, and tumors).
- AIS algorithms evolve populations of artificial antibodies and antigens, where antibodies represent solutions to a problem and antigens represent problem instances.
- AIS is used in anomaly detection, pattern recognition, and optimization problems, particularly in dynamic and uncertain environments.

Swarm Intelligence.

- This encompasses various algorithms influenced by social insects' collective behavior or other animals, like bees, termites, and birds.
- Examples include bee colony optimization, firefly algorithm, and bat algorithm, each mimicking specific behaviors observed in natural swarms or flocks.
- Swarm intelligence algorithms are used in optimization problems, data clustering, and routing in communication networks.

These bio-inspired algorithms often provide benefits like robustness, adaptability, and scalability, making them effective for solving complex optimization problems in diverse domains.

1.1 Bio-Inspired Algorithms and Their Relevance to Healthcare

In order to address difficult optimization and decision-making problems, bio-inspired algorithms take inspiration from the concepts and mechanisms seen in natural systems. These algorithms imitate biological organisms or processes systems' behavior, such as evolutionary processes, swarm behavior, or genetic mechanisms, to find solutions to various real-world problems. Within the structure of healthcare, bio-inspired algorithms offer innovative approaches to address challenges in diagnosis, treatment planning, patient monitoring, and biomedical research. These algorithms use natural principles to provide resilient and flexible solutions to challenging healthcare issues. Their relevance lies in their ability to effectively tackle optimization, decision-making, and pattern recognition tasks, ultimately leading to improved healthcare outcomes **(Victor Hugo C. de Albuquerque et al., 2020)**, more efficient healthcare systems, and advancements in biomedical research. Here's an overview of some prominent bio-inspired algorithms and their relevance to healthcare.

Genetic Algorithms (GAs). In healthcare, GAs are applied in tasks such as feature selection for medical diagnosis, parameter optimization for treatment planning, and drug discovery through molecular structure optimization.
Particle Swarm Optimization (PSO). PSO has been used in healthcare for tasks such as optimizing treatment plans in radiation therapy, parameter tuning in machine learning models for disease prediction, and optimizing neural network architectures for medical image analysis.
Ant Colony Optimization (ACO). In healthcare, ACO has been applied in tasks such as optimizing hospital scheduling and resource allocation, as well as in medical image registration and segmentation.
Artificial Immune Systems (AIS). AIS has been used in healthcare for tasks such as anomaly detection in medical data, classification of biological data, and optimization of vaccination strategies.
Swarm Intelligence. In healthcare, swarm intelligence algorithms have been employed for tasks such as optimizing hospital workflows, clustering patient data for personalized medicine, and designing adaptive healthcare systems.

1.2 Importance of Deep Learning and Machine Learning in Healthcare

Machine learning (ML) and deep learning (DL) have become indispensable tools in healthcare **(Abdel-Jaber et al., 2022**), offering transformative capabilities that enhance various aspects of medical practice, research, and patient care. Their importance lies in their ability to enhance diagnostic accuracy, enable predictive analytics, facilitate personalized medicine, revolutionize medical imaging analysis, optimize healthcare operations, provide clinical decision support, and drive research and discovery **(Kumari et al., 2023).** Through the utilization of data-driven algorithms, machine learning and DL hold the promise of transforming delivery of healthcare, enhancement of patient outcomes, and progression of medical science **(Rubinger et al., 2023**).

In Figure 1 below let us see some key reasons highlighting their importance.

Figure 1. Importance of machine learning and deep learning in healthcare

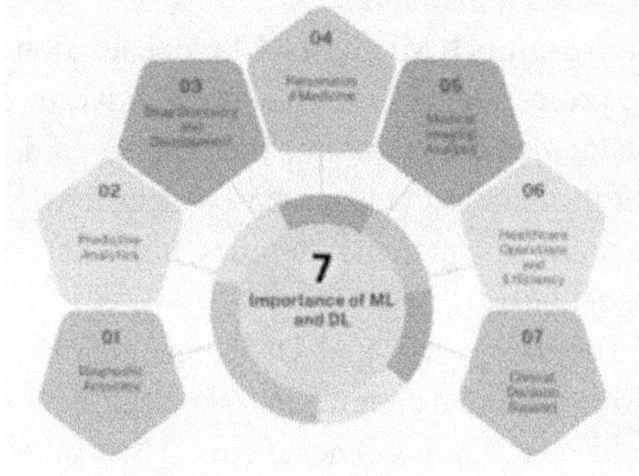

Diagnostic Accuracy. ML and DL Complex medical data, like medical imaging, is an excellent subject for algorithms to analyze genomic sequences, and electronic health records (EHRs), to aid in disease diagnosis. They are able to identify patterns, anomalies, and subtle features that could not be apparent to human observers, thereby improving diagnostic accuracy and reducing errors.
Predictive Analytics. By leveraging historical patient data, ML and DL models can predict disease risk, progression, and treatment outcomes. These predictive analytics empower clinicians to proactively identify at-risk patients, personalize treatment plans, and optimize resource allocation within healthcare systems.
Drug Discovery and Development. ML and DL techniques are transforming the pharmaceutical discovery process by accelerating compound screening, predicting drug-target interactions, and optimizing molecular designs. This expedites the identification of novel therapeutic compounds and streamlines the development of new drugs, potentially leading to more effective treatments for various diseases.
Personalized Medicine. ML and DL enable the analysis of enormous volumes of patient data, including genetic data, clinical indicators, and lifestyle factors, in order to customize medical treatments for specific individuals. Personalized medicine seeks to maximize therapeutic efficacy, reduce side effects, and enhance patient outcomes by taking into account each patient's particular traits and medical history.
Medical Imaging Analysis. Medical image analysis has been transformed by the use of convolutional neural networks (CNNs), in particular, which automate processes like object detection, disease categorization, and image segmentation. Large-scale medical imaging datasets-trained deep learning models can help

radiologists analyze scans more quickly and accurately, which helps speed up diagnosis and treatment planning.

Clinical Decision Support. ML-based Clinical decision support technologies give healthcare professionals instantaneous guidance. providers by synthesizing patient data, medical literature, and clinical guidelines. These systems assist clinicians in making evidence-based decisions, selecting appropriate treatments, and predicting patient outcomes, ultimately improving the quality and safety of patient care.

Research and Discovery. ML and DL techniques facilitate biomedical research by uncovering hidden patterns in large-scale datasets, identifying novel biomarkers, and generating hypotheses for further investigation. These tools accelerate scientific discovery, fuel innovation in healthcare, and contribute to a deeper understanding of disease mechanisms.

1.3 Motivation for Integrating Bio-Inspired Approaches into Existing Models

The motivation for integrating bio-inspired approaches into existing models **(Yang & Karamanoglu, 2013)** stems from their robustness, adaptability, ability to explore novel solutions, parallelism, scalability, biological relevance, cross-disciplinary insights, and adaptive learning capabilities. By harnessing the principles of nature, we can enhance the capabilities of machine learning and deep learning models and unlock new opportunities for innovation in healthcare and beyond. Some of the key reasons for incorporating bio-inspired approaches can be seen in Figure 2 below

Figure 2. Key reasons for incorporating bio-inspired approaches

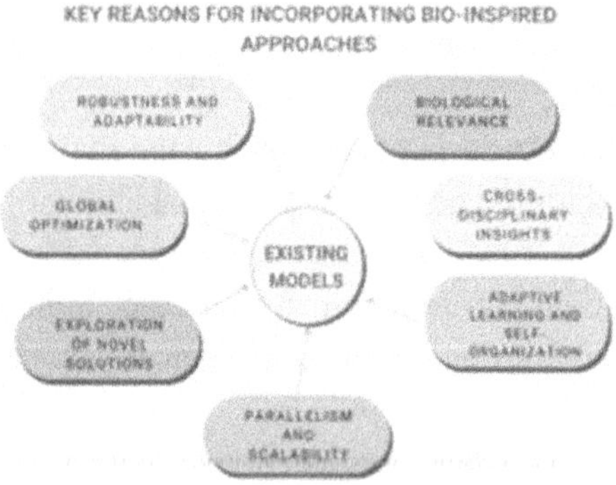

Robustness and Adaptability. Bio-inspired algorithms, such as genetic algorithms and swarm intelligence, have evolved over millions of years to effectively resolve challenging issues in dynamic and uncertain environments. By integrating these algorithms into existing models, we can imbue them with robustness and adaptability, allowing them to handle noisy or incomplete data and adapt to changing conditions in healthcare settings.

Global Optimization. Bio-inspired algorithms excel at finding globally optimal solutions to optimization problems by exploring a vast search space efficiently. In healthcare, where the search for optimal treatment plans, diagnostic criteria, and biomarkers can be highly complex, bio-inspired approaches offer the potential to identify superior solutions that may elude traditional optimization techniques.

Exploration of Novel Solutions. Nature-inspired algorithms often generate diverse and innovative solutions by mimicking the evolutionary processes observed in biological systems. Integrating bio-inspired approaches into existing models enables the exploration of novel solutions that may not be evident through conventional methods, leading to breakthroughs in disease diagnosis, treatment, and biomedical research.

Parallelism and Scalability. Many bio-inspired algorithms exhibit inherent parallelism and scalability, making them well-suited for high-performance computing environments and distributed systems. By harnessing the parallel processing capabilities of bio-inspired approaches, we can accelerate the optimization process and handle large-scale healthcare datasets more efficiently.

Biological Relevance. The foundation of bio-inspired algorithms is developed from concepts found in natural systems, including immunological reactions, social behavior, and genetic evolution. By leveraging these biological analogies, we can develop models that better capture the underlying mechanisms of complex biological processes, leading to more interpretable and physiologically relevant solutions in healthcare.

Cross-Disciplinary Insights. Integrating bio-inspired approaches into existing models fosters interdisciplinary collaboration between computer science, biology, medicine, and other fields. This cross-disciplinary synergy enables researchers to draw inspiration from diverse domains, uncover new insights, and develop innovative solutions that address multifaceted healthcare challenges from holistic perspectives.

Adaptive Learning and Self-Organization. Algorithms that draw inspiration from biological structures and functions, such neural networks, demonstrate the ability to learn adaptively and self-organize. By incorporating these principles into machine learning and deep learning models, we can create systems that learn from experience, adapt to changing environments, and autonomously improve their performance over time, mirroring the resilience and flexibility observed in living organisms.

2. BIO-INSPIRED ALGORITHMS IN HEALTHCARE

Algorithms inspired by biology are essential for addressing many constantly changing problems in healthcare. By providing powerful computational tools that leverage the principles of nature to optimize models, handle complex data, and enable personalized and innovative approaches to patient care.

Bio-inspired algorithms are needed in healthcare **(Gill & Buyya, 2019)** for several reasons.

Complexity of Healthcare Problems. Healthcare problems are often complex, multifaceted, and dynamic. Bio-inspired algorithms offer innovative solutions by mimicking the adaptive and problem-solving mechanisms found in nature, which can help tackle these complex challenges more effectively.

Optimization of Models and Processes. In healthcare, there is a continuous need for optimization, whether it's optimizing machine learning models for diagnostic accuracy or optimizing clinical processes for efficiency. Bio-inspired algorithms, such as particle swarm optimization and genetic algorithms are well-suited for optimization tasks, offering robust and efficient solutions.

Handling Noisy and Incomplete Data. Healthcare data can be noisy, heterogeneous, and incomplete, posing challenges for traditional machine learning approaches. Bio-inspired algorithms, inspired by the robustness and adaptability of biological systems, can help mitigate these challenges by providing mechanisms for handling noisy and incomplete data effectively.

Personalized Medicine and Treatment Planning. Bio-inspired algorithms can facilitate the development of personalized medicine approaches by analyzing large-scale biological and clinical data to identify individualized treatment plans. These algorithms can model complex biological interactions and patterns, resulting in more accurate and successful treatment plans that are customized to each patient's particular needs.

Exploration of Novel Solutions. By drawing inspiration from natural processes, bio-inspired algorithms offer a fresh perspective for exploring novel solutions to healthcare problems. They can uncover unconventional approaches that may not have been considered using traditional computational methods, resulting in innovations in the diagnosis, treatment, and provision of healthcare.

2.1 Review of Prominent Bio-Inspired Algorithms and Their Applications in Healthcare

Ant colony optimization, particle swarm optimization, and genetic algorithms are prominent bio-inspired algorithms with diverse applications in healthcare which can be seen in Figure 3 below. These algorithms offer innovative solutions to complex optimization problems **(Binitha & Sathya, 2012),** ranging from feature selection and treatment optimization to medical image analysis and disease surveillance. By leveraging the principles of nature, bio-inspired algorithms contribute to advancements in healthcare delivery, biomedical research, and patient care.

Figure 3. Bio-inspired algorithms and their applications in healthcare

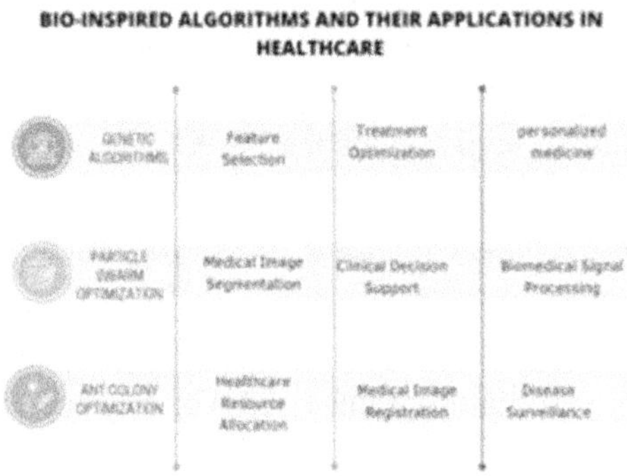

Here, we review three prominent bio-inspired algorithms—Ant colony optimization, particle swarm optimization, and genetic algorithms and their specific applications in healthcare.

Genetic Algorithms (GAs). Genetic algorithms are methods of optimization that draw inspiration from natural selection and genetics. They work by imitating the processes of biological evolution by developing a population of candidate solutions through crossover, mutation, and selection.

Applications in Healthcare.

Feature Selection. In the medical field, genetic algorithms (GAs) are frequently employed for feature selection. This is especially true when analyzing high-dimensional medical data, including gene expression profiles or imaging data. GAs assist in enhancing the performance of machine learning models for illness prognosis and diagnosis by choosing pertinent features.

Treatment Optimization. GAs are applied to optimize treatment plans in various healthcare domains, including radiation therapy, chemotherapy scheduling **(Yaqoob et al., 2023),** and personalized medicine. By exploring the space of possible treatment options, GAs help identify optimal dosing regimens and treatment strategies tailored to individual patient characteristics.

Particle Swarm Optimization (PSO). Particle swarm optimization draws inspiration from the collective behavior of fish schools and bird flocks, where individuals modify their placements based on group knowledge as well as personal experience.

Applications in Healthcare.

Medical Image Segmentation. PSO has been employed for image segmentation tasks in diagnostic imaging procedures including CT and MRI scans. By optimizing segmentation parameters, PSO helps delineate anatomical structures and identify regions of interest for diagnosis and treatment planning.

Clinical Decision Support. Clinical decision support systems, which help medical professionals make evidence-based decisions, are optimized using PSO-based algorithms. By synthesizing patient data and clinical guidelines, PSO helps prioritize diagnostic tests, recommend treatment options, and predict patient outcomes.

Biomedical Signal Processing. PSO is applied in the analysis of biomedical signals, such as electrocardiograms (ECG) or electroencephalograms (EEG). By optimizing signal processing algorithms, PSO enhances the detection of abnormal patterns and physiological events, aiding in the diagnosis of cardiovascular diseases and neurological disorders.

Ant Colony Optimization (ACO). The shortest route to a food source is determined by ants using pheromone trails to communicate with one another, a process that serves as the model for ant colony optimization.

Applications in Healthcare.

Healthcare Resource Allocation. ACO-based algorithms are used to optimize healthcare resource allocation, including hospital scheduling, staff assignment, and patient flow management. By finding efficient routes and schedules, ACO helps minimize waiting times, reduce resource bottlenecks, and improve patient access to care.

Medical Image Registration. ACO is employed for image registration tasks in medical imaging, where multiple images need to be aligned and fused for diagnosis or treatment planning. By optimizing registration parameters, ACO enhances the spatial alignment of images and improves the accuracy of subsequent analysis.

Disease Surveillance. Disease surveillance systems use ACO-based algorithms to track the spread of infectious diseases and pinpoint outbreak hotspots.. By optimizing surveillance strategies and resource allocation, ACO helps public health authorities monitor disease trends, implement targeted interventions, and mitigate the spread of contagious pathogens.

2.2 How These Algorithms Mimic Biological Processes to Optimize Model Performance and Decision-Making in Healthcare Tasks

Despite their distinct mechanisms, these algorithms share common principles in mimicking biological processes as seen in Figure 4 below to optimize model performance and decision-making in healthcare tasks and offer innovative solutions to complex optimization problems, contributing to advancements in healthcare delivery, biomedical research, and patient care.

Figure 4. How BIA's mimic biological processes

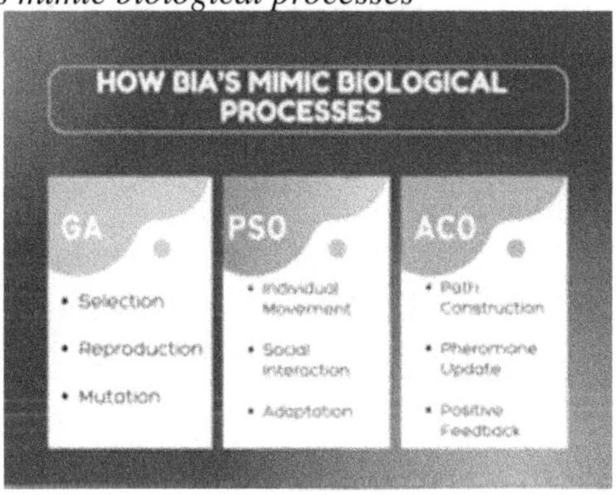

Genetic Algorithms (GAs).

They work by mimicking how a population of potential solutions to an issue evolves, with individuals (solutions) undergoing selection, reproduction (crossover), and mutation to produce offspring that inherit beneficial traits from their parents.

In healthcare tasks, GAs mimic biological evolution by.

Selection. Individuals with higher fitness scores, representing better solutions to the problem at hand (e.g., accurate disease diagnosis or GAs are inspired by genetics and the process of natural selection. They function by imitating the way a population of possible answers to a problem develops.
Reproduction (Crossover). GAs performs crossover operations to exchange genetic information between selected individuals, leading to the creation of new candidate solutions that combine features from multiple parents.
Mutation. Occasionally, GAs introduce random changes (mutations) to candidate solutions, allowing for investigating fresh areas in the solution space and preventing premature convergence to suboptimal solutions.
Particle Swarm Optimization (PSO).

In PSO model every particle serves as a potential answer to the optimization problem, and the swarm as a whole works together to find the optimal answer. In healthcare tasks, PSO mimics collective behavior by **(Yang & Karamanoglu, 2013)**.

Individual Movement. Particles alter their positions in the solution space according to their velocities, which are determined by their individual best-known positions (personal best) and the swarm's best-known positions (global best).
Social Interaction. PSO allows particles to communicate and share information with each other, enabling them to collectively converge towards promising regions of the solution space.
Adaptation. PSO facilitates effective exploration and exploitation of the solution space by dynamically adjusting the velocity and movement of particles based on their own experience and the collective knowledge of the swarm.
Ant Colony Optimization (ACO).

In ACO, artificial ants construct solutions to optimization problems by iteratively building and improving paths based on pheromone concentrations and heuristic information. In healthcare tasks, ACO mimics ant foraging behavior by.

Path Construction. Artificial ants construct solutions by iteratively selecting paths through the solution space, aided by heuristic data and pheromone trails related to the problem domain.
Pheromone Update. Ants leave behind pheromone on the pathways they follow; the quantity left behind is correlated with the caliber of the solution. Pheromone trails are modified over time in response to the caliber of solutions that particular ants find.

Positive Feedback. High-quality solutions attract more ants to explore the corresponding paths, reinforcing positive feedback loops and leading to the discovery of optimal or near-optimal solutions.

3. ENHANCING MACHINE LEARNING MODELS WITH BIO-INSPIRED ALGORITHMS

Improving machine learning models with bio-inspired algorithms represents a frontier in computational intelligence, leveraging nature's optimization strategies to tackle complex problems. Bio-inspired algorithms, drawing from evolutionary processes, social behaviors, and immune system dynamics, offer novel approaches for optimizing model parameters, feature selection, and architecture design. For instance, genetic algorithms mimic natural selection to evolve solutions over generations, while ant colony optimization simulates ants' foraging behavior to find optimal paths. Integrating these algorithms with machine learning models not only enhances their performance but also imbues them with adaptability and robustness across various domains, from healthcare to finance and beyond.

In healthcare, the integration of BIA's with machine learning holds particular promise, offering innovative solutions for analysis of medical image, disease diagnosis, and treatment plans personalized for patients **(Kumari et al., 2023)**. By harnessing the principles of nature, these algorithms optimize model performance and facilitate the discovery of insights from complex biological and clinical data. However, challenges such as scalability, interpretability, and ethical considerations persist, requiring further R and D efforts. As the field continues to evolve, the synergy between algorithms that is bio-inspired and machine learning models promises to revolutionize problem-solving approaches, paving the way for more efficient and effective solutions to real-world challenges.

3.1 How Bio-Inspired Algorithms Can Improve Traditional Machine Learning Models for Tasks Such as Disease Diagnosis, Prognosis, and Treatment Planning

Bio-inspired algorithms offer innovative approaches to improve traditional machine learning models for disease diagnosis, prognosis, and treatment planning in healthcare. By optimizing model parameters, enhancing feature selection, and facilitating personalized risk assessment, bio-inspired algorithms contribute to more accurate, interpretable, and adaptive healthcare decision-making processes.

Here's an exploration of how bio-inspired algorithms can improve these traditional models in each of these areas.

Disease Diagnosis.

They can enhance traditional machine learning models for disease diagnosis by improving feature selection, model optimization, and interpretability.

Feature Selection. Genetic algorithms (GAs) can be used to optimize feature selection in SVMs or decision trees by identifying the most relevant features from large-scale biomedical datasets. GAs iteratively selects subsets of features that maximize classification performance, improving the discriminative power of the models.
Model Optimization. Particle swarm optimization (PSO) can optimize hyper parameters in SVMs or decision trees, such as kernel parameters or tree depth, to maximize classification accuracy. PSO iteratively adjusts model parameters based on performance feedback, leading to better generalization and robustness.
Interpretability. Ant colony optimization (ACO) can improve the interpretability of decision trees by guiding the selection of optimal splitting criteria at each node. ACO prioritizes features that contribute most to classification accuracy, resulting in decision trees that are easier to interpret and understand by healthcare practitioners.
Prognosis.

They can aid in prognosis by enhancing the predictive capabilities of traditional machine learning models and facilitating personalized risk assessment.

Model Fusion. Genetic algorithms can be used to fuse multiple traditional models, such as SVMs or decision trees, into ensemble models that capture diverse aspects of patient data. By combining complementary predictive models, genetic algorithms improve the robustness and accuracy of prognosis predictions.
Temporal Modeling. Particle swarm optimization can optimize temporal modeling techniques, like hidden Markov models (HMMs) or recurrent neural networks (RNNs) for disease progression prediction. PSO adjusts model parameters to capture temporal dependencies in longitudinal patient data, enabling more accurate prognosis over time.
Personalized Risk Assessment. Ant colony optimization can tailor prognostic models to individual patient characteristics by optimizing personalized risk scores. ACO optimizes model parameters based on patient-specific features, allowing for personalized risk assessment and treatment planning.
Treatment Planning.

They can optimize treatment planning by automating decision-making processes and identifying optimal intervention strategies.

Treatment Optimization. Genetic algorithms can optimize treatment parameters, such as drug dosages or radiation schedules, to maximize treatment efficacy while minimizing adverse effects. GAs search for optimal treatment plans by iteratively adjusting dosage levels or treatment modalities based on patient responses and clinical outcomes.
Adaptive Learning. Particle swarm optimization can facilitate adaptive treatment planning by optimizing dynamic treatment strategies over time. PSO adjusts treatment parameters based on real-time patient data and feedback, enabling adaptive learning and personalized treatment optimization.
Resource Allocation. Ant colony optimization can optimize resource allocation in healthcare systems by identifying optimal patient scheduling and resource utilization strategies. ACO guides the allocation of healthcare resources, such as hospital beds or medical equipment, to maximize patient outcomes and minimize costs.

3.2 Case Studies Demonstrating the Effectiveness of Integrating Bio-Inspired Approaches Into Healthcare Applications

They highlight the effectiveness of integrating bio-inspired approaches into healthcare applications, resulting in more effective healthcare delivery, better treatment outcomes, and greater patient care (**Al-Tawil et al., 2023**). Here are a few case studies demonstrating the effectiveness of integrating bio-inspired approaches into healthcare applications.

Optimization of Radiation Therapy Treatment Plans using Genetic Algorithms.
Case Study. Researchers at a cancer treatment center integrated genetic algorithms (GAs) into the optimization process for radiation therapy treatment planning.
Approach. GAs were used to optimize the distribution of radiation dosage across tumor volumes while minimizing the exposure of healthy tissues to radiation.
Results. The integration of GAs significantly improved the quality of treatment plans, leading to better tumor coverage and reduced toxicity to surrounding organs. Patients receiving optimized treatment plans experienced fewer side effects and better treatment outcomes.
Personalized Drug Dosage Optimization using Particle Swarm Optimization.

Case Study. A pharmaceutical company utilized particle swarm optimization (PSO) to optimize personalized drug dosages for patients with chronic conditions.

Approach. PSO was applied to determine optimal drug dosages based on patient-specific factors such as age, weight, genetic markers, and disease progression.

Results. The integration of PSO led to the development of personalized dosing regimens that maximized drug efficacy while minimizing adverse effects. Patients receiving optimized dosages showed improved treatment adherence and better symptom management compared to standard dosing protocols.

Medical Image Segmentation using Ant Colony Optimization.

Case Study. Researchers in medical imaging applied ant colony optimization (ACO) to improve the brain tumor segmentation using magnetic resonance imaging (MRI) scans.

Approach. ACO was used to optimize the parameters of image segmentation algorithms, such as thresholding and region growing, to accurately delineate tumor boundaries.

Results. The integration of ACO resulted in more precise and reproducible tumor segmentations, facilitating treatment planning and monitoring for patients with brain tumors. Clinicians benefited from enhanced visualization of tumor morphology and better quantitative assessment of tumor size and growth over time.

Healthcare Resource Allocation using Swarm Intelligence.

Case Study. A hospital management system implemented swarm intelligence techniques to optimize resource allocation and patient scheduling.

Approach. Swarm intelligence algorithms were employed to allocate hospital beds, operating rooms, and staff resources efficiently.

Results. By integrating swarm intelligence into resource allocation algorithms, the hospital achieved better patient flow, reduced waiting times, and improved staff utilization. Healthcare providers were able to deliver timely and high-quality care, leading to enhanced patient satisfaction and operational efficiency.

4. DEEP LEARNING AND BIO-INSPIRED ALGORITHMS FUSION

It describes the fusion of optimization methods based on biological processes seen in nature with deep learning techniques. Deep learning enables tasks like picture recognition, audio recognition, and natural language processing by using multiple-layered artificial neural networks to learn complicated representations from data. In contrast, bio-inspired algorithms utilize biological processes like immune systems, swarm behavior, and evolution as models for solving optimization and search issues.

The fusion of deep learning and BIA's involves leveraging the strengths of both approaches to address challenges in various domains **(Reddy et al., 2024).** For example, bio-inspired algorithms can be applied to enhance the standards of deep neural networks, improving their training efficiency and generalization performance. They can also enhance the robustness of deep learning models by providing mechanisms for adaptation and exploration in complex and dynamic environments.

4.1 Investigation into the Fusion of Deep Learning Architectures With Bio-Inspired Algorithms to Enhance Feature Extraction, Model Optimization, and Interpretability in Healthcare

Convolutional and recurrent neural networks (RNNs) are examples of deep learning architectures that have shown remarkable performance in various healthcare applications, including medical imaging analysis, disease diagnosis, and predictive modeling. By leveraging the complementary strengths of deep learning's representation learning capabilities and bio-inspired algorithms' optimization prowess, this fusion approach has the potential to improve feature extraction, model optimization, and interpretability in healthcare applications which can be seen in **Table 1** below. This investigation delves into the potential benefits and challenges of such fusion, exploring its implications for advancing healthcare AI systems.

Table 1. Fusion of deep learning architectures with bio-inspired algorithms

Aspect	Description	Example
Feature Extraction	Incorporating bio-inspired algorithms into deep learning architectures to enhance feature extraction from complex healthcare data.	Using genetic algorithms to optimize the feature selection process in convolutional neural networks for medical image analysis.
Model Optimization	Combining bio-inspired algorithms with deep learning models to optimize model parameters, improve convergence, and enhance performance on healthcare tasks.	Applying particle swarm optimization to tune hyper parameters of recurrent neural networks for time-series analysis in healthcare.
Interpretability	Utilizing bio-inspired algorithms to enhance the interpretability of deep learning models in healthcare, making predictions more understandable and trustworthy for clinicians.	Employing ant colony optimization to identify relevant features and explain the decision-making process of deep neural networks in diagnosing diseases.

4.2 Challenges and Opportunities in Integrating Complex Deep Learning Models With Bio-Inspired Optimization Techniques

Figure 5. Challenges and opportunities in integrating complex deep learning models with bio-inspired optimization techniques

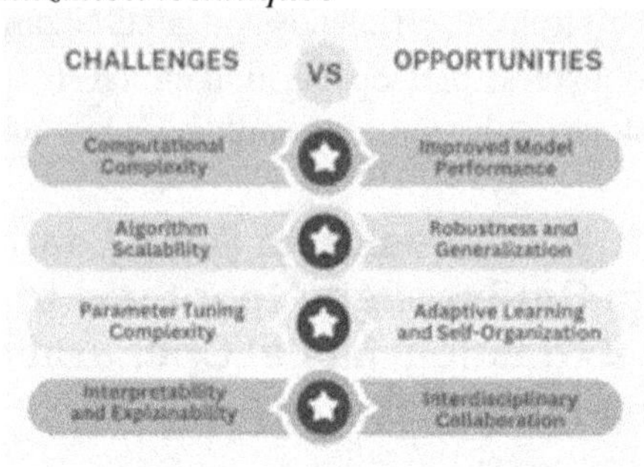

Integrating complex deep learning models with bio-inspired optimization techniques presents both challenges and opportunities in healthcare and other domains. From the above Figure 5 let us explore some of these challenges and opportunities.

Challenges.

Computational Complexity. Deep learning models, especially complex designs such as recurrent neural networks (RNNs) or deep convolutional neural networks (CNNs), often require significant computational resources for training and inference. Integrating bio-inspired optimization techniques may further increase computational complexity, particularly when optimizing hyper parameters or model architectures.

Algorithm Scalability. Optimization methods impacted by nature, include genetic algorithms and particle swarm optimization, may face scalability challenges when applied to large-scale deep learning models and datasets. Scaling these algorithms to handle high-dimensional parameter spaces and massive datasets efficiently is a non-trivial task.

Parameter Tuning Complexity. Deep learning models often have numerous hyper parameters that require careful tuning for optimal performance. Integrating bio-inspired optimization techniques adds another layer of complexity to the parameter tuning process, requiring researchers to navigate a complex search space effectively.

Interpretability and Explain ability. DL models are often criticized for their lack of interpretability and explain ability, which can be further exacerbated when combined with bio-inspired optimization techniques. Understanding the reasoning behind model decisions and interpreting the learned representations becomes challenging in such complex systems.

Opportunities.

Improved Model Performance. Integrating bio-inspired optimization techniques with complex deep learning models can lead to improved model performance, as these techniques are adept at exploring high-dimensional search spaces and finding optimal solutions to complex optimization problems.

Robustness and Generalization. Bio-inspired optimization strategies can improve deep learning models resilience and generalization powers by optimizing hyper parameters and regularization strategies. This can help mitigate overfitting and improve model performance on unseen data.

Feature Representation Learning. Deep learning models excel at learning hierarchical representations from raw data, while bio-inspired optimization techniques can enhance feature extraction and representation learning. This combination can lead to more informative and discriminative feature representations in healthcare applications, such as medical imaging analysis or patient data classification.

Adaptive Learning and Self-Organization. Optimization methods that draw motivation from nature, like swarm intelligence and evolutionary algorithms, exhibit adaptive learning and self-organization capabilities. Integrating these techniques with deep learning models can enable adaptive model optimization, where the model dynamically adjusts its parameters in response to changing data distributions or task requirements.

Interdisciplinary Collaboration. The integration of complex deep learning models with bio-inspired optimization techniques fosters interdisciplinary collaboration between computer science, biology, and healthcare domains. This collaboration enables researchers to leverage insights from both fields and develop innovative solutions to complex healthcare challenges.

5. APPLICATIONS OF BIO-INSPIRED ALGORITHMS IN HEALTHCARE

Bio-inspired algorithms have emerged like powerful tools in revolutionizing healthcare across multiple fronts. Leveraging principles from nature, these algorithms find applications in medical imaging, aiding in the precise detection and characterization of abnormalities from complex scans **(Mandave & Patil, 2023)**. They also facilitate drug discovery, optimizing molecular structures to accelerate the development of novel therapeutics. In disease diagnosis and prognosis, bio-inspired algorithms analyze vast patient datasets to enhance accuracy and personalize treatment plans. Additionally, they streamline healthcare resource management, optimizing processes like ambulance routing and staff scheduling for efficient care delivery. From predictive modeling to personalized medicine, these algorithms pave the way for innovative solutions that enhance patient outcomes and drive progress in healthcare.

5.1 Specific Healthcare Domains Where Bio-Inspired Algorithms Have Shown Promise

They play a critical role in advancing healthcare by optimizing medical processes, improving diagnostic and treatment outcomes, and facilitating personalized and efficient patient care. Their versatility and adaptability make them valuable tools for addressing complex healthcare challenges and driving innovation in biomedical research and clinical practice. Some of the Healthcare domains where bio-inspired algorithms have shown promise can be seen from Figure 6below.

Figure 6. Healthcare domains where bio-inspired algorithms have shown promise

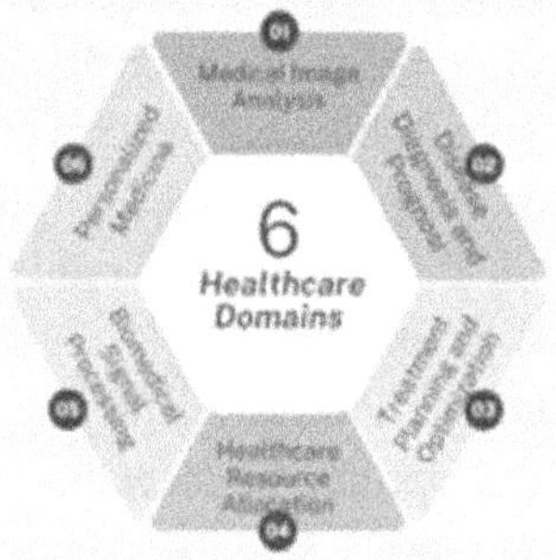

Here are some key applications of bio-inspired algorithms in healthcare.

Medical Image Analysis.
Segmentation. Bio-inspired algorithms such as particle swarm optimization and genetic algorithms are used to optimize image segmentation techniques, enabling accurate delineation of anatomical structures and lesions in medical images (e.g., MRI, CT scans).
Registration. Simulated annealing and ant colony optimization and algorithms facilitate the registration of multiple medical images, aligning them spatially for comparison, fusion, and treatment planning.
Disease Diagnosis and Prognosis.

Feature Selection. Genetic algorithms assist in selecting relevant features from high-dimensional medical datasets, improving the accuracy of machine learning models for disease diagnosis and prognosis.

Predictive Modeling. Swarm intelligence algorithms optimize predictive models for disease progression, treatment response, and patient outcomes by fine-tuning model parameters and ensemble learning techniques.

Treatment Planning and Optimization.

Radiation Therapy. Genetic algorithms optimize treatment plans in radiation therapy, determining optimal radiation dosages and beam angles to reduce harm to the surrounding healthy tissues and optimize tumor control.

Drug Discovery. Artificial immune system algorithms aid in drug discovery by simulating the immune system's recognition of foreign agents, facilitating the identification of novel drug targets and lead compounds.

Healthcare Resource Allocation.

Hospital Scheduling. Ant colony optimization algorithms optimize hospital scheduling by efficiently allocating resources such as operating rooms, medical staff, and patient appointments, reducing patient wait times and improving resource utilization.

Supply Chain Management. Swarm intelligence algorithms optimize supply chain logistics in healthcare, ensuring the timely delivery of medical supplies, medications, and equipment to healthcare facilities.

Biomedical Signal Processing.

Electrocardiogram (ECG) Analysis. Bio-inspired algorithms aid in the analysis of biomedical signals such as ECGs, detecting abnormalities, arrhythmias, and cardiac events with high accuracy and reliability.

Brain-Computer Interfaces. For brain-computer interfaces, evolutionary algorithms optimize feature extraction and classification techniques, allowing people with motor limitations to control assistive equipment in real time.

Personalized Medicine.

Treatment Optimization. By maximizing the effectiveness of treatment and reducing side effects, genetic algorithms enhance individualized treatment plans based on unique patient features such as genetic profiles, biomarkers, and clinical histories.

Disease Risk Assessment. Swarm intelligence algorithms analyze multi-omics data to assess individualized disease risk profiles, enabling early detection, prevention, and intervention strategies.

Healthcare System Optimization.

Workflow Optimization. Ant colony optimization algorithms optimize healthcare workflows by streamlining administrative processes, reducing bottlenecks, and improving overall efficiency in healthcare delivery.

Patient Monitoring. Bio-inspired algorithms facilitate real-time monitoring of patient vital signs, predicting adverse events and enabling early interventions to prevent complications.

5.2 Examination of Real-World Implementations and Their Impact on Clinical Practice and Patient Outcomes

Real-world implementations of bio-inspired algorithms have had a profound influence on patient outcomes and clinical practice in a variety of healthcare domains **(Reddy et al., 2024)**. By leveraging principles from nature, these algorithms have optimized medical processes, improved diagnostic accuracy, personalized treatment approaches, and enhanced overall quality of care, ultimately leading to better health outcomes for patients **(Kumari et al., 2023)**. Figure 7 below helps us in understanding the impacts of real world implementations.

Figure 7. Examination of real-world implementations and their impact on clinical practice and patient outcomes

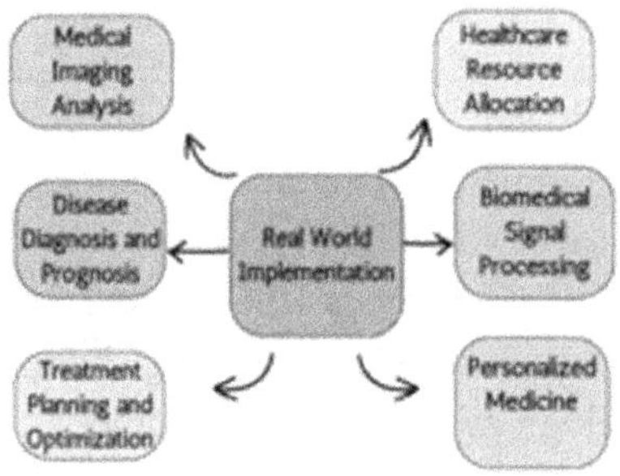

Let's examine some examples.

Medical Imaging Analysis.
Impact. They have been used to optimize image segmentation and registration techniques, leading to more accurate and efficient interpretation of medical images.
Outcome. Improved segmentation and registration algorithms have enhanced the precision of treatment planning in radiation therapy and surgery, resulting in better tumor targeting and reduced damage to surrounding healthy tissues.
Disease Diagnosis and Prognosis.

Impact. Bio-inspired algorithms have facilitated the development of predictive models for disease diagnosis and prognosis, leveraging genetic algorithms, swarm intelligence, and other techniques.
Outcome. These models have made it possible to identify illnesses like cancer, heart problems, and neurological abnormalities early, which has improved patient outcomes and allowed for prompt interventions.
Treatment Planning and Optimization.
Impact. They have been applied to optimize treatment plans in radiation therapy, chemotherapy scheduling, and personalized medicine.
Outcome. Optimized treatment plans have resulted in more precise and effective delivery of therapies, maximizing tumor control while minimizing adverse effects on patients, thereby improving treatment outcomes and quality of life.
Healthcare Resource Allocation.
Impact. Bio-inspired optimization techniques have been deployed to optimize resource allocation in healthcare facilities, including hospital scheduling, staff assignment, and inventory management.
Outcome. Streamlined workflows and efficient resource utilization have reduced patient wait times, enhanced patient throughput, and improved overall quality of care in clinical settings.
Biomedical Signal Processing.
Impact. Bio-inspired algorithms have enhanced the analysis of biomedical signals such as electrocardiograms (ECGs), electroencephalograms (EEGs), and respiratory waveforms.
Outcome. Improved signal processing techniques have enabled more accurate diagnosis of cardiac arrhythmias, seizure disorders, and sleep apnea, resulting in prompt interventions and improved patient condition management.
Personalized Medicine.
Impact. They have made it easier to create individualized treatment plans that take into account the unique traits of each patient as well as their biomarker profiles.
Outcome. Tailored treatment approaches have optimized therapeutic outcomes, minimized adverse effects, and enhanced patient compliance with treatment regimens, leading to better long-term prognosis and disease management.
Healthcare System Optimization.
Impact. Bio-inspired optimization algorithms have optimized healthcare workflows, patient scheduling, and resource allocation in healthcare systems.
Outcome. Enhanced operational efficiency and resource utilization have increased patient happiness with healthcare services, decreased healthcare costs, and better access to care.

6. CHALLENGES AND FUTURE DIRECTIONS

Implementing bio-inspired algorithms within machine learning (ML) and deep learning (DL) frameworks presents a fascinating yet complex landscape ripe with challenges and promising future directions **(Theodosiou & Read, 2023)**. One key challenge lies in bridging the gap between the intricate dynamics of biological systems and the abstraction necessary for computational modeling. Achieving this demands a multidisciplinary approach, combining expertise from biology, computer science, and mathematics. Additionally, scalability remains a hurdle, as bio-inspired algorithms often find it difficult to manage big databases effectively. Furthermore, the interpretability of these algorithms in complex ML and DL models poses significant challenges. Despite these obstacles, the future holds exciting prospects. Advancements in hardware, such as neuromorphic computing, could revolutionize the efficiency and scalability of bio-inspired algorithms. Moreover, integrating principles from evolutionary biology, neuroscience, and ecology can result in the creation of AI systems that are more resilient and adaptable.. Embracing these challenges with interdisciplinary collaboration promises to unlock the full potential of bio-inspired algorithms in shaping the future of ML and DL.

6.1 Identification of Challenges and Limitations in the Adoption of Bio-Inspired Algorithms in Healthcare With Existing Healthcare Systems

Challenges and limitations in the adoption of bio-inspired algorithms in healthcare with existing healthcare systems can be seen in Figure 8 below.

Figure 8. Challenges and limitations in the adoption of bio-inspired algorithms in healthcare with existing healthcare systems

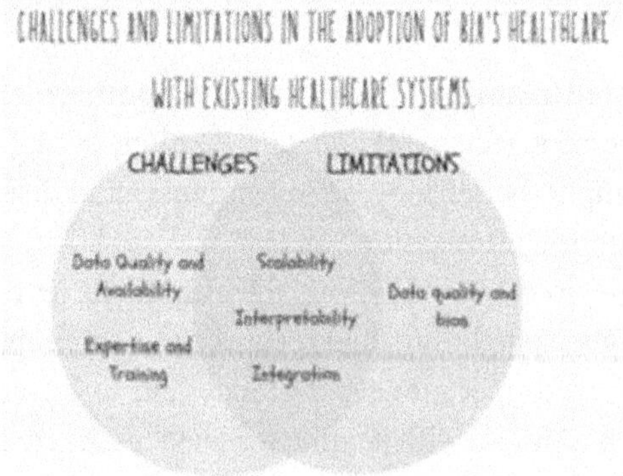

The **challenges f**aced in the adoption of bio-inspired algorithms in healthcare.

Scalability.
Challenge. These algorithms may struggle to handle the scale of healthcare data, which is often vast, heterogeneous, and high-dimensional. Scaling algorithms to process large datasets efficiently while maintaining computational tractability can be challenging.
Impact. Limited scalability can hinder the practical application of bio-inspired algorithms in healthcare settings, restricting their ability to analyze large-scale patient data, optimize complex models, or support real-time decision-making.
Interpretability.
Challenge. These algorithms, such as evolutionary algorithms or swarm intelligence, often produce solutions that are difficult to interpret or explain. Understanding the rationale behind algorithmic decisions is crucial in healthcare, where transparency and interpretability are essential for clinical acceptance.
Impact. Lack of interpretability can undermine trust in algorithmic predictions, impede clinical adoption, and pose challenges for regulatory compliance and patient safety. Healthcare professionals may be hesitant to rely on algorithms they cannot understand or trust.
Integration with Existing Healthcare Systems.
Challenge. Integrating bio-inspired algorithms into existing healthcare systems poses technical, organizational, and regulatory challenges. Compatibility issues, It is necessary to address data interoperability as well as adherence to healthcare standards and laws (such as HIPΛΛ).
Impact. Incompatible systems, data silos, and regulatory barriers can impede the smooth incorporation of algorithms into clinical workflows, hindering their adoption and utility in real-world healthcare settings.
Data Quality and Availability.
Challenge. Bio-inspired algorithms require high-quality, well-curated data for training and validation. However, healthcare data are often noisy, incomplete, and subject to biases, posing challenges for algorithm development and performance evaluation.
Impact. Poor data quality and availability can undermine the reliability and generalizability of bio-inspired algorithms, leading to suboptimal performance, biased predictions, and limited applicability in diverse patient populations or clinical contexts.
Expertise and Training.

Challenge. Leveraging bio-inspired algorithms effectively in healthcare requires expertise in both computational techniques and domain-specific knowledge. Healthcare professionals may lack the necessary training and skills to implement, interpret, and validate these algorithms effectively.
Impact. Insufficient expertise and training can impede the adoption and utilization of bio-inspired algorithms in healthcare, limiting their potential to address clinical challenges, improve patient outcomes, and drive innovation in healthcare delivery and research.

The adoption of bio-inspired algorithms in healthcare indeed presents several challenges and **limitations**. Here's an overview.

Scalability. Many bio-inspired algorithms, such as artificial neural networks or genetic algorithms can be computationally intensive and may struggle to scale effectively when applied to large datasets or complex problems. In healthcare, where vast amounts of data are generated daily, scalability becomes a crucial issue. Processing real-time patient data or large-scale population health data can strain computational resources and increase processing time, limiting the practicality of these algorithms in certain healthcare applications.
Interpretability. Bio-inspired algorithms often operate like "black-box" models, meaning that they provide predictions or solutions without clear explanations of how they arrived at those outcomes. In healthcare, where transparency and interpretability are essential for clinical decision-making, this lack of explain ability can be a significant barrier to adoption. Clinicians may be hesitant to trust algorithmic recommendations if they cannot understand the underlying rationale or if the decision-making process is not transparent. Interpretability issues also raise concerns about accountability and potential biases embedded in the algorithms.
Integration with existing healthcare systems. Healthcare systems are highly complex and diverse, with a wide range of medical devices, electronic health record (EHR) systems, and data standards in use. Integrating bio-inspired algorithms seamlessly into existing healthcare infrastructure poses technical challenges related to data interoperability, compatibility with different software systems, and adherence with regulatory standards (e.g., HIPAA in the United States). Moreover, healthcare providers may face resistance or reluctance to adopt new technologies due to concerns about disruptions to workflow, training requirements, and perceived risks to patient privacy and safety.
Data quality and bias. The effectiveness of bio-inspired algorithms in healthcare relies on a strong emphasis on quality, completeness, and representativeness of the data used for training and validation. However, healthcare data are often

heterogeneous, messy, and subject to various biases (e.g., demographic bias, selection bias, or measurement bias). Biases in the data can result in biased predictions or recommendations, exacerbating health disparities and undermining the fairness and equity of healthcare delivery. Addressing data quality issues and mitigating biases require careful data preprocessing, algorithmic fairness assessments, and ongoing monitoring and evaluation of algorithm performance in diverse patient populations.

6.2 Proposal of Future Research Directions to Address These Challenges and Unlock the Full Potential of Bio-Inspired Approaches in Healthcare

To unlock the complete capability of bio-inspired approaches in healthcare and address the challenges outlined earlier, several future research directions can be proposed.

Scalable and Efficient Algorithms. Develop novel bio-inspired algorithms that are specifically tailored for scalability and efficiency in healthcare applications. This may involve designing algorithms that can leverage distributed computing resources, parallel processing, and optimized hardware architectures (e.g., GPUs, TPUs). Research should also focus on developing techniques for reducing the computational complexity of bio-inspired algorithms without compromising accuracy, such as algorithmic optimizations, feature selection, and dimensionality reduction methods.

Interpretable and Explainable Models. Enhance the interpretability and explain ability of bio-inspired algorithms to improve trust and acceptance among healthcare professionals. Research should explore techniques for generating human-interpretable explanations of algorithmic predictions or decisions, such as model-agnostic interpretability methods, visualization techniques, and post-hoc explanation approaches. Additionally, investigating ways to incorporate domain knowledge and clinical guidelines into algorithmic decision-making processes can improve transparency and alignment with clinical practice standards.

Integration with Healthcare Systems. Develop interoperable and user-friendly solutions for integrating bio-inspired algorithms into existing healthcare systems and workflows. Research efforts should focus on developing standardized interfaces, data exchange protocols, and integration frameworks that enable seamless communication between bio-inspired algorithms and medical devices, electronic health record (EHR) systems, and other healthcare IT infrastructure. Collaboration with healthcare providers, IT vendors, and regulatory agencies is essential to ensure compatibility, compliance, and usability of integrated solutions.

Data Quality Assurance and Bias Mitigation. Address data quality issues and mitigate biases in healthcare datasets to improve the reliability and fairness of bio-inspired algorithms. Future research should explore techniques for data preprocessing, anomaly detection, and bias correction to enhance the quality, wholeness, and representativeness of training using medical data and validation. Additionally, developing robust methods for evaluating algorithmic fairness, identifying and mitigating biases, and ensuring equitable healthcare delivery across diverse patient populations is crucial for promoting health equity and reducing disparities.

Ethical and Regulatory Considerations. Investigate ethical and regulatory frameworks for the responsible development, deployment, and evaluation of bio-inspired algorithms in healthcare. Research needs to take ethical issues into account related to patient privacy, consent, autonomy, and the potential impact of algorithmic decision-making on healthcare outcomes and patient well-being. Collaboration between researchers, policymakers, ethicists, and healthcare stakeholders is necessary to develop guidelines, standards, and best practices for ensuring ethical AI in healthcare and safeguarding patient rights and interests.

By pursuing these future research directions, we can conquer difficulties and attain your fullest potential of bio-inspired approaches in healthcare, ultimately leading to improved patient care, clinical outcomes, and healthcare system efficiency.

6.3 Discussion on the Ethical and Regulatory Considerations Associated With Deploying Bio-Inspired Algorithms in Clinical Settings

Deploying bio-inspired algorithms in clinical settings raises several ethical and regulatory considerations that need careful attention to ensure patient safety, privacy, autonomy, and equitable healthcare delivery. Here's a discussion on some of the key considerations.

Patient Safety and Benefit. One of the primary ethical imperatives in healthcare is to prioritize patient safety and well-being. Before deploying bio-inspired algorithms in clinical settings, thorough validation and evaluation are necessary to ensure that the algorithms are accurate, reliable, and clinically beneficial. Ethical considerations include assessing the potential risks and benefits of algorithmic decision-making compared to standard care practices and ensuring that algorithmic recommendations do not harm patients or compromise their health outcomes.

Transparency and Explain ability. Healthcare providers and patients have the right to understand how algorithmic decisions are made and the rationale behind them. Ensuring transparency and explain ability of bio-inspired algorithms is essential for building trust, fostering shared decision-making between clinicians and patients, and enabling accountability for algorithmic recommendations. Ethical considerations include providing clear explanations of algorithmic predictions or decisions in a language that is understandable to patients and clinicians and disclosing any limitations or uncertainties associated with algorithmic outputs.

Patient Privacy and Data Security. Personal health information is contained in extremely sensitive and confidential healthcare data, which must be shielded from misuse, disclosure, and unauthorized access. Ensuring bio-inspired algorithms adhere to relevant privacy laws (such as HIPAA in the US) and data security guidelines are one way to address ethical concerns. Strong data encryption, audit trails, anonymization methods, and access controls can all be used to protect patient privacy and lessen the chance of data breaches or illegal access.

Algorithmic Bias and Fairness. Bio-inspired algorithms trained on healthcare data may inadvertently perpetuate or exacerbate existing biases and disparities in healthcare delivery. Ethical considerations include assessing and mitigating biases in algorithmic decision-making processes to ensure fairness, equity, and non-discrimination across diverse patient populations. To mitigate algorithmic bias and advance health equity, strategies such as implementing fairness-aware machine learning approaches, carrying out bias audits, and utilizing varied and representative training data should be used.

Informed Consent and Autonomy. Patients have the right to make informed decisions about their healthcare and to understand the implications of algorithmic decision-making on their treatment options and outcomes. Ethical considerations include obtaining informed consent from patients before using bio-inspired algorithms to guide clinical decision-making, ensuring that patients are adequately informed about the nature, purpose, risks, and benefits of algorithmic interventions, and respecting patients' autonomy to accept or decline algorithmic recommendations.

Regulatory Compliance and Oversight. Healthcare regulatory agencies play a crucial role in ensuring the safety, effectiveness, and ethical use of bio-inspired algorithms in clinical practice. Ethical considerations include complying with regulatory requirements for the development, validation, deployment, and monitoring of healthcare AI systems, obtaining regulatory approvals or clearances where necessary, and establishing mechanisms for ongoing oversight, evaluation, and reporting of algorithm performance and outcomes.

Addressing these ethical and regulatory considerations requires collaboration and engagement among healthcare professionals, researchers, policymakers, ethicists, patients, and other stakeholders. By adopting a multi-stakeholder approach and We can utilize the potential of bio-inspired algorithms to enhance patient care, clinical outcomes, and healthcare system efficiency while maintaining ethical standards and safeguarding patient rights. These can be achieved by upholding ethical principles of beneficence, non-malfeasance, justice, and respect for patient autonomy.

7. CONCLUSION

7.1 Summary of Key Findings and Insights

Bio-inspired algorithms have enormous potential to transform the way that healthcare is provided, enhance patient outcomes, and progress medical research. For bio-inspired algorithms to reach their full potential and to tackle the intricate problems that face contemporary healthcare systems, further research and innovation in this area are necessary **(Anisha et al., 2022)** .Key findings and insights regarding the application of bio-inspired algorithms in healthcare include.

Improved Diagnostic Accuracy. When it comes to identifying different medical diseases, bio-inspired algorithms—like artificial neural networks and genetic algorithms—have proven to be more effective than conventional techniques. These algorithms can identify diseases earlier and more accurately by analyzing intricate patterns in medical data, such as genetic sequences or imaging scans.

Personalized Treatment Planning. Personalized treatment plans that are based on the genetic composition, medical history, and lifestyle aspects of each patient can be created thanks to bio-inspired algorithms. These algorithms are able to enhance treatment plans and enhance patient outcomes through the analysis of big datasets and the identification of biomarkers or risk factors unique to each patient.

Drug Discovery and Development. Due to their ability to anticipate drug-target interactions, identify promising therapeutic compounds, and optimize drug design, bio-inspired algorithms are a key component in the acceleration of drug discovery and development processes. With the aid of these algorithms, scientists may more effectively comb through a large chemical space and identify potential new medication candidates.

Healthcare Resource Optimization. Bio-inspired algorithms can optimize healthcare resource allocation, such as hospital bed management, staff scheduling, and inventory management, to improve operational efficiency and patient flow. By analyzing real-time data and predicting demand patterns, these algorithms help healthcare facilities optimize resource utilization and enhance patient access to care.

Disease Surveillance and Epidemiological Modeling. Bio-inspired algorithms contribute to disease surveillance efforts and epidemiological modeling by analyzing large-scale healthcare data, such as electronic health records and public health databases. These algorithms can identify disease outbreaks, predict disease spread, and inform public health interventions to mitigate the impact of infectious diseases and other health threats.

Challenges and Limitations. Despite their potential benefits, the adoption of bio-inspired algorithms in healthcare faces several challenges, including scalability, interpretability, integration with existing healthcare systems, data quality assurance, and ethical and regulatory considerations. Addressing these challenges requires interdisciplinary collaboration, methodological innovations, and ethical guidelines to ensure the responsible and effective use of bio-inspired algorithms in clinical practice.

7.2 Call to Action for Researchers to Continue Exploring Bio-Inspired Algorithms for Advancing Machine Learning and Deep Learning Models in Healthcare

Bio-inspired algorithms have enormous promise to advance deep learning and machine learning models in the healthcare industry. The promise of bio-inspired algorithms to revolutionize healthcare delivery and enhance patient outcomes can be fully realized by researchers through interdisciplinary collaboration, practical application of ethical principles, exchange of knowledge and best practices, and real-world healthcare difficulties and outcomes addressed **(Reddy et al., 2024)**.The following is a request that scientists carry out more study on bio-inspired algorithms in the field of medicine.

Push the Boundaries of Innovation. Researchers should strive to push the boundaries of innovation by exploring novel bio-inspired algorithms and adapting existing algorithms to address specific challenges in healthcare. This could involve developing hybrid algorithms that combine elements of multiple bio-inspired approaches or designing algorithms that mimic complex biological processes more accurately.

Collaborate Across Disciplines. Collaboration across disciplines is essential for advancing the field of bio-inspired algorithms in healthcare. Researchers from computer science, biomedical engineering, medicine, and other relevant fields should collaborate to leverage their diverse expertise and perspectives. Interdisciplinary research teams can tackle complex healthcare problems more effectively and develop innovative solutions that bridge the gap between theory and practice.

Address Real-World Healthcare Challenges. Researchers should focus on addressing real-world healthcare challenges through the application of bio-inspired algorithms. This includes developing algorithms for clinical decision support, disease diagnosis and prognosis, personalized medicine, drug discovery, healthcare resource optimization, and epidemiological modeling. By targeting high-impact healthcare applications, researchers can demonstrate the tangible benefits of bio-inspired algorithms and drive adoption in clinical practice.

Ensure Ethical and Responsible AI. Ethical considerations should be at the forefront of research efforts in applying bio-inspired algorithms to healthcare. Researchers must prioritize patient safety, privacy, autonomy, and fairness throughout the algorithm development lifecycle. This includes addressing algorithmic biases, ensuring transparency and interpretability, obtaining informed consent, and complying with regulatory requirements. By upholding ethical principles and promoting responsible AI practices, researchers can build trust with healthcare stakeholders and facilitate the responsible deployment of bio-inspired algorithms in clinical settings.

Share Knowledge and Best Practices. Researchers should actively share their findings, methodologies, and best practices with the broader scientific community. This includes publishing research papers, participating in conferences and workshops, contributing to open-source projects, and collaborating with industry partners. By fostering a culture of knowledge sharing and collaboration, researchers can accelerate progress in the field of bio-inspired algorithms and drive innovation in healthcare.

7.3 Outlook on the Transformative Impact of Bio-Inspired Approaches on the Future of Healthcare Delivery and Patient Care

Precision medicine, early disease detection and prevention, optimal resource allocation, improved clinical decision support, and population health management are just a few of the revolutionary effects of bio-inspired approaches that will reshape the entire healthcare landscape. Healthcare systems have the potential to transform into more individualized, proactive, and efficient care models that emphasize patient outcomes, enhance community health, and stimulate innovation in healthcare delivery by utilizing computational intelligence and biological inspiration.

Here's an outlook on how these approaches will shape the future.

Precision and Personalized Medicine. The creation of extremely precise and customized treatment plans based on each patient's unique genetic composition, medical background, lifestyle choices, and environmental influences is made possible by bio-inspired algorithms. Through the process of scrutinizing copious quantities of patient data and discerning nuanced patterns and associations, these algorithms enable medical professionals to provide focused interventions that optimize treatment effectiveness while mitigating harmful consequences.

Early Disease Detection and Prevention. Bio-inspired algorithms have the potential to revolutionize disease detection and prevention efforts by enabling earlier and more accurate diagnosis of various medical conditions. By analyzing multi-modal data from sources such as medical imaging, genomic sequencing, wearable sensors, and electronic health records, these algorithms can identify disease biomarkers and risk factors at the earliest stages, facilitating timely interventions and preventive measures to improve patient outcomes and reduce healthcare costs.

Optimized Healthcare Delivery and Resource Allocation. Bio-inspired algorithms optimize healthcare delivery and resource allocation by predicting patient demand, optimizing scheduling and staffing, and streamlining workflow processes. These algorithms make it possible for medical facilities to better manage their resources, shorten wait times, and increase patient access to care.. By maximizing the utilization of healthcare resources, bio-inspired algorithms enhance healthcare system efficiency and resilience, particularly in the face of growing demand and resource constraints.

Enhanced Clinical Decision Support. Bio-inspired algorithms provide valuable decision support tools for healthcare providers, enabling them to make more informed and evidence-based clinical decisions. These algorithms analyze complex patient data, medical literature, and clinical guidelines to generate personalized treatment recommendations, assist in diagnosis and prognosis, and identify optimal care pathways. By augmenting clinical expertise with computational intelligence, bio-inspired algorithms empower healthcare providers to deliver high-quality, patient-centered care and improve clinical outcomes.

Public Health Surveillance and Population Health Management. Bio-inspired algorithms contribute to public health surveillance efforts and population health management by analyzing large-scale healthcare data to monitor disease trends, detect outbreaks, and inform public health interventions. These algorithms enable proactive and data-driven approaches to disease prevention and control, facilitating early intervention strategies, targeted health promotion initiatives, and resource allocation decisions. By leveraging real-time data analytics and predictive modeling, bio-inspired algorithms support population-level health monitoring and improve health outcomes across diverse communities.

REFERENCES

Abdel-Jaber, H., Devassy, D., Al Salam, A., Hidaytallah, L., & El-Amir, M. (2022). A review of deep learning algorithms and their applications in healthcare. *Algorithms*, 15(2), 71. DOI: 10.3390/a15020071

Al-Tawil, M., Mahafzah, B. A., Al Tawil, A., & Aljarah, I. (2023). Bio-inspired machine learning approach to Type 2 Diabetes Detection. *Symmetry*, 15(3), 764. DOI: 10.3390/sym15030764

Anisha, P. R., Reddy, C. K. K., Nguyen, N. G., Bhushan, M., Kumar, A., & Hanafiah, M. M. (Eds.). (2022). *Intelligent Systems and Machine Learning for Industry: Advancements, Challenges, and Practices.* CRC Press.

Binitha, S., & Sathya, S. S. (2012). A survey of bio inspired optimization algorithms. *International journal of soft computing and engineering,* 2(2), 137-151.

de Albuquerque, V. H. C., Gupta, D., De Falco, I., Sannino, G., & Bouguila, N. (2020). Special issue on Bio-inspired optimization techniques for Biomedical Data Analysis: Methods and applications. *Applied Soft Computing*, 95, 106672.

Dorigo, M., & Blum, C. (2005) Ant colony optimization theory. A survey,Theoretical Computer Science,Vol 344, Issues 2–3,Pages 243-278,ISSN 0304-3975.

Gandomi, A. H., Gun, J. Y., Yang, X.-S., & Talatahari, S. (2013),Chaos-enhanced accelerated particle swarm optimization,Communications in Nonlinear Science and Numerical Simulation,Volume 18, Issue 2,Pages 327-340,ISSN 1007-5704.

Gill, S. S., & Buyya, R. (2019). Bio-inspired algorithms for big data analytics. a survey, taxonomy, and open challenges. In *Big data analytics for intelligent healthcare management* (pp. 1–17). Academic Press. DOI: 10.1016/B978-0-12-818146-1.00001-5

Jiang, F., Jiang, Y., Zhi, H., Dong, Y., Li, H., Ma, S., Wang, Y., Dong, Q., Shen, H., & Wang, Y. (2017). Artificial intelligence in healthcare. past, present and future. *Stroke and Vascular Neurology*, 2(4), 230–243. Advance online publication. DOI: 10.1136/svn-2017-000101 PMID: 29507784

Kar, A. K. (2016). Bio inspired computing–a review of algorithms and scope of applications. *Expert Systems with Applications*, 59, 20–32. DOI: 10.1016/j.eswa.2016.04.018

Kumari, J., Kumar, E., & Kumar, D. (2023). A structured analysis to study the role of machine learning and deep learning in the healthcare sector with big data analytics. *Archives of Computational Methods in Engineering*, 30(6), 3673–3701. DOI: 10.1007/s11831-023-09915-y PMID: 37359744

Mandave, D. D., & Patil, L. V. (2023). Bio-inspired computing algorithms in dementia diagnosis–a application-oriented review. *Results in Control and Optimization*, 12, 100276.

Mittal, M., Goyal, L. M., Kaur, S., Kaur, I., Verma, A., & Hemanth, D. J. (2019), Deep learning based enhanced tumor segmentation approach for MR brain images,Applied Soft Computing,Volume 78,Pages 346-354,ISSN 1568-4946.

Reddy, C. K. K., Anisha, P. R., Khan, S., Hanafiah, M. M., Pamulaparty, L., & Mohana, R. M. (Eds.). (2024). *Sustainability in Industry 5.0: Theory and Applications*. CRC Press.

Reddy, P. Y., Reddy, C. K. K., & Sithole, N. T. (2024). A study and analysis on diagnosis of melanoma cancer with deep learning: A case study. In *Federated Learning and AI for Healthcare 5.0* (pp. 203–218). IGI Global Scientific Publishing.

Rubinger, L., Gazendam, A., Ekhtiari, S., & Bhandari, M. (2023),Machine learning and artificial intelligence in research and healthcare,Volume 54, Supplement 3,Pages S69-S73,ISSN 0020-1383,Injury.

Shamshirband, S., Fathi, M., Dehzangi, A., Chronopoulos, A. T., & Alinejad-Rokny, H. (2021), A review on deep learning approaches in healthcare systems. Taxonomies, challenges, and open issues,Journal of Biomedical Informatics,Volume 113,103627,ISSN 1532-0464.

Yang, X.-S., & Karamanoglu, M. (2013). *Swarm Intelligence and Bio-Inspired Computation. An Overview*. Elsevier.

Yaqoob, A., Aziz, R. M., Verma, N. K., Lalwani, P., Makrariya, A., & Kumar, P. (2023). A review on nature-inspired algorithms for cancer disease prediction and classification. *Mathematics*, 11(5), 1081. DOI: 10.3390/math11051081

Yu, KH., Beam, A.L. & Kohane, I.S. (2018) Artificial intelligence in healthcare. *Nat Biomed Eng* 2, 719–731 . https.//.DOI: 10.1038/s41551-018-0305-z

Chapter 3

Current Neuroinnovative Techniques With Machine Learning Algorithms in the Diagnosis and Classification of Neurodegenerative Diseases

Esra Demır Unal
https://orcid.org/0000-0002-1752-9619
Medical Faculty, Yenimahalle Training and Research Hospital, Ankara Yıldırım Beyazıt University, Turkey

ABSTRACT

Neurodegenerative diseases are one of the most debilitating diseases associated with progressive neuronal dysfunction and protein accumulation. The increasing prevalence of Alzheimer's disease (AD), Parkinson's disease (PD), and amyotrophic lateral sclerosis (ALS) in population is leading to the development of neuroinnovative techniques for early recognition, classification and treatment. Various computer-aided software programs based on machine learning (ML) studies have been developed in recent decades in medical scope. ML is a branch of artificial intelligence that recognizes and models different data sets and algorithms. Different disciplines such as neural networks, support vector machines, Decision Trees, Random Forests, logistic regression, k-Nearest Neighbor techniques such as multi-layer perceptron, Neural Networks, Gaussian Mixture Models, and Boosting methods have been introduced

DOI: 10.4018/979-8-3373-1399-3.ch003

in neurology. This section will discuss neuropathological and neuroradiological evaluation of AD, PD, and ALS, and ML-based modeling algorithms used in diagnosing and classification so far.

INTRODUCTION

Artificial intelligence (AI) is a branch of science based on computer-based activations in different scientific fields, and it has become widespread, especially in the last decade. AI has revolutionized the early diagnosis, classification, and treatment of diseases by creating various algorithms within the framework of acquired learning instructions (Oliver et al., 2021). Machine learning (ML) is a subdiscipline of AI and identifies different data sets by learning and checking them with acquired learning methods (Mosavi et al., 2019). In this respect, the most frequently used algorithms so far are neural networks, support vector machines (SVM), decision trees (DT), random forest (RF), logistic regression, and many more. Also, some others are subsections of the neural network, such as Goodfellow's generative adversarial network (GAN) (Goodfellow et al., 2020). Neurodegenerative diseases (NDDs) are characterized by pathological protein aggregation in the central and/or peripheral nervous system that causes synaptic and neuronal network dysfunction, abnormal proteostasis, cytoskeletal abnormalities, altered energy metabolism, DNA and RNA defects, neuronal cell death, and damage the structure of neural networks. The most common neurodegenerative diseases in society include Alzheimer's disease (AD), Parkinson's disease (PD), and amyotrophic lateral sclerosis (ALS). ML is involved in groundbreaking innovations in neurology, as in many medical disciplines, especially in computer-aided identification, monitoring, and management of symptoms associated with neurodegenerative movement disorders.

AD is one of the most common neurodegenerative diseases in society, and its main etiopathogenesis is the abnormal accumulation of amyloid-beta (Aβ) and tau in intra/extra neuronal structures, which are amyloid precursors. Extracellular Aβ plaques and hyper-phosphorylated tau forms intracellular NFT, which cause synaptic degeneration, and neuronal death follow, driving a neurodegenerative progression through various brain regions (Blennow et al., 2006; Wakhloo et al., 2022). ML is used to detect disease progression and differential diagnosis, for which various ML algorithms such as support vector machines (SVM), decision trees (DT), and ensemble models have been used so far (Dara et al., 2023).

PD is characterized by the pathogenic accumulation of alpha-synuclein, resulting in the loss of dopaminergic neurons in the substantia nigra pars compacta (Raina et al., 2020; Psol et al., 2021; Garg et al., 2022). It is known that progressive neuronal loss occurs long before clinical findings (Mahajani et al., 2021; Raina et al., 2021), and the disease can be diagnosed with early detection. However, although it is the second most common neurodegenerative disorder, its detection and definitive treatment method in the premotor stages of PD have not yet been developed (DeMaagd & Phillip, 2015). Today, the most frequently used ML algorithms for PD are on wearable devices, thus contributing to reducing disease progression. In addition, different ML algorithms such as SVM, Random Forest (RF), DT, k-Nearest Neighbor (k-NN), and Multi-Layer Perceptron (MLP) can be used to distinguish PD patients from healthy individuals (Alshammri et al., 2023).

ALS is a neurodegenerative disease that occurs at a relatively earlier stage with high mortality. Although it is still in development, data sets obtained through ML algorithms help diagnose diseases and make differential diagnoses. In addition, patient classification and staging according to prognostic categories is thus possible. For this purpose, RF (Jahandideh et al., 2017; Pfohl et al., 2018; Querin et al., 2018), SVM (Bandini et al., 2018; D'hulst et al., 2018), Neural Networks (NN) (Beaulieu-Jones & Greene, 2016; van der Burgh et al., 2017), Gaussian Mixture Models (GMM) (Huang et al., 2017), Boosting methods (Jahandideh et al., 2017; Ong et al., 2017), k-NN (Beaulieu-Jones & Greene, 2016; Bandini et al., 2018) was used. Generalized linear regression models are also commonly used (Gordon et al., 2009; Taylor et al. et al., 2016; Huang et al., 2017; Li et al., 2018; Pfohl et al., 2018).

In this section, the possible etiopathogenetic processes that have been explained so far in AD, PD, and ALS, which present clinically with progressive clinical disability by causing irreversible dysfunction in the structure and function of neurons, will be evaluated from neuroradiological and neuropathological perspectives, and the role of artificial intelligence-based algorithmic modeling in early recognition, classification, and treatment of neurodegenerative diseases will be analyzed.

NEURORADIOLOGICAL AND NEUROPATHOLOGICAL PROCESSES IN AD

AD occurs in individuals over the age of 65, with an insidious onset and a slowly progressive course, in which subjective memory complaints turn into an isolated progressive memory disorder that can be detected objectively; over time, disorders in other cognitive areas are added, and finally, the severity of the mental disorder impedes independence in activities of daily living (ADLs) and impairs functionality. When the functional impairment threshold is exceeded, the AD dementia (ADD) stage, which is divided into sub-stages as mild, moderate, and severe, is reached. The course up to this threshold is called mild cognitive impairment (MCI) or prodromal AD. Isolated progressive memory impairment, which is a typical feature of prodromal AD, is called amnestic MCI (aMCI). AD has an initial phase called preclinical AD, which lasts for many years before presenting its clinical symptoms, during which only the pathological features of the disease evolve. Subjective memory complaints are still considered the last stages of the preclinical phase and are called subjective cognitive impairment (SCI). Preclinical and clinical stages are also called pathological AD (AD-P) and clinical AD (AD-C). The early-onset presentation is called early-onset AD (EOAD) and has a predominantly genetic causation. The typical late-onset presentation is called late-onset AD (LOAD), and this class consists mainly of sporadic AD (Sahin, 2009; Chen et al., 2017). AD is a progressive neurodegenerative disease that occurs with the loss of neurons and synapses in some parts of the central nervous system (CNS). Genetic predisposition is one of the most critical factors in AD development. It is a complex disease that manifests itself with disorders at many levels in more than one pathway. Two basic pathological findings related to the disease have been defined. These are amyloid plaques and NFT (Ozpak et al., 2017). The main genes found to be associated with AD are amyloid precursor protein (APP), presenilin 1 (PS1), presenilin 2 (PS2), and ApoE. The genes responsible for AD, called EOAD, which occurs before age 65, are APP, PS1, and PS2. The gene responsible for LOAD (after age 65) has been identified as ApoE. It has been observed that mutation of the APP genes located on chromosome 21 and PS2 genes located on chromosome 1 increases amyloid-β peptide levels in AD. These mutations cause faulty cleavage of the APP gene, production of toxic amyloidβ, hyper-phosphorylation of Tau proteins, and NFT formation. 2-3% of early-onset familial AD occurs due to APP, 20% due to PS2, and 70-80% due to PS1 gene mutation (Ozkay et al., 2011). Molecular studies have shown that the main component of amyloid plaques is amyloid beta (Aβ), and NFT are tau protein (Meraz-Rios et al., 2010). With the contribution of studies conducted in experimental animals, it has been shown that excess Aβ, both oligomeric and aggregate-forming Aβ, has a vital role in the pathogenesis of the disease (Haass & Selkoe, 2007). Studies on the genetics of AD have supported these observations. The mutations detected in familial AD are in genes related to Aβ formation, and even genetic changes that may predispose to the disease are closely related to Aβ biology. It is known that tau hyperphosphorylation not only disrupts microtubule stability and function but also acquires toxic function in the pathogenesis of AD; for example, tau aggregates induce apoptosis. However, similar to Aβ, it is also thought that the tau oligomers are associated with neurodegeneration and memory impairment (Meraz-Rios et al., 2010). In addition to parenchymal lesions, cerebral amyloid angiopathy may accompany the pathological process. Neuron and synapse loss are the main pathological findings of the disease.

AD is a neurodegenerative disease characterized by the progressive, irreversible loss of neurons and synapses in cortical and subcortical areas, causing cerebral gross atrophy (including the temporal lobe, parietal lobe, and parts of the frontal cortex and cingulate gyrus) (Brasure, 2018). The primary imaging methods used are commonly detectable by using radiological and nuclear neuroimaging methods such as magnetic resonance imaging (MRI), computed tomography (CT), positron emission tomography (PET), and single-photon emission computed tomography (SPECT). In addition, a number of different modalities (in particular, structural volumetric, diffusion, perfusion, functional, and magnetic metabolic resonance methods) are available to determine diagnostic algorithms in AD. Neuroimaging methods aim to identify specific structural hallmarks based on neurological disease and to make a differential diagnosis from another neurodegenerative disease that may affect treatment efficacy. According to Braak staging (Braak & Braak, 1991), the involvement of cortical structures is as follows: early involvement of the entorhinal cortex, followed by the limbic system and hippocampus, spreading to the cortex (precuneus in particular) with temporal lobes (Patel et al., 2020). However, there are also atypical cases. Appropriate neurological and morphological evaluation with currently used MRI modalities and technique subtypes (volumetric (MR-volumetry), functional MRI (fMRI)), structure (*perfusion* (arterial spin labeling (ASL), dynamic susceptibility contrast (DSC) MRI, dynamic contrast-enhanced (DCE) MRI)), *diffusion* (diffusion-weighted imaging (DWI), diffusion tensor imaging (DTI)), *metabolism* (MR spectroscopy (MRS)) can be performed. PET and SPECT are perfusion imaging methods that feature cranial morphometric mapping via radiotracers. During the dissociation process, isotopes emit gamma rays that are collected in detectors and produce 3D scans that reflect brain functions such as regional blood flow, glucose metabolism, abnormal protein accumulation, and neurotransmitter deficiencies (Rowley et al., 2020; Lu & Yuan, 2015). Neuroimaging with SPECT and PET contributes significantly to the elucidation of functional and molecular pathological processes in the premorbid phase, and both methods are used in the differential diagnosis of other stroke, epilepsy, and neoplastic (bone diseases) diseases (Cri an et al., 2022).

The Neuroradiodiagnostic Importance of ML in The Light of Current Neuroinnovative Developments in AD

ML techniques are among the frequently used innovative methods in medical practice and have been used so far in diagnosing and classifying different disease groups (Table 1). Many open-source databases for AD continue to research AD classification, early diagnosis, and premorbid stages (Clifford et al., 2015; Michael et al., 2017). The most frequently used databases for this purpose are ADNI (Dallas et al., 2018) (adni.loni.usc.edu), AIBL (Ellis et al., 2009) (aibl.csiro.au), and OASIS (Marcus et al., 2007) (www.oasis-brains.org). The critical challenge is scanning MRI data and determining the characteristic features quickly. In this respect, open-question data sets, such as Statistical Parametric Mapping (SPM) developed by the Wellcome Center for Human Neuroimaging, were used (Flandin & Friston, 2008). SVM, ANN, and deep learning (Elaheh et al., 2015; Enrico et al., 2018; Termenon et al., 2013). In 2017, Litjens et al. (Litjens et al., 2017) took an important step toward synthesizing medical imaging analyses with deep learning and mentioned some statistical techniques in order to bring the existing uncertainty to predictable levels in the synthesis of these methods. Shen et al. (2017) conducted a survey on behalf of AD regarding the feasibility of AI and supported this fact. In 2018, José et al. conducted a study on the applicability of neuroimaging techniques on brain diseases and obtained significant statistical results for ML techniques (José et al., 2018). Pellegrini et al. (2018) discussed the machine learning techniques used for dementia and cognitive impairment from 2006 to 2016 in 111 papers. It was emphasized that new ML techniques could be developed for AD. Additionally, Rathore et al. (2017) reviewed feature extraction to classify AD and its prodromal stages. Image zoning techniques of ML are also among the methods used to diagnose AD. It consists of methods for image partitioning that consist of thresholding, clustering, machine learning, and deep learning. Statistical methods try to classify images and determine the diagnosis of defects (Ban et al., 2022). While clustering methods, such as K-means and fuzzy clustering, are effective in detection and identification, they do not have acquired learning capabilities (Coraggio et al., 2023). In addition, in cases where the number of clusters cannot be determined clearly, significant erroneous results may be obtained. The clustering area loses sensitivity when the cluster center cannot be determined. Group intelligence increases execution time and helps improve clustering methods (Akinola et al., 2022).

MRI is a commonly used neuroimaging method for diagnosing and classifying neurodegenerative diseases, as in many neurological diseases. Methods that can be applied with the help of MRI, such as image processing, machine learning, and deep learning, have been utilized to diagnose diseases such as Alzheimer's and Alzheimer's plaques. Segmentation techniques are among the most frequently used. These techniques are needed in imaging to detect cerebral amyloid plaques using MRI in AD patients. The other one is Zoning techniques that aim to distinguish damaged brain tissue from healthy ones; this technique has been applied in multiple stages (Oh et al., 2019). Using MRI data as input and normalization are the first two stages and distinguishing the essential characters of the image is the third step after the standard image zoning process. Functional MRI (fMRI is an important piece of equipment in ML methods; in one study, tosa-straw resting-state fMRI was used, and spontaneous activities consolidated into several distinct resting-state networks (RSNs) that exhibit comparable temporal characteristics were measured (Moussa et al., 2012). Similarly, several studies use structural MRI (sMRI) based on the synthesis model of medical photographs (Moussa et al., 2012; Wolz et al., 2011). Data obtained from sMRI include the collective regulation of structural adaptations based on neurons or synapses. Current AI studies focus on early AD detection to detect cranial changes long before clinical findings in individuals with high susceptibility to AD (i.e., based on family history or genetics). For this purpose, MRI scans in various brain regions have been used to detect connections and interconnections at the cellular level in the brain (Eskildsen et al., 2014).

AD is the most common progressively debilitating neurodegenerative disorder and may be accompanied by dozens of different etiological predisposing factors. Although an obvious therapeutic cure has not yet been found in this process, diagnostic studies that can provide possible early treatment by taking advantage of current technological opportunities are on the agenda. For this purpose, AI-based studies have accelerated the diagnosis, differential diagnosis, and classification of the disease at stages where it can benefit from early treatment.

Table 1. Overview of the current ML algorithms studies in AD

Paper	Algorithms	Study Population	Dataset/Population	Location
				United States and Pakistan
Rathore et al., 2017	Feature Extraction, Feature Selection, Dimensionality Reduction, And Feature-Based Classification	AD	409	United States and Canada
Dallas et al., 2018	ADNI (Alzheimer's Disease Neuroimaging Initiative)	AD	1442	United Kingdom and European Union
Pellegrini et al., 2018	QUADAS (Quality Assessment of studies of Diagnostic Accuracy in Systematic reviews)	AD vs CN	111	United States
Marcus et al., 2007	OASIS (Open Access Series of Imaging Studies)	AD, MCI, PD and orher neurodegenerative diseases vs CN	416	Australia
Ellis et al., 2009	AIBL (Australian Imaging, Biomarkers & Lifestyle Flagship Study of Ageing)	MCI, AD, CN	1166 (AD (211), MCI (133) and CN (768)	United States

continued on following page

Table 1. Continued

Paper	Algorithms	Study Population	Dataset/Population	Location
Eskildsen et al., 2014	ADNI (Alzheimer's Disease Neuroimaging Initiative)	MCI	862	Korea
Oh et al., 2019	Volumetric Convolutional Neural Network and Transfer Learning	AD, MCI vs NC	694 (AD (198), NC (230), MCI (267)	

Neuroradiological and Neuropathological Processes in PD

The main clinical disorders that determine PD characteristics are bradykinesia, tremor, rigidity, postural reflex disorder, and dopaminergic dysfunction in the substantia nigra (SN). There are ongoing studies for determining which structures are responsible for the dysfunction in PD. The reflex time, motor cortical preparation potential recording (Bereitschafts potential (BP)), emission tomographies (Positron Emission Tomography (PET) and Single Photon Emission Tomography (SPECT)), fMRI, transcranial magnetic stimulation (TMS) techniques have been investigated. The progressive neurodegenerative pathology in SN is the main dysfunction in PD, and various MRI biomarkers have been used to detect neurodegenerative changes (Du et al., 2011; Pe ran et al., 2010). Decreased fractional isotropy (FA) (Du et al., 2011) and modifications in axial diameter (AD) (Vaillancourt et al., 2009), radial diameter (RD) (Du et al., 2011) or mean diffusivity (Schwarz et al., 2013) via DTI have been reported in the SN with some variability across studies. Additionally, iron accumulation in the SN can be demonstrated by R2* relaxometry (Du et al., 2011) or susceptibility imaging (Barbosa et al., 2015). Neuromelanin, the dopamine product that accumulates in dopaminergic neurons, has high signaling in T1-weighted spin-echo images due to its paramagnetic property (Sasaki et al., 2006), and it has been shown that neuromelanin volume or signaling in the SN decreases in PD (Schwarz et al., 2011). The basal ganglia is the main cerebral area where the primary pathology begins and progresses in PD, and topographic imaging is important for early diagnosis. With the PET technique, it has been determined that dopamine transporter (DAT) is significantly reduced in the basal ganglia (especially in the putamen), and 18 Fdopa uptake is reduced in the putaminal region in early PD (Ordidge et al., 1994; Sasaki et al., 2006). In different studies, the decrease in the amount of neuromelanin, one of the main findings of PD, was investigated. Sasaki et al. (2006) reported that the signal increase in the locus cereleus (LC) and substantia nigra pars compacta (SNc) in the 3T MRI T1 sequence is related to the amount of neurolemanin. Kane et al. (2008) showed that when the signals of the background brain region were suppressed, and the signal increase of the SNc and LC containing neuromelanin was increased, the contrast difference in these regions was related to the neuromelanin they contained. Otsuka et al. (2013) found a significant decrease in PH contrast ratio and signal increase in neuromelanin-sensitive imaging in 3T MRI T1 sequence in SNc compared to the control group. A decrease in neuromelanin begins years before the clinical symptoms of the disease, and its detection at an early stage is essential in determining the effective treatment of the disease progression. The pathophysiology of PD is characterized by the selective and progressive loss of dopamine (DA)-producing neurons containing melanin pigment in the SNc. In this process, intracytoplasmic α-synuclein positive Lewy bodies are formed in the subcortical nuclei and hippocampus, less frequently in the cerebral cortex, and dystrophic Lewy neurites are observed in these regions (Hurtig et al., 2010; Pahwa et al., 2003; Pollanen et al., 1993). Glial cell proliferation reactive to these changes can be seen (Hunot & Hirsch, 2003). "In vivo" isotope radiographic brain imaging and postmortem brain tissue examinations have shown that pathological changes in PD begin years before clinical symptoms, and dopaminergic neuron loss in the basal ganglia reaches 50%-75% in the subclinical or preclinical phase of the degenerative process. The clinical spectrum occurs when there is an 80% decrease in striatal DA content (Lester et al., 2010). In the akinetic rigid form, one of the two major clinical subtypes of PD, more severe involvement is observed in the ventrolateral part of the SN. In contrast, in the tremor-predominant clinical form, more severe neuronal loss is observed in the medial part of the SNc and the retroviral area (Jellinger, 2014). Neuron degeneration has also been observed in the ventral tegmental section of the mesencephalon in PD, and it is known that the axons of the cells in this area project to frontal cortical areas such as entorhinal, hippocampal, anterior cingulate, and limbic areas such as nucleus accumbens, hypothalamus, and amygdaloid nucleus. In addition, mesocortical and mesolimbic projections contribute to the normal function of behavior and cognition phenomena; in PD, there is a mild-moderate DA loss in the ventral tegmental-basal forebrain-cortical loop, and this causes cognition and behavioral disorders (Hayes, 2019). PD, striatal dopaminergic deafferentation constitutes the main pathology of clinical symptoms, and neurodegeneration progressively diffuses from dopamine-producing cells to neuromelanin-containing neurons in the SNc. Detection of the neurodegenerative process at an early stage, both through neuroimaging and neuropathology, is essential for preventing progression and early intervention.

The Neuroradiodiagnostic Importance of ML in The Light of Current Neuroinnovative Developments in PD

In many studies, methods for classifying PD with the help of AI are being developed, and these studies provide concrete evidence of the solutions that ML creates against the difficulties encountered in PD subclassification, risk assessment, and diagnosis (Table 2) (Senturk, 2020). A neural network, SVM, and classification and regression trees were used to categorize patients in PD, and performance comparisons showed that SVM with Recursive Feature Elimination shows 93.84% accuracy compared to other performers and seems more suitable in quantity and quality. In a methodological study based on artificial neural networks and SVM conducted by Gil and Manuel (2009), these two methods showed high performance with an accuracy of about 90% in diagnosing PD. Dias (2010) compared various classification techniques to make an accurate PD diagnosis. In the study by Alshammri et al., the accuracy performances of DT, Regression, NN, and DM neural classification were compared, and the NN's overall classification performance was 92.9%. In a different study by Al-Fatlawi et al., a deep belief network (DBN) was used, and it was reported that it significantly contributed to the diagnostic evaluation of PD (Al-Fatlawi et al., 2016). Also, Rasheed et al. (2020) proposed two different methods based on voice measurement to clarify the PD classification. In the artificial neural network created with a variable adaptive moment-based backpropagation algorithm called BPVAM, dimensionality reduction methods such as principal component analysis (PCA) in conjunction with BPVAM were used. Diagnostic effectiveness in the early stages of PD was investigated, and the best results were obtained by BPVAM and BPVAM PCA (97.50%), followed by ANN with Levenberg-Marquardt (95.89%) (Kadam & Jadhav, 2019). As a result, they proposed an ensemble learning method based on sparse autoencoders (specificity of 97.28% and 90%, respectively).

Several ongoing studies continuously determine the motor characteristics of basic irregular movements in PD and distinguish between normal and abnormal movements. Researchers implemented classifier-based supervised ML algorithms on data derived from patients given horizontal drawing tasks (Hoq et al., 2021). To see the superiority of wearable technologies and to make an objective motor evaluation, four inertial devices (two SensHand and two SensFoot) and related processing algorithms for extracting parameters from limb motion were compared, and kinematic parameters from four limbs were measured. As a result, the best performance has been observed in linear SVM and Gaussian SVM (Rovini et al., 2020). In another study, data from supporting software using patients' personal computers showed significant similarities with clinicians' scoring scores (Nair et al., 2020). In this respect, motor function analyses that can be performed via smartphones and tablets are practical due to their accessibility and cost-effectiveness (Woodzinski et al., 2019).

The ML-based hardware systems, Parkinson's KinetiGraph and the Kinesia system, which are based on the specific determination of dyskinesia and bradykinesia in PD, were used (Power et al., 2019) and tried to determine the superiority of data over each other using supervised ML algorithms. Another critical development in PD is gait analysis algorithms that were found to show essential data for the early detection and definition of motor diseases (Heldman et al., 2017). Heldman's study (2017) applied recurrent neural networks and long short-term memory using the records obtained to classify gait analysis and stride-to-stride footfall times. It was proven that these algorithms provide 95% to 100% accuracy in diagnosing PD, distinguishing it from other motor diseases, and making the correct analysis and diagnosis. Existing algorithms can also be improved to optimize performance (Heldman et al., 2017; Belić et al., 2019). In a study conducted by Latourelle et al., patients (312) and healthy (117) individuals participated in detecting progression in PD and worked on the principle of detecting motor progression based on molecular and genetic diagnoses. As a result of the study, progression modeling found a relationship between higher baseline motor scores, old age, and male sex and rapid motor progression, and also identified new predictors such as genetic variation and biomarkers (Latourelle et al., 2017).

Thanks to technological developments, thousands of data and PD risk factors can be compared quickly in light of current data. Cost-effectively, disciplines such as ML and wearable technology can be used as predictive tools for the early diagnosis of PD. Early diagnosis of cognitive decline in the presymptomatic phase at the cellular level is now possible. By using wearable sensors and technological remote control devices, it is possible to detect the disability and improve mobility through gait analysis and monitor and treat the patient in his/her comfort via telemedicine.

Table 2. Overview of the current ML algorithms in PD

Paper	Algorithms	Study Population	Dataset	Location
				Iraq
Al-Fatlawi et al., 2016	DBN (Deep Belief Network)	PD vs NC	31 (PD (23), NC (8))	United Kingdom
Latoureller et al., 2018	Bayesian Multivariate Predictive Inference	PD	429 (PD(312), NC(117))	Spain
Gil-Martin et al., 2019	CNN (Convolutional Neural Network)	PD vs NC	77 (PD (62), NC (15))	Serbia

continued on following page

Table 2. Continued

Paper	Algorithms	Study Population	Dataset	Location
Belić et al., 2019	Support Vector Machine, Support Vector Regression, Naïve Bayes, Logistic Regression, Artificial Neural Network, Extreme Machine Learning, Dynamic Neural Networks, k-Nearest Neighbours, Linear Discriminant Analysis, Tree-based algorithms, Ada Boost DT, Hidden Markov Models	PD	48	India
Kadam & Jadhav, 2019	Deep Neural Network, · Ensemble Learning	PD vs NC	31 (PD (23), NC (8))	Italy
Rovini et al., 2020	Support vector machine, Random Forest, Naïve Bayes	PD vs NC	80 (PD (40), NC(40))	Turkey
Rasheed et al., 2020	BPVAM (Variable Adaptive Moment-Based Backpropagation Algorithm), PCA (Principal Component Analysis)	PD vs NC	31 (PD (23), NC (8))	Turkey
Senturk, 2020	Feature selection	PD vs NC	31 (PD (23), NC (8))	

Neuroradiological and Neuropathological Processes in ALS

Amyotrophic lateral sclerosis (ALS) is a progressive neurodegenerative disease that affects motor neurons in the cerebral cortex, brainstem, and spinal cord (Brooks et al., 1994). Most cases die within 2-3 years, and the 5-year survival rate is reported as 10-20% (Traynor et al., 1994). In ALS, where most cases are sporadic, familial transmission is 5-10%, and in approximately 10-20% of these, the cause is a mutation in the Cu/Zn superoxide dismutase-1 (SOD1) enzyme. There are three members of the Superoxide dismutase enzyme family known to be involved in the pathophysiology of ALS, and the SOD1 (Cu-Zn SOD / cytosolic SOD) enzyme is the one associated with familial ALS. The familial form of ALS, which has an autosomal dominant SOD enzyme, creates an oxidant environment by converting superoxide into hydrogen peroxide with its antioxidant effect and causes neurotoxicity. The SOD1 enzyme mutated in familial ALS is cytotoxic, and one of the proposed mechanisms for this cytotoxicity is the apoptotic process initiated by SOD1 function loss (Rothstein et al., 1994; Greenlund et al., 1995). Another important mechanism in the pathogenesis of this form is the nitration of neurofilaments (Beckman, 2009; Bogdanov et al., 1998). Motor neurons contain many interconnecting neurofilament proteins due to their long axons. When the assembly of neurofilaments is prevented, caused by the peroxynitrite produced by the mutant SOD1 enzyme, many neurofilament subunits become vulnerable to nitration (Crow et al., 1997). A small nitrated neurofilament subunit disrupts uninitiated neurofilament networks, and aberrant neurofilament aggregates are formed in degenerated motor neurons' soma and proximal axons (Bogdanov et., 1998). Another mediator that plays a role in the pathophysiology of ALS is glutamate. AMPA/kainate receptors have the most critical role among glutamate receptors, and studies show that NMDA receptors are especially responsible for acute glutamate toxicity (Askmark et al., 1993; Blin et al., 1996). On the other hand, three environmental toxins that cause motor neuron damage in humans act on AMPA receptors (Weiss et al., 2000). While AMPA receptor agonists cause disease (Carriedo et al., 1996; Ikonomidou et al., 1996), their antagonists stop motor neuron damage (Canton et al., 2001; Van Damme et al., 2003), and AMPA/kainate receptors affect ALS in pathophysiology (Rao & Weiss, 2004). Additionally, while most AMPA/kainate receptors are resistant to calcium migration thanks to their GluR2 AMPA subunits, it has been reported that the absence of this subunit in motor neurons causes selective motor neuron death (Vandenberghe et al., 2000; Van Damme et al., 2002). Another disorder that lays the groundwork for the neurotoxicity that occurs in ALS is mitochondrial dysfunction. Increased mitochondrial volume was detected in muscle biopsies of patients with ALS, and it was determined that the increased calcium level in the mitochondria causes the neurodegenerative system activation (S. Klos et al., 1996). Additionally, errors have been detected in patients' complexes 1 and 4 of the mitochondrial respiratory chain (Wiedemann et al., 1998; Borthwick et al., 1999).

To summarize the basic pathophysiology, the increase in glutamate level secondary to hypoxia, environmental toxins, trauma, inflammation, infection, mutant SOD1 or EAAT2 deficiency ultimately acts through AMPA/kainate receptors in motor neurons that do not have the calcium permeable-GluR2 subunit or do not work well. As a result, it initiates neurotoxicity by causing high levels of calcium influx into the motor neuron. The oxidative stress increased by the damaged mitochondria of the motor neuron and the participation of neighboring microglia and astrocytes, and the toxic effects created by the activated microglia cells compose the main pathophysiologic basis of the disease

There is no definitive neuroradiologic diagnostic tool for ALS diagnoses, but detecting the signals from the upper and lower motor neurons via electroneuromyography (EMG) can be helpful in the diagnosis course. This method is a unique technique that records the electrical activity of muscles. The exact results of EMG are effective in diagnosing the patient with ALS. Another testing method is measuring nerve conduction velocity (NCV), which can be sensitive to degenerative processes in the peripheric nerve system. Abnormality in the results of this test may indicate that a patient is suspected of having ALS and/or peripheral neuropathy (damage to the peripheral nervous system) or myopathy (muscle disease). Although MRI may be a sensitive method, its specificity must be clarified.

The Neuroradiodiagnostic Importance of ML in The Light of Current Neuroinnovative Developments in ALS

Today, thanks to the models developed with ML, various classifiers and predictive algorithms have been developed in ALS (Table 3) (Bede, 2017). Current ML models have the potential to best approximate and replace current diagnostic criteria that use the El Escorial criteria based on clinical evaluation. RF is one of the most frequently used modeling methods in ALS (Hothorn & Jung, 2014; Ko et al., 2014; Pfohl et al., 2018), and in the light of the reviews made so far, it is one of the best in terms of performance (Beaulieu-Jones & Greene, 2016; Taylor et al. et al., 2016; Huang et al., 2017; Pfohl et al., 2018). Bosting is another ensemble method, and Jahandideh et al. (2017) and Ong et al. (2017) conducted a study on this method. Another model also used in ALS is regression models (including generalized linear and nonlinear models) (Taylor et al., 2016; Pfohl et al., 2018; Westeneng et al., 2018). Despite the assumptions made in regression models, this method has great potential for clinical use (Westeneng et al., 2018; Seibold et al., 2017). This model used the innovative radio frequency ML approach to establish the impact of Riluzole therapy and explained that the model revealed predictive algorithms for functional decline and survival. Prognosis in ALS is typically addressed as a classification or regression problem. In studies using the classification approach, categories are defined based on functional decline (Gomeni & Fava, 2013; Ko et al., 2014; Ong et al., 2017; Westeneng et al., 2018), survival (Schuster et al., 2017; Pfohl et al., 2018), or disease phase (Bandini et al., 2018). In studies based on a regression approach, survival (Huang et al., 2017; van der Burgh et al., 2017; Pfohl et al., 2018), Riluzole effect (Seibold et al., 2017), functional decline (Hothorn & Jung, 2014; Taylor et al. et al., 2016), or respiratory function (Jahandideh et al., 2017). An SFRS-r emphasized that this approach could be the gold standard in monitoring disease progression. Modeling on radiological data in ALS is also applicable. In a small-scale study by Schuster et al. (2017) and van der Burgh et al. (2017), modeling showed that they could be a reliable method for diagnosing ALS with an accuracy of over 79%. Objective performance evaluations, robust missing data management, and adherence to reporting guidelines using standard measurements in ALS are essential requirements for future ML-based ALS studies.

Table 3. Overview of the current ML algorithms studies in ALS

Paper	Algorithms	Study Population	Dataset/Population	Location
				France, United States
Gomeni & Fava, 2013	Longitudinal Model, Probabilistic Model	ALS	338	Switzerland
Hothorn &Jung, 2014	Random Forest Model	ALS	1197	United States
Ko et al., 2014	Random Forest Model	ALS	8635	United States
Beaulieu-Jones & Greene, 2016	DAPS Model (Denoising Autoencoders for Phenotype Stratification)	ALS vs NC	10000 (ALS (5000), NC (5000)	United States
Huang et al., 2017	GuanRank Model	ALS	8,635	United States
Jahandideh et al., 2017	GBM Model (Gradient-Boosting Machine Model)	ALS	4346	Singapore, United Kingdom
Ong et al., 2017	RUS/Ada Boost, Naïve Bayes, Decision Tree, Random Forest Models	ALS	6355	Switzerland, Austria
Seibold et al., 2017	Random Forest, Split Procedure, Personalised Models	ALS	73	Netherlands
van der Burgh et al., 2017	Artificial Neural Network	ALS	135	United States
Pfohl et al., 2018	Generalized Linear and Random Forest Models	ALS	1,585	Canada, United States
Bandini et al., 2018	Support Vector Machine, K-Nearest Neighbor	ALS	64	Netherlands, Ireland, Italy, United Kingdom, Portugal, Germany,, Switzerland, France, Belgium
Westeneng et al., 2018	Royston-Parmar Model	ALS	1936	

CONCLUSION

Neurodegenerative disorders are a group of diseases of which prevalence in society is increasing day by day, and etiopathogenesis and curative treatment methodology have not yet been elucidated. The synthesis of current neuroradiological and neuropathological diagnostic tools with AI technology has broken new ground in terms of early diagnosis, complication detection, clinical morbidity determination and development of treatment methods in this disease group. In AD, many open source databases such as ADNI, AIBL, OASIS have guided the studies carried out in terms of both diagnosis and classification. In current studies with AI-based algorithms in AD, it has been proven that Segmentation and Zoning techniques, SVM, ANN, and DL have high reliability in neuroimaging, identification, differentiation and diagnosis, while feature extraction in classification and K-means, fuzzy clustering and image zoning techniques of ML in detection, identification and diagnosis. It is known that AI technology is most frequently used for subclassification, risk assessment, and diagnosis in PD. It was found to have high accuracy of SVM, classification and regression trees and DBN algorithms in subtype categorization and diagnosis, NN in classification, BPVAM and BPVAM PCA in early PD diagnosis, and classifier-based supervised ML in distinguishing abnormal movements in PD. In addition, various wearable technologies are being developed for extrapyramidal movement disorder, which is the main clinical pathology in PD, and for this purpose, the high accuracy and reliability of SVM, Gaussian SVM, Parkinson's KinetiGraph algorithms are used, also in gait analysis, the high reliability of NN and long short-term memory algorithms has been reported. ALS is a mortal pathology that progresses at a younger age and has a high death rate than other neurodegenerative diseases, so current studies aim at early diagnosis and classification. While RF and Bostingmethods have been proven to have highis accuracy especially for diagnostic purposes, regression models have been used in terms of treatment effectiveness, functional decline, respiratory function, survival and disease phase. Also ANN algorithms have been reported to have high reliability in neuroimaging-based studies. New ML technologies have been developing day by day in the diagnosis, treatment and follow-up of neurodegenerative diseases, and in this chapter, the radiological and pathological aspects of neurodegenerative diseases, each of which come from different pathogenic processes, are evaluated and information is given about how they are synthesized with current ML techniques from a clinico-radiopathological perspective.

REFERENCES

Akinola, O. O., Ezugwu, A. E., Agushaka, J. O., Zitar, R. A., & Abualigah, L. (2022). Multiclass feature selection with metaheuristic optimization algorithms: A review. *Neural Computing & Applications*, 34(22), 19751–19790. DOI: 10.1007/s00521-022-07705-4 PMID: 36060097

Al-Fatlawi, A. H., Jabardi, M. H., & Ling, S. H. (2016). "Efficient diagnosis system for parkinson's disease using deep belief network," in 2016 IEEE Congress on evolutionary computation (CEC) (Vancouver, BC, Canada: IEEE), 1324–1330. DOI: 10.1109/CEC.2016.7743941

Alshammri, R., Alharbi, G., Alharbi, E., & Almubark, I. (2023). Machine learning approaches to identify Parkinson's disease using voice signal features. *Frontiers in Artificial Intelligence*, 6, 1084001. DOI: 10.3389/frai.2023.1084001 PMID: 37056913

Askmark, H., Aquilonius, S. M., Gillberg, P. G., Liedholm, L. J., Stålberg, E., & Wuopio, R. (1993). A pilot trial of dextromethorphan in amyotrophic lateral sclerosis. *Journal of Neurology, Neurosurgery, and Psychiatry*, 56(2), 197–200. DOI: 10.1136/jnnp.56.2.197 PMID: 8437010

Ban, Y., Zhang, X., & Lao, H. (2022). Diagnosis of Alzheimer's disease using structure highlighting key slice stacking and transfer learning. *Medical Physics*, 49(9), 5855–5869. DOI: 10.1002/mp.15888 PMID: 35894542

Bandini, A., Green, J. R., Wang, J., Campbell, T. F., Zinman, L., and Yunusova, Y. (2018). Kinematic features of jaw and lips distinguish symptomatic from presymptomatic stages of bulbar decline in amyotrophic lateral sclerosis. J. Speech Lang. Hear. Res. 61:1118. doi: jslhr-s-17-0262DOI: 10.1044/2018

Barbosa, J. H. O., Santos, A. C., Tumas, V., Liu, M., Zheng, W., Haacke, E. M., & Salmon, C. E. G. (2015). Quantifying brain iron deposition in patients with Parkinson's disease using quantitative susceptibility mapping, R2 and R2. *Magnetic Resonance Imaging*, 33(5), 559–565. DOI: 10.1016/j.mri.2015.02.021 PMID: 25721997

Beaulieu-Jones, B. K., & Greene, C. S. (2016). Semi-supervised learning of the electronic health record for phenotype stratification. *Journal of Biomedical Informatics*, 64, 168–178. DOI: 10.1016/j.jbi.2016.10.007 PMID: 27744022

Beckman, J. S. (2009). Understanding peroxynitrite biochemistry and its potential for treating human diseases. *Archives of Biochemistry and Biophysics*, 484(2), 114–116. DOI: 10.1016/j.abb.2009.03.013 PMID: 19393377

Bede, P. (2017). From qualitative radiological cues to machine learning: MRI-based diagnosis in neurodegeneration. *Future Neurology*, 12(1), 5–8. DOI: 10.2217/fnl-2016-0029

Belić, M., Bobić, V., Badža, M., Šolaja, N., Đurić-Jovičić, M., & Kostić, V. S. (2019). Artificial intelligence for assisting diagnostics and assessment of Parkinson's disease—A review. *Clinical Neurology and Neurosurgery*, 184, 105442. DOI: 10.1016/j.clineuro.2019.105442 PMID: 31351213

Blennow, K., de Leon, M. J., & Zetterberg, H. (2006). Alzheimer's disease. *Lancet*, 368(9533), 387–403. DOI: 10.1016/S0140-6736(06)69113-7 PMID: 16876668

Blin, O., Azulay, J. P., Desnuelle, C., Bille-Turc, F., Braguer, D., Besse, D., Branger, E., Crevat, A., Serratrice, G., & Pouget, J. Y. (1996). A controlled one-year trial of dextromethorphan in amyotrophic lateral sclerosis. *Clinical Neuropharmacology*, 19(2), 189–192. DOI: 10.1097/00002826-199619020-00009 PMID: 8777774

Bogdanov, M. B., Ramos, L. E., Xu, Z., & Beal, M. F. (1998). Elevated "hydroxyl radical" generation in vivo in an animal model of amyotrophic lateral sclerosis. *Journal of Neurochemistry*, 71(3), 1321–1324. DOI: 10.1046/j.1471-4159.1998.71031321.x PMID: 9721759

Borthwick, G. M., Johnson, M. A., Ince, P. G., Shaw, P. J., & Turnbull, D. M. (1999). Mitochondrial enzyme activity in amyotrophic lateral sclerosis: Implications for the role of mitochondria in neuronal cell death. *Annals of Neurology*, 46(5), 787–790. DOI: 10.1002/1531-8249(199911)46:5<787::AID-ANA17>3.0.CO;2-8 PMID: 10553999

Braak, H., & Braak, E. (1991). Neuropathological stageing of Alzheimer-related changes. *Acta Neuropathologica*, 82(4), 239–259. DOI: 10.1007/BF00308809 PMID: 1759558

Brasure, M., Desai, P., Davila, H., Nelson, V. A., Calvert, C., Jutkowitz, E., Butler, M., Fink, H. A., Ratner, E., Hemmy, L. S., McCarten, J. R., Barclay, T. R., & Kane, R. L. (2018). Physical activity interventions in preventing cognitive decline and Alzheimer-type dementia: A systematic review. *Annals of Internal Medicine*, 168(1), 30–38. DOI: 10.7326/M17-1528 PMID: 29255839

Brooks, B. R. (1994). El Escorial World Federation of Neurology criteria for the diagnosis of amyotrophic lateral sclerosis. Subcommittee on Motor Neuron Diseases/ Amyotrophic Lateral Sclerosis of the World Federation of Neurology Research Group on Neuromuscular Diseases and the El Escorial" Clinical limits of amyotrophic lateral sclerosis" workshop contributors. *Journal of the Neurological Sciences*, 124, 96–107. DOI: 10.1016/0022-510X(94)90191-0 PMID: 7807156

Canton, T., Böhme, G. A., Boireau, A., Bordier, F., Mignani, S., Jimonet, P., Jahn, G., Alavijeh, M., Stygall, J., Roberts, S., Brealey, C., Vuilhorgne, M., Debono, M.-W., Le Guern, S., Laville, M., Briet, D., Roux, M., Stutzmann, J.-M., & Pratt, J. (2001). RPR 119990, a novel α-amino-3-hydroxy-5-methyl-4-isoxazolepropionic acid antagonist: Synthesis, pharmacological properties, and activity in an animal model of amyotrophic lateral sclerosis. *The Journal of Pharmacology and Experimental Therapeutics*, 299(1), 314–322. DOI: 10.1016/S0022-3565(24)29332-0 PMID: 11561094

Carriedo, S. G., Yin, H. Z., & Weiss, J. H. (1996). Motor neurons are selectively vulnerable to AMPA/kainate receptor-mediated injury in vitro. *The Journal of Neuroscience : The Official Journal of the Society for Neuroscience*, 16(13), 4069–4079. DOI: 10.1523/JNEUROSCI.16-13-04069.1996 PMID: 8753869

Chen, Y., Sillaire, AR., Dallongeville, J., Skrobala, E., Wallon, D., Dubois, B. Hannequin, D., Pasquier, F. (2017). Low Prevalence and Clinical Effect of Vascular Risk Factors in Early-Onset Alzheimer's Disease. Journal of Alzhemier's Disease, 60, 1045-1054. 15.

Coraggio, L., & Coretto, P. (2023). Selecting the number of clusters, clustering models, and algorithms. A unifying approach based on the quadratic discriminant score. *Journal of Multivariate Analysis*, 196, 105181. DOI: 10.1016/j.jmva.2023.105181

Cri an, G., Moldovean-Cioroianu, N. S., Timaru, D. G., Andrie , G., Căinap, C., & Chi , V. (2022). Radiopharmaceuticals for PET and SPECT imaging: A literature review over the last decade. *International Journal of Molecular Sciences*, 23(9), 5023. DOI: 10.3390/ijms23095023 PMID: 35563414

Crow, J. P., Ye, Y. Z., Strong, M., Kirk, M., Barnes, S., & Beckman, J. S. (1997). Superoxide dismutase catalyzes nitration of tyrosines by peroxynitrite in the rod and head domains of neurofilament-L. *Journal of Neurochemistry*, 69(5), 1945–1953. DOI: 10.1046/j.1471-4159.1997.69051945.x PMID: 9349539

D'hulst, L., Weehaeghe, D. V., Chiò, A., Calvo, A., Moglia, C., & Canosa, A.. (2018). Multicenter validation of [18f]-FDG PET and support-vector machine discriminant analysis in automatically classifying patients with amyotrophic lateral sclerosis versus controls. *Amyotrophic Lateral Sclerosis & Frontotemporal Degeneration*, 19(7-8), 570–577. Advance online publication. DOI: 10.1080/21678421.2018.1476548 PMID: 29862846

Dara, O. A., Lopez-Guede, J. M., Raheem, H. I., Rahebi, J., Zulueta, E., & Fernandez-Gamiz, U. (2023). Alzheimer's Disease Diagnosis Using Machine Learning: A Survey. *Applied Sciences (Basel, Switzerland)*, 13(14), 8298. DOI: 10.3390/app13148298

Das, R. (2010). A comparison of multiple classification methods for diagnosis of parkinson disease. *Expert Systems with Applications*, 37(2), 1568–1572. DOI: 10.1016/j.eswa.2009.06.040

DeMaagd, G., & Philip, A. (2015). Parkinson's disease and its management: part 1: disease entity, risk factors, pathophysiology, clinical presentation, and diagnosis. *P&T*, 40, 504–532. PMID: 26236139

Du, G., Lewis, M. M., Styner, M., Shaffer, M. L., Sen, S., Yang, Q. X., & Huang, X. (2011). Combined R2* and diffusion tensor imaging changes in the substantia nigra in Parkinson's disease. *Movement Disorders*, 26(9), 1627–1632. DOI: 10.1002/mds.23643 PMID: 21618607

Ellis, K. A., Bush, A. I., Darby, D., De Fazio, D., Foster, J., Hudson, P., Lautenschlager, N. T., Lenzo, N., Martins, R. N., Maruff, P., Masters, C., Milner, A., Pike, K., Rowe, C., Savage, G., Szoeke, C., Taddei, K., Villemagne, V., Woodward, M., & Ames, D. (2009). The Australian Imaging, Biomarkers and Lifestyle (AIBL) study of aging: Methodology and baseline characteristics of 1112 individuals recruited for a longitudinal study of Alzheimer's disease. *International Psychogeriatrics*, 21(4), 672–687. DOI: 10.1017/S1041610209009405 PMID: 19470201

Eskildsen, S. F., Coupé, P., Fonov, V., & Collins, D. L. (2014, September). Detecting Alzheimer's disease by morphological MRI using hippocampal grading and cortical thickness. In *Proc MICCAI Workshop Challenge on Computer-Aided Diagnosis of Dementia Based on Structural MRI Data* (pp. 38-47).

Garg, P., Maass, F., Sundaram, S. M., Mollenhauer, B., Mahajani, S., van Riesen, C., Kügler, S., & Bähr, M. (2022). The relevance of synuclein autoantibodies as a biomarker for Parkinson's disease. *Molecular and Cellular Neurosciences*, 121, 103746. DOI: 10.1016/j.mcn.2022.103746 PMID: 35660088

Gil, D., & Manuel, D. J. (2009). Diagnosing parkinson by using artificial neural networks and support vector machines. *Glob. J. Comput. Sci. Technol.*, 9, 63–71.

Gomeni, R., & Fava, M. (2013). Amyotrophic lateral sclerosis disease progression model. *Amyotrophic Lateral Sclerosis & Frontotemporal Degeneration*, 15(1-2), 119–129. DOI: 10.3109/21678421.2013.838970 PMID: 24070404

Goodfellow, I., Pouget-Abadie, J., Mirza, M., Xu, B., Warde-Farley, D., Ozair, S., Courville, A., & Bengio, Y. (2020). Generative adversarial networks. *Communications of the ACM*, 63(11), 139–144. DOI: 10.1145/3422622

Gordon, P. H., Cheng, B., Katz, I. B., Mitsumoto, H., & Rowland, L. P. (2009). Clinical features that distinguish PLS, upper motor neuron-dominant ALS, and typical ALS. *Neurology*, 72(22), 1948–1952. DOI: 10.1212/WNL.0b013e3181a8269b PMID: 19487653

Greenlund, L. J., Deckwerth, T. L., & Johnson, E. M.Jr. (1995). Superoxide dismutase delays neuronal apoptosis: A role for reactive oxygen species in programmed neuronal death. *Neuron*, 14(2), 303–315. DOI: 10.1016/0896-6273(95)90287-2 PMID: 7857640

Haass, C., & Selkoe, D. J. (2007). Soluble protein oligomers in neurodegeneration: Lessons from the Alzheimer's amyloid β-peptide. *Nature Reviews. Molecular Cell Biology*, 8(2), 101–112. DOI: 10.1038/nrm2101 PMID: 17245412

Hayes, M. T. (2019). Parkinson's disease and parkinsonism. *The American Journal of Medicine*, 132(7), 802–807. DOI: 10.1016/j.amjmed.2019.03.001 PMID: 30890425

Heldman, D. A., Harris, D. A., Felong, T., Andrzejewski, K. L., Dorsey, E. R., Giuffrida, J. P., Goldberg, B., & Burack, M. A. (2017). Telehealth management of Parkinson's disease using wearable sensors: An exploratory study. *Digital Biomarkers*, 1(1), 43–51. DOI: 10.1159/000475801 PMID: 29725667

Heuvel, M. P. (2017). Deep learning predictions of survival based on MRI in amyotrophic lateral sclerosis. *NeuroImage. Clinical*, 13, 361–369. DOI: 10.1016/j.nicl.2016.10.008 PMID: 28070484

Hoq, M., Uddin, M. N., & Park, S. B. (2021). Vocal feature extraction-based artificial intelligent model for Parkinson's disease detection. *Diagnostics (Basel)*, 11(6), 1076. DOI: 10.3390/diagnostics11061076 PMID: 34208330

Hothorn, T., & Jung, H. H. (2014). RandomForest4life: A random forest for predicting ALS disease progression. *Amyotrophic Lateral Sclerosis & Frontotemporal Degeneration*, 15(5-6), 444–452. DOI: 10.3109/21678421.2014.893361 PMID: 25141076

Huang, Z., Zhang, H., Boss, J., Goutman, S. A., Mukherjee, B., Dinov, I. D., & Guan, Y. (2017). Complete hazard ranking to analyze rightcensored data: An ALS survival study. *PLoS Computational Biology*, 13(12), e1005887. DOI: 10.1371/journal.pcbi.1005887 PMID: 29253881

Hunot, S., & Hirsch, E. C. (2003). Neuroinflammatory processes in Parkinson's disease. *Annals of Neurology*, 53(S3), S49–S60. DOI: 10.1002/ana.10481 PMID: 12666098

Hurtig, H. (2000). What is Parkinson's disease? Neuropathology, neurochemistry, and pathophysiology. In *Parkinson's disease and movement disorders: Diagnosis and treatment guidelines for the practicing physician* (pp. 57–70). Humana Press.

Ikonomidou, C., Qin, Y. Q., Labruyere, J., & Olney, J. W. (1996). Motor neuron degeneration induced by excitotoxin agonists has features in common with those seen in the SOD-1 transgenic mouse model of amyotrophic lateral sclerosis. *Journal of Neuropathology and Experimental Neurology*, 55(2), 211–224. DOI: 10.1097/00005072-199602000-00010 PMID: 8786380

Jack, C. R.Jr, Barnes, J., Bernstein, M. A., Borowski, B. J., Brewer, J., Clegg, S., Dale, A. M., Carmichael, O., Ching, C., DeCarli, C., Desikan, R. S., Fennema-Notestine, C., Fjell, A. M., Fletcher, E., Fox, N. C., Gunter, J., Gutman, B. A., Holland, D., Hua, X., & Weiner, M. (2015). Magnetic resonance imaging in Alzheimer's disease neuroimaging initiative 2. *Alzheimer's & Dementia*, 11(7), 740–756. DOI: 10.1016/j.jalz.2015.05.002 PMID: 26194310

Jahandideh, S., Taylor, A. A., Beaulieu, D., Keymer, M., Meng, L., Bian, A., Atassi, N., Andrews, J., & Ennist, D. L. (2017). Longitudinal modeling to predict vital capacity in amyotrophic lateral sclerosis. *Amyotrophic Lateral Sclerosis & Frontotemporal Degeneration*, 19(3-4), 294–302. DOI: 10.1080/21678421.2017.1418003 PMID: 29260584

Jellinger, K. A. (2014). Neuropathology of Parkinson's disease. In *Inflammation in Parkinson's disease: scientific and clinical aspects* (pp. 25–74). Springer International Publishing. DOI: 10.1007/978-3-319-08046-8_2

Kadam, V. J., & Jadhav, S. M. (2019). In Iyer, B., Nalbalwar, S., & Pathak, N. (Eds.), *"Feature ensemble learning based on sparse autoencoders for diagnosis of parkinson's disease," in Computing, Communication and Signal Processing. Advances in Intelligent Systems and Computing* (Vol. 810, pp. 567–581). Springer., DOI: 10.1007/978-981-13-1513-8_58

Ko, K. D., El-Ghazawi, T., Kim, D., & Morizono, H. (2014). "Predicting theseverity of motor neuron disease progression using electronic health record data with a cloud computing big data approach," in *2014 IEEE Conference on Computational Intelligence in Bioinformatics and Computational Biology* (Honolulu, HI: IEEE), 1–6.

Latourelle, J. C., Beste, M. T., Hadzi, T. C., Miller, R. E., Oppenheim, J. N., Valko, M. P., Wuest, D. M., Church, B. W., Khalil, I. G., Hayete, B., & Venuto, C. S. (2017). Large-scale identification of clinical and genetic predictors of motor progression in patients with newly diagnosed Parkinson's disease: A longitudinal cohort study and validation. *Lancet Neurology*, 16(11), 908–916. DOI: 10.1016/S1474-4422(17)30328-9 PMID: 28958801

Lester, D. B., Rogers, T. D., & Blaha, C. D. (2010). Acetylcholine–dopamine interactions in the pathophysiology and treatment of CNS disorders. *CNS Neuroscience & Therapeutics*, 16(3), 137–162. DOI: 10.1111/j.1755-5949.2010.00142.x PMID: 20370804

Lhommée, E., Batir, A., Quesada, J. L., Ardouin, C., Fraix, V., Seigneuret, E., & Krack, P. (2014). Dopamine and the biology of creativity: Lessons from Parkinson's disease. *Frontiers in Neurology*, 5, 55. DOI: 10.3389/fneur.2014.00055 PMID: 24795692

Li, T., Howells, J., Lin, C., Garg, N., Kiernan, M., & Park, S. (2018). 8. Predicting motor disorders from nerve excitability studies. *Clinical Neurophysiology*, 129(4), e4. DOI: 10.1016/j.clinph.2017.12.021

Litjens, G., Kooi, T., Bejnordi, B. E., Setio, A. A. A., Ciompi, F., Ghafoorian, M., van der Laak, J. A. W. M., van Ginneken, B., & Sánchez, C. I. (2017). A survey on deep learning in medical image analysis. *Medical Image Analysis*, 42, 60–88. DOI: 10.1016/j.media.2017.07.005 PMID: 28778026

Lu, F. M., & Yuan, Z. (2015). PET/SPECT molecular imaging in clinical neuroscience: Recent advances in the investigation of CNS diseases. *Quantitative Imaging in Medicine and Surgery*, 5(3), 433. PMID: 26029646

Mahajani, S., Bähr, M., & Kügler, S. (2021). Patterning inconsistencies restrict the true potential of dopaminergic neurons derived from human induced pluripotent stem cells. *Neural Regeneration Research*, 16(4), 692–693. DOI: 10.4103/1673-5374.295316 PMID: 33063729

Marcus, D. S., Wang, T. H., Parker, J., Csernansky, J. G., Morris, J. C., & Buckner, R. L. (2007). Open Access Series of Imaging Studies (OASIS): Cross-sectional MRI data in young, middle aged, nondemented, and demented older adults. *Journal of Cognitive Neuroscience*, 19(9), 1498–1507. DOI: 10.1162/jocn.2007.19.9.1498 PMID: 17714011

Mateos-Pérez, J. M., Dadar, M., Lacalle-Aurioles, M., Iturria-Medina, Y., Zeighami, Y., & Evans, A. C. (2018). Structural neuroimaging as clinical predictor: A review of machine learning applications. *NeuroImage. Clinical*, 20, 506–522. DOI: 10.1016/j.nicl.2018.08.019 PMID: 30167371

Meraz-Ríos, M. A., Lira-De León, K. I., Campos-Peña, V., De Anda-Hernández, M. A., & Mena-López, R. (2010). Tau oligomers and aggregation in Alzheimer's disease. *Journal of Neurochemistry*, 112(6), 1353–1367. DOI: 10.1111/j.1471-4159.2009.06511.x PMID: 19943854

Moradi, E., Pepe, A., Gaser, C., Huttunen, H., & Tohka, J. (2015). Machine learning framework for early MRI-based Alzheimer's conversion prediction in MCI subjects. *NeuroImage*, 104, 398–412. DOI: 10.1016/j.neuroimage.2014.10.002 PMID: 25312773

Mosavi, A., Salimi, M., Ardabili, S. F., Rabczuk, T., Shamshirband, S., & Varkonyi-Koczy, A. R. (2019). State of the Art of Machine Learning Models in Energy Systems, a Systematic Review. *Energies*, 12(7), 1301. DOI: 10.3390/en12071301

Moussa, M. N., Steen, M. R., Laurienti, P. J., & Hayasaka, S. (2012). Consistency of network modules in resting-state FMRI connectome data, 7(8):e44428. DOI: 10.1371/journal.pone.0044428

Nair, P., Trisno, R., Baghini, M. S., Pendharkar, G., & Chung, H. (2020). Predicting early stage drug induced Parkinsonism using unsupervised and supervised machine learning. *Annual International Conference of the IEEE Engineering in Medicine and Biology Society*, 2020, 776–779. DOI: 10.1109/EMBC44109.2020.9175343 PMID: 33018101

Nakane, T., Nihashi, T., Kawai, H., & Naganawa, S. (2008). Visualization of neuromelanin in the Substantia nigra and locus ceruleus at 1.5 T using a 3D-gradient echo sequence with magnetization transfer contrast. *Magnetic Resonance in Medical Sciences*, 7(4), 205–210. DOI: 10.2463/mrms.7.205 PMID: 19110515

Niso, G., Botvinik-Nezer, R., Appelhoff, S., De La Vega, A., Esteban, O., Etzel, J. A., Finc, K., Ganz, M., Gau, R., Halchenko, Y. O., Herholz, P., Karakuzu, A., Keator, D. B., Markiewicz, C. J., Maumet, C., Pernet, C. R., Pestilli, F., Queder, N., Schmitt, T., & Rieger, J. W. (2022). Open and reproducible neuroimaging: From study inception to publication. *NeuroImage*, 263, 119623. DOI: 10.1016/j.neuroimage.2022.119623 PMID: 36100172

Oh, K., Chung, Y. C., Kim, K. W., Kim, W. S., & Oh, I. S. (2019). Classification and visualization of Alzheimer's disease using volumetric convolutional neural network and transfer learning. *Scientific Reports*, 9(1), 18150. DOI: 10.1038/s41598-019-54548-6 PMID: 31796817

Ohtsuka, C., Sasaki, M., Konno, K., Koide, M., Kato, K., Takahashi, J., Takahashi, S., Kudo, K., Yamashita, F., & Terayama, Y. (2013). Changes in substantia nigra and locus coeruleus in patients with early-stage Parkinson's disease using neuromelanin-sensitive MR imaging. *Neuroscience Letters*, 541, 93–98. DOI: 10.1016/j.neulet.2013.02.012 PMID: 23428505

Oliver, J. M., Esteban, M. D., López-Gutiérrez, J.-S., Negro, V., & Neves, M. G. (2021). Optimizing Wave Overtopping Energy Converters by ANN Modelling: Evaluating the Overtopping Rate Forecasting as the First Step. *Sustainability (Basel)*, 13(3), 1483. DOI: 10.3390/su13031483

Ong, M.-L., Tan, P. F., & Holbrook, J. D. (2017). Predicting functional decline and survival amyotrophic lateral sclerosis. *PLoS One*, 12(4), e0174925. DOI: 10.1371/journal.pone.0174925 PMID: 28406915

Ordidge, R. J., Gorell, J. M., Deniau, J. C., Knight, R. A., & Helpern, J. A. (1994). Assessment of relative brain iron concentrations using T2-weighted and T2*-weighted MRI at 3 Tesla. *Magnetic Resonance in Medicine*, 32(3), 335–341. DOI: 10.1002/mrm.1910320309 PMID: 7984066

Ozkay, U. D., Oztürk, Y., & Can, O. D. (2011). The disease of the aging world: Alzheimer's disease. S.D.U. *Faculty of Medicine Journal*, 18(1), 35–42.

Ozpak, L., Pazarbası, A., & Keser, N. (2017). Genetics and Epigenetics of Alzheimer's Disease. *Journal of Archive Literature Review*, 26(1), 34–49.

Patel, K. P., Wymer, D. T., Bhatia, V. K., Duara, R., & Rajadhyaksha, C. D. (2020). Multimodality imaging of dementia: Clinical importance and role of integrated anatomic and molecular imaging. *Radiographics*, 40(1), 200–222. DOI: 10.1148/rg.2020190070 PMID: 31917652

Pellegrini, E., Ballerini, L., Hernandez, M. D. C. V., Chappell, F. M., González-Castro, V., Anblagan, D., Danso, S., Muñoz-Maniega, S., Job, D., Pernet, C., Mair, G., MacGillivray, T. J., Trucco, E., & Wardlaw, J. M. (2018). Machine learning of neuroimaging for assisted diagnosis of cognitive impairment and dementia: A systematic review. *Alzheimer's & Dementia: Diagnosis, Assessment & Disease Monitoring*, 10(1), 519–535. DOI: 10.1016/j.dadm.2018.07.004 PMID: 30364671

Péran, P., Cherubini, A., Assogna, F., Piras, F., Quattrocchi, C., Peppe, A., Celsis, P., Rascol, O., Démonet, J.-F., Stefani, A., Pierantozzi, M., Pontieri, F. E., Caltagirone, C., Spalletta, G., & Sabatini, U. (2010). Magnetic resonance imaging markers of Parkinson's disease nigrostriatal signature. *Brain*, 133(11), 3423–3433. DOI: 10.1093/brain/awq212 PMID: 20736190

Pfohl, S. R., Kim, R. B., Coan, G. S., & Mitchell, C. S. (2018). Unraveling the complexity of amyotrophic lateral sclerosis survival prediction. *Frontiers in Neuroinformatics*, 12, 36. DOI: 10.3389/fninf.2018.00036 PMID: 29962944

Pollanen, M. S., Dickson, D. W., & Bergeron, C. (1993). Pathology and biology of the Lewy body. Journal of Neuropathology & Experimental Neurology, 52(3), 183-191. 74.

Powers, R., Etezadi-Amoli, M., Arnold, E. M., Kianian, S., Mance, I., Gibiansky, M., Trietsch, D., Alvarado, A. S., Kretlow, J. D., Herrington, T. M., Brillman, S., Huang, N. C., Lin, P., Pham, H. A., & Ullal, A. (2021). Smartwatch inertial sensors continuously monitor real-world motor fluctuations in Parkinson's disease. *Science Translational Medicine*, 13(579), eabd7865. DOI: 10.1126/scitranslmed.abd7865 PMID: 33536284

Psol, M., Darvas, S. G., Leite, K., Mahajani, S. U., Bähr, M., & Kügler, S. (2021). Dementia with Lewy bodies-associated β-synuclein mutations V70M and P123H cause mutation-specific neuropathological lesions. *Human Molecular Genetics*, 30(3-4), 247–264. DOI: 10.1093/hmg/ddab036 PMID: 33760043

Querin, G., Mendili, M.-M. E., Bede, P., Delphine, S., & Lenglet, T. (2018). Multimodal spinal cord MRI offers accurate diagnostic classification in ALS. *Journal of Neurology, Neurosurgery, and Psychiatry*, 89(11), 1220–1221. DOI: 10.1136/jnnp-2017-317214 PMID: 29353238

Raina, A., Leite, K., Guerin, S., Mahajani, S. U., Chakrabarti, K. S., Voll, D., Becker, S., Griesinger, C., Bähr, M., & Kügler, S. (2021). Dopamine promotes the neurodegenerative potential of β-synuclein. *Journal of Neurochemistry*, 156(5), 674–691. DOI: 10.1111/jnc.15134 PMID: 32730640

Raina, A., Mahajani, S., Bähr, M., & Kügler, S. (2020). Neuronal trans-differentiation by transcription factors Ascl1 and Nurr1: Induction of a dopaminergic neurotransmitter phenotype in cortical GABAergic neurons. *Molecular Neurobiology*, 57(1), 249–260. DOI: 10.1007/s12035-019-01701-x PMID: 31317490

Rao, S. D., & Weiss, J. H. (2004). Excitotoxic and oxidative cross-talk between motor neurons and glia in ALS pathogenesis. *Trends in Neurosciences*, 27(1), 17–23. DOI: 10.1016/j.tins.2003.11.001 PMID: 14698606

Rasheed, J., Hameed, A. A., Ajlouni, N., Jamil, A., Ozyavas, A., & Orman, Z. (2020). “Application of adaptive back-propagation neural networks for parkinson’s disease prediction,” in *2020 International Conference on Data Analytics for Business and Industry: Way Towards a Sustainable Economy (ICDABI)* (IEEE), 1–5. DOI: 10.1109/ICDABI51230.2020.9325709

Rathore, S., Habes, M., Iftikhar, M. A., Shacklett, A., & Davatzikos, C. (2017). A review on neuroimaging-based classification studies and associated feature extraction methods for Alzheimer’s disease and its prodromal stages. *NeuroImage*, 155, 530–548. DOI: 10.1016/j.neuroimage.2017.03.057 PMID: 28414186

Rothstein, J. D., Bristol, L. A., Hosler, B., Brown, R. H.Jr, & Kuncl, R. W. (1994). Chronic inhibition of superoxide dismutase produces apoptotic death of spinal neurons. *Proceedings of the National Academy of Sciences of the United States of America*, 91(10), 4155–4159. DOI: 10.1073/pnas.91.10.4155 PMID: 7910402

Rovini, E., Maremmani, C., & Cavallo, F. (2020). A wearable system to objectify assessment of motor tasks for supporting parkinson’s disease diagnosis. *Sensors (Basel)*, 20(9), 2630. DOI: 10.3390/s20092630 PMID: 32380675

Rowley, P. A., Samsonov, A. A., Betthauser, T. J., Pirasteh, A., Johnson, S. C., & Eisenmenger, L. B. (2020). Amyloid and Tau PET imaging of Alzheimer disease and other neurodegenerative conditions. In *Seminars in Ultrasound, CT and MRI. 41(6): 572-583.* WB Saunders. DOI: 10.1053/j.sult.2020.08.011

Sahin, H. A. (2009). Alzheimer hastalığını klinik belirtileri ve seyri. *Türkiye Klinikleri*, 2(1), 3135.

Sasaki, M., Shibata, E., Tohyama, K., Takahashi, J., Otsuka, K., Tsuchiya, K., Takahashi, S., Ehara, S., Terayama, Y., & Sakai, A. (2006). Neuromelanin magnetic resonance imaging of locus ceruleus and substantia nigra in Parkinson’s disease. *Neuroreport*, 17(11), 1215–1218. DOI: 10.1097/01.wnr.0000227984.84927.a7 PMID: 16837857

Schuster, C., Hardiman, O., & Bede, P. (2017). Survival prediction in amyotrophic lateral sclerosis based on MRI measures and clinical characteristics. *BMC Neurology*, 17(1), 73. DOI: 10.1186/s12883-017-0854-x PMID: 28412941

Schwarz, S. T., Abaei, M., Gontu, V., Morgan, P. S., Bajaj, N., & Auer, D. P. (2013). Diffusion tensor imaging of nigral degeneration in Parkinson’s disease: A region-of-interest and voxel-based study at 3 T and systematic review with meta-analysis. *NeuroImage. Clinical*, 3, 481–488. DOI: 10.1016/j.nicl.2013.10.006 PMID: 24273730

Schwarz, S. T., Rittman, T., Gontu, V., Morgan, P. S., Bajaj, N., & Auer, D. P. (2011). T1-weighted MRI shows stage-dependent substantia nigra signal loss in Parkinson's disease. *Movement Disorders*, 26(9), 1633–1638. DOI: 10.1002/mds.23722 PMID: 21491489

Seibold, H., Zeileis, A., & Hothorn, T. (2017). Individual treatment effect prediction for amyotrophic lateral sclerosis patients. *Statistical Methods in Medical Research*, 27(10), 3104–3125. DOI: 10.1177/0962280217693034 PMID: 29298618

Senturk, Z. K. (2020). Early diagnosis of parkinson's disease using machine learning algorithms. Medical Hypoth. 138, 109603. doi:. 109603DOI: 10.1016/j.mehy.2020

Shen, D., Wu, G., & Suk, H. I. (2017). Deep learning in medical image analysis. *Annual Review of Biomedical Engineering*, 19(1), 221–248. DOI: 10.1146/annurev-bioeng-071516-044442 PMID: 28301734

Siklós, L., Engelhardt, J., Harati, Y., Smith, R. G., Joó, F., & Appel, S. H. (1996). Ultrastructural evidence for altered calcium in motor nerve terminals in amyotrophc lateral sclerosis. *Annals of Neurology*, 39(2), 203–216. DOI: 10.1002/ana.410390210 PMID: 8967752

Taylor, J. P., Brown, R. H., and Cleveland, D. W. (2016). Decoding ALS: from genes to mechanism. Nature 539, 197–206. doi: 20413DOI: 10.1038/nature

Taylor, J. P., Brown, R. H., and Cleveland, D. W. (2016). Decoding ALS: from genes to mechanism. Nature 539, 197–206. doi: 20413DOI: 10.1038/nature

Termenon, M., Grana, M., Besga, A., Echeveste, J., & Gonzalez-Pinto, A. (2013). Lattice independent component analysis feature selection on diffusion weighted imaging for Alzheimer's disease classification. *Neurocomputing*, 114, 132–141. DOI: 10.1016/j.neucom.2012.08.044

Traynor, B. J., Codd, M. B., Corr, B., Forde, C., Frost, E., & Hardiman, O. M. (2000). Clinical features of amyotrophic lateral sclerosis according to the El Escorial and Airlie House diagnostic criteria: A population-based study. *Archives of Neurology*, 57(8), 1171–1176. DOI: 10.1001/archneur.57.8.1171 PMID: 10927797

Vaillancourt, D. E., Spraker, M. B., Prodoehl, J., Abraham, I., Corcos, D. M., Zhou, X. J., Comella, C. L., & Little, D. M. (2009). High-resolution diffusion tensor imaging in the substantia nigra of de novo Parkinson disease. *Neurology*, 72(16), 1378–1384. DOI: 10.1212/01.wnl.0000340982.01727.6e PMID: 19129507

Van Damme, P., Leyssen, M., Callewaert, G., Robberecht, W., & Van Den Bosch, L. (2003). The AMPA receptor antagonist NBQX prolongs survival in a transgenic mouse model of amyotrophic lateral sclerosis. *Neuroscience Letters*, 343(2), 81–84. DOI: 10.1016/S0304-3940(03)00314-8 PMID: 12759169

Van Damme, P., Van den Bosch, L., Van Houtte, E., Callewaert, G., & Robberecht, W. (2002). GluR2-dependent properties of AMPA receptors determine the selective vulnerability of motor neurons to excitotoxicity. *Journal of Neurophysiology*, 88(3), 1279–1287. DOI: 10.1152/jn.2002.88.3.1279 PMID: 12205149

van der Burgh, H. K., Schmidt, R., Westeneng, H.-J., de Reus, M. A., van den Berg, L. H., & van den Heuvel, M. P. (2017). Deep learning predictions of survival based on MRI in amyotrophic lateral sclerosis. *NeuroImage. Clinical*, 13, 361–369. DOI: 10.1016/j.nicl.2016.10.008 PMID: 28070484

Vandenberghe, W., Robberecht, W., & Brorson, J. R. (2000). AMPA receptor calcium permeability, GluR2 expression, and selective motoneuron vulnerability. *The Journal of Neuroscience : The Official Journal of the Society for Neuroscience*, 20(1), 123–132. DOI: 10.1523/JNEUROSCI.20-01-00123.2000 PMID: 10627588

Veitch, D. P., Weiner, M. W., Aisen, P. S., Beckett, L. A., Cairns, N. J., Green, R. C., Harvey, D., Jack, C. R.Jr, Jagust, W., Morris, J. C., Petersen, R. C., Saykin, A. J., Shaw, L. M., Toga, A. W., & Trojanowski, J. Q. (2019). Understanding disease progression and improving Alzheimer's disease clinical trials: Recent highlights from the Alzheimer's Disease Neuroimaging Initiative. *Alzheimer's & Dementia*, 15(1), 106–152. DOI: 10.1016/j.jalz.2018.08.005 PMID: 30321505

Wakhloo, D., Oberhauser, J., Madira, A., & Mahajani, S. (2022). From cradle to grave: Neurogenesis, neuroregeneration and neurodegeneration in Alzheimer's and Parkinson's diseases. *Neural Regeneration Research*, 17(12), 2606–2614. DOI: 10.4103/1673-5374.336138 PMID: 35662189

Weiner, M. W., Veitch, D. P., Aisen, P. S., Beckett, L. A., Cairns, N. J., Green, R. C., Harvey, D., Jack, C. R.Jr, Jagust, W., Morris, J. C., Petersen, R. C., Saykin, A. J., Shaw, L. M., Toga, A. W., & Trojanowski, J. Q. (2017). Recent publications from the Alzheimer's Disease Neuroimaging Initiative: Reviewing progress toward improved AD clinical trials. *Alzheimer's & Dementia*, 13(4), e1–e85. DOI: 10.1016/j.jalz.2016.11.007 PMID: 28342697

Weiss, J. H., & Sensi, S. L. (2000). Ca2+–Zn2+ permeable AMPA or kainate receptors: Possible key factors in selective neurodegeneration. *Trends in Neurosciences*, 23(8), 365–371. DOI: 10.1016/S0166-2236(00)01610-6 PMID: 10906800

Westeneng, H.-J., Debray, T. P. A., Visser, A. E., van Eijk, R. P. A., Rooney, J. P. K., Calvo, A., Martin, S., McDermott, C. J., Thompson, A. G., Pinto, S., Kobeleva, X., Rosenbohm, A., Stubendorff, B., Sommer, H., Middelkoop, B. M., Dekker, A. M., van Vugt, J. J. F. A., van Rheenen, W., Vajda, A., & van den Berg, L. H. (2018). Prognosis for patients with amyotrophic lateral sclerosis: Development and validation of a personalised prediction model. *Lancet Neurology*, 17(5), 423–433. DOI: 10.1016/S1474-4422(18)30089-9 PMID: 29598923

Wiedemann, F. R., Winkler, K., Kuznetsov, A. V., Bartels, C., Vielhaber, S., Feistner, H., & Kunz, W. S. (1998). Impairment of mitochondrial function in skeletal muscle of patients with amyotrophic lateral sclerosis. *Journal of the Neurological Sciences*, 156(1), 65–72. DOI: 10.1016/S0022-510X(98)00008-2 PMID: 9559989

Wodzinski, M., Skalski, A., Hemmerling, D., Orozco-Arroyave, J. R., & Nöth, E. (2019, July). Deep learning approach to Parkinson's disease detection using voice recordings and convolutional neural network dedicated to image classification. In 2019 41st annual international conference of the IEEE engineering in medicine and biology society (EMBC) (pp. 717-720). IEEE. DOI: 10.1109/EMBC.2019.8856972

Wolz, R., Julkunen, V., Koikkalainen, J., Niskanen, E., Zhang, D. P., Rueckert, D., Soininen, H., & Lötjönen, J. (2011). Multi-method analysis of MRI images in early diagnostics of Alzheimer's disease. *PLoS One*, 6(10), e25446. DOI: 10.1371/journal.pone.0025446 PMID: 22022397

Chapter 4
Advanced Polyp Segmentation Using U-Net Architecture:
A Review

Oumnia Sabri
Chouaib Doukkali University, Morocco

El mehdi El Aroussi
Chouaib Doukkali University, Morocco

Karim Abouelmehdi
Chouaib Doukkali University, Morocco

ABSTRACT

Early detection of polyps in colonoscopy images is crucial for preventing and treating colorectal cancer, a leading cause of cancer deaths worldwide. Accurate segmentation, which isolates polyps within the image, is essential for this detection process. This paper reviews the application of deep learning, specifically the U-Net architecture, for polyp segmentation. Thanks to its encoder-decoder architecture, U-Net can efficiently collect contextual information while maintaining spatial features, making it suitable for such use. Benchmark dataset evaluation confirms the effectiveness of U-Net models in accurately segmenting polyps within colonoscopy images. The presented technique is a viable option for computer-aided polyp detection, with the potential to improve early cancer diagnosis.

DOI: 10.4018/979-8-3373-1399-3.ch004

I. INTRODUCTION

Medical imaging has become an essential tool in modern healthcare. Magnetic Resonance Imaging (MRI) and Computed Tomography (CT) scans provide highly detailed views of internal organs and structures, aiding in accurate diagnoses, treatment planning, and surgical guiding. However, because of their complex architecture and overlapping properties, deciphering these images can be challenging.

Medical image segmentation (Prasantha et al., 2010) emerges as an effective solution to this issue. It allows for the identification of Regions of Interest (ROIs) within various 2D/3D imaging modalities, like MRI, CT or X-Rays. These ROIs could include organs, tissues, or even specific illnesses like cancer. Isolating these zones enables medical staff to take more precise measurements, perform quantitative analyses, and extract important information for better clinical decision-making.

One critical use of medical image segmentation is in gastroenterology, notably in the analysis of colonoscopy images to locate polyps. Colonoscopy is important in early colorectal cancer (CRC) screening because it allows for visual examination of the colon's inner lining as well as the identification and removal of polyps, which are precancerous growths that can progress to cancer. However, reliable polyp diagnosis during a colonoscopy can be difficult due to differences in polyp size, shape, and color, as well as potential limits in human visual perception.

Focusing on deep learning, a powerful branch of AI, this study investigates its potential to overcome this obstacle. This review explores the potential of U-Net, a powerful deep learning network for semantic segmentation, in tackling this challenge. The review looks at how U-Net can be applied to segment polyps in colonoscopy images, potentially aiding in improved polyp detection and ultimately earlier diagnosis.

The U-Net (Ronneberger et al., 2015) architecture is very useful for this application because to its unique design. This architecture leverages an encoder-decoder structure. The encoder extracts contextual information from the input image, which the decoder then utilizes to generate a pixel-by-pixel segmentation map. This enables the U-Net to not only detect the existence of polyps, but also precisely define their borders.

The remainder of this paper is organized as follows. Section II summarizes the various datasets used in polyp segmentation and the evaluation metrics employed to analyze the performance of segmentation algorithms. Section III discusses the structure of the U-Net architecture. An overview of optimization methods for polyp segmentation is given in Section IV. Section V examines the deep learning approaches based on U-Net used in polyp segmentation. The concluding section (Section VI) summarizes the paper's key findings and highlights potential directions for future research in this field.

II. DATASETS AND EVALUATION METRICS

A. Datasets

Colorectal polyps are small growths in the colon's inner lining. While most are benign, some can progress to colorectal cancer over time. Colonoscopy is the gold standard for finding these polyps, however precise diagnosis during the operation can be difficult. This is where colorectal polyp segmentation comes in, Table 1 provides a summary of some colonoscopy datasets.

Kvasir-SEG is a freely available dataset specifically designed for polyp segmentation in medical image analysis. It offers gastrointestinal polyp images with corresponding segmentation masks. These masks were meticulously created and validated by a medical professional and a gastroenterologist to ensure high-quality annotations for accurate analysis.

Kvasir-Sessile is a dataset of images containing sessile polyps from the Kvasir-SEG dataset.

CVC-ClinicDB is a collection of colonoscopy videos containing frames with polyps. 612 still images from 29 distinct sequences.

CVC-ColonDB 300 frames from 15 distinct videos. Polyps are given binary ground truth.

ETIS-LaribPolypDB is a dataset including 196 still images derived from 34 colonoscopy videos. There are 44 instances of distinct polyps displayed in various sizes and perspectives. The images have a HD resolution of 1225 x 966 pixels.

The **CVC-300** dataset includes 300 colonoscopy images. Each image is likely from a colonoscopy video and includes comments identifying the polyps.

CVC-ClinicVideoDB is a completely public video annotated database that contains 18 distinct standard definition video sequences, all of which feature a polyp. CVC-ClinicVideoDB has 10924 frames (768 × 576), with 9221 containing polyps.

Table 1. Summary of publicly available colonoscopy datasets

Dataset	Image Size	Number of patients	Number of images	Format
Kvasir-SEG	From 332x487 to 1920x1072	-	1000	Image
Kvasir-Sessile	From 401 × 415 to 1348 × 1070	-	196	Image
CVC-ClinicDB	288 × 384	23	612	Image
CVC-ColonDB	500 × 574	13	300	Image
ETIS-LaribPolypDB	966 × 1225	-	196	Image
CVC-300	500 × 574	-	60	Image
CVC-ClinicVideoDB	768 x 576	10924	18	video

B. Evaluation Metrics

Medical image segmentation is a significant activity in healthcare since it helps with diagnosis, therapy planning, and illness monitoring. Various metrics can be employed to assess the efficacy of machine learning algorithms used in medical image segmentation.

To guarantee that these machine learning algorithms perform optimally, it is critical to employ appropriate evaluation metrics. These metrics quantify the segmentation's accuracy by measuring the agreement between the algorithm's generated segmentation and a previously defined, ideal segmentation (ground truth). The suitability and reliability of different segmentation algorithms depends on various factors, and a robust evaluation involves multiple metrics to fully assess their performance. Common metrics include similarity coefficients, overlap measures, distance metrics, and volumetric differences. However, picking appropriate metrics is a delicate process, since the most informative ones are determined by the specific context of medical imaging and the properties of the datasets in question.

All provided metrics are based on the generation of the confusion matrix (Table 2) for a binary segmentation mask, which includes.

- True Positive (TP). Pixels that are truly part of the foreground and are correctly identified by the algorithm as foreground.
- False Positive (FP). Background pixels that are incorrectly classified as foreground, leading to over-segmentation.
- True Negative (TN). Pixels that are truly part of the background and are correctly identified by the algorithm as background.
- False Negative (FN). Foreground pixels that are incorrectly classified as background, leading to under-segmentation.

The values of all presented measures range from zero (worst) to one (best).

Accuracy, as shown in Equation (1), reflects the overall proportion of pixels in the input image that were correctly classified. The Dice Similarity Coefficient (DSC), also known as the F1 score (Equation (3)), measures the overlap between the model's predictions and the ground truth. Equations (4) and (5) detail the mathematical formulations for recall and precision, respectively. Recall, sometimes referred to as sensitivity or true positive rate, emphasizes the model's ability to identify true foreground pixels. Finally, the Jaccard score (Equation (6)), also known as Intersection-over-Union (IoU), quantifies the area of overlap between the predicted segmentation and the ground truth.

Table 2. The confusion matrix (Predicted vs. Ground Truth)

		Ground truth	
		Foreground	*Background*
Prediction	***Foreground***	TP	FP
	Background	FN	TN

$$A\chi\chi = (T\Pi+TN) / (T\Pi+TN+\Phi\Pi+\Phi N) \qquad (1)$$

$$\Delta\Sigma X = T\Pi / (2.T\Pi + \Phi\Pi + \Phi N)$$
(2)

$$\Phi 1-\sigma\chi o\rho\varepsilon = (2.\Pi\rho\varepsilon.P\varepsilon\chi) / (\Pi\rho\varepsilon + P\varepsilon\chi)$$
(3)

$$P\varepsilon\chi = \Sigma\varepsilon\nu = T\Pi / (T\Pi + \Phi N)$$
(4)

$$\Pi\rho\varepsilon = T\Pi / (T\Pi + \Phi\Pi)$$
(5)

$$Io\Upsilon = \vartheta I = T\Pi / (T\Pi + \Phi\Pi + \Phi N) \qquad (6)$$

III. OVERVIEW OF U-NET ARCHITECTURE

The U-Net (Ronneberger et al., 2015) architecture, developed by Ronneberger et al. which marks a significant improvement in medical image segmentation. Deep convolutional neural networks have revolutionized various visual identification tasks in recent years, surpassing traditional methods in terms of efficiency and accuracy. However, using neural networks to biomedical image interpretation presents distinct obstacles due to both the scarcity of labeled training data and the intricacy of biological structures.

One of the U-Net architecture's main advantages is its capacity to effectively utilize data augmentation. By including numerous transformations into the training data, such as rotations, flips, and scaling, the network may learn from a wide variety of instances and generalize more well to previously unknown data. This approach is especially useful in biological imaging, where gathering huge annotated datasets is either difficult or costly.

The U-Net architecture is presented in Figure 1, here is an overview of how the U-Net works.

Contracting Path (Encoder). The contracting path uses max pooling for downsampling, ReLU activation, and repeated 3x3 convolutions to capture context. With each downsampling step along this process, the number of feature channels is increased.

Expansive Path (Decoder). Unlike the shrinking contracting path, the expansive path in U-Net expands features back to their original size. It cleverly combines this upsampled information with details from the contracting path to achieve precise localization.

Skip Connections. The U-Net architecture includes skip connections that link the layers in the contracting path to the equivalent layers in the expanding path. These connections enhance segmentation accuracy and aid in the preservation of spatial information throughout the up-sampling procedure.

Overlap-Tile Strategy. U-Net can handle very large images by using an "overlap-tile" approach. Here, the image is broken down into smaller overlapping sections, each predicted individually. Missing information at the edges is filled in by mirroring the existing image data. This allows U-Net to seamlessly segment even massive images.

Training Efficiency. U-Net's ability to take use of data augmentation techniques, which artificially enlarge the training dataset, gives rise to its efficacy even with a small number of annotated images. This efficient training strategy, along with the network's design, helps it succeed in biomedical image segmentation tasks.

Figure 1. U-net architecture

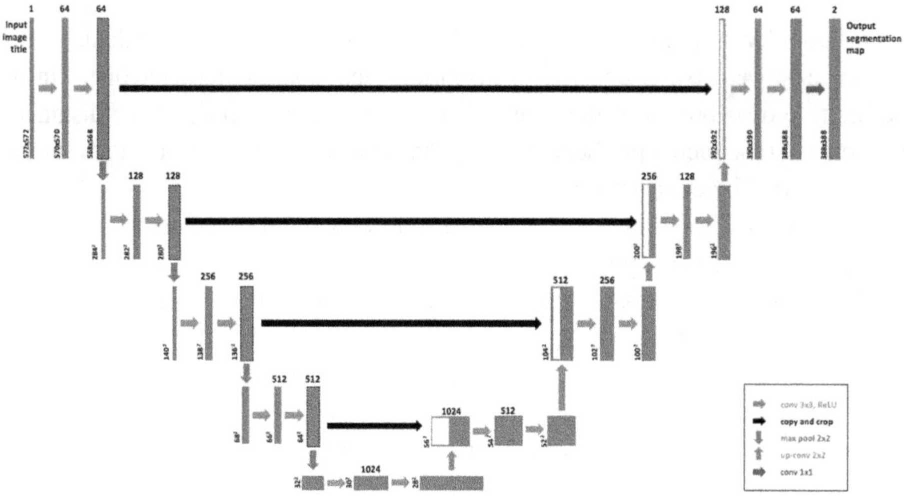

Overall, the U-Net architecture's unique design, with its contracting and expansive paths, skip connections, and overlap-tile strategy, is an enormous progress forward in biomedical image analysis, providing a strong framework for accurate and fast segmentation of complex biological structures. U-Net has established itself as a leader in biomedical image segmentation by combining novel design ideas and successful training methodologies.

IV. OPTIMIZATION TECHNIQUES

A. Loss Functions

A loss function (Azad et al., 2023), also known as a cost function or objective function, is a key idea in neural networks and machine learning that is essential to model training. It measures the disparity between a model's predicted output and the actual target values. Essentially, it offers a numerical gauge of the model's performance, indicating how well or poorly it is functioning. Through its ability to indicate how the model's predictions differ from the expected outcome, the loss function serves as a guidance during the learning process. During training, the model continuously adjusts its internal parameters (weights and biases) to minimize this loss. A lower loss signifies better performance, while a higher loss indicates that the model's predictions deviate significantly from the target values.

1) Cross Entropy

One popular loss function in machine learning is cross-entropy. It calculates the difference between two probability distributions. the model's predicted distribution and the true distribution of the labels, or the actual target values. The "distance" between what the model predicts and what the actual distribution of the data should be is quantified by cross-entropy.

In binary classification tasks, such as image segmentation, where each pixel or instance needs to be categorized into one of two categories, Binary Cross-Entropy (BCE) and Weighted Cross-Entropy are two of the loss functions that are utilized (e.g., foreground vs. background).

A loss function called binary cross-entropy calculates the difference between a binary classification model's anticipated probability and the actual binary labels. It is frequently applied when a model yields a probability score (from 0 to 1) that represents the chance that a given input is a member of a particular class. It is given by.

$$\mathbf{BCE} = -[y.\log(p)+(1-y).\log(1-p)]$$

where the real binary label, y, is either 0 or 1 and p represents the expected likelihood that the input is a member of the class denoted by the number 1.

To address class imbalance, the usual Binary Cross-Entropy Loss is modified to use Weighted Cross-Entropy. In Weighted Cross-Entropy, various weights are assigned to the classes in the loss function to give more importance to the minority class. This guarantees that the model focuses more on accurately predicting the less frequent class, thereby preventing bias towards the majority class. We obtain the Weighted Cross-Entropy Loss as follows.

$$\mathbf{WCE} = -[\alpha.y.\log(p)+\beta.(1-y).\log(1-p)]$$

Where weights α and β are assigned to the positive class (designated as 1) and the negative class (designated as 0), respectively.

In medical image segmentation, when the regions of interest (such as tumors) may constitute a relatively tiny amount of the image relative to the background, Weighted Cross-Entropy is very helpful. The model is motivated to concentrate more on identifying these critical areas by assigning the minority class more weights, which enhances segmentation performance overall.

2) Dice Loss

Dice Loss is a loss function particularly made for testing and fine-tuning models in medical image segmentation tasks, where the objective is to precisely identify regions of interest, such organs or tumors, inside an image. It is based on the Dice Similarity Coefficient (DSC), a statistical comparison of two sets of data (in this example, the anticipated segmentation and the ground truth segmentation) to see how similar they are.

Achieving exact boundaries is essential in medical image segmentation since even little inaccuracies can have serious clinical consequences. By explicitly optimizing the overlap between the predicted and true segmentation masks, the Dice Loss aids in resolving this. It has the following definition.

Dice Loss $= 1 - (2 \times |A \cap B|) / (|A|+|B|)$

where A denotes the set of pixels in the predicted segmentation and B denotes the set of pixels in the ground truth segmentation. The numerator $2 \times |A \cap B|$ represents the intersection between the two sets, while the denominator $|A|+|B|$ reflects their combined sizes. This formulation guarantees that the Dice Loss ranges from 0 to 1, where 0 signifies perfect overlap and 1 signifies no overlap.

The model is motivated to maximize the overlap between the predicted and actual segments by reducing Dice Loss, which enhances the reliability and precision of medical image segmentation tasks.

3) Focal Loss

By adding a modulating factor $(1 - p_t)^\gamma$ to the loss, Focal Loss (Lin et al., 2017) alters the standard Cross-Entropy Loss in a way that decreases the contribution of easy examples (those the model correctly classifies with high confidence) and increases the emphasis on hard examples (those the model misclassifies or classifies with low confidence). For a single example, the Focal Loss is defined as.

Focal Loss $= -\alpha_t.(1 - p_t)^\gamma.\log(p_t)$

Where the predicted probability of the correct class is denoted by pt. To be more precise, if the real class label is y = 1, then $p_t = p$; otherwise, $p_t = 1\text{-}p$. αt is a weighting factor that equalizes the significance of instances that are positive and negative. This is helpful in cases when there is an important class imbalance. The focusing parameter, γ, controls how quickly simple samples are down-weighted. The emphasis on challenging cases is increased with a greater value of γ (usually 2).

B. Optimizers

The application of optimization techniques is essential to AI model optimization as it contributes to better training and increased model performance. In order to train a neural network model with its highest possible performance and accuracy, the optimizer plays a vital part. In deep learning, the optimizer is crucial in modifying the model's parameters to reduce the loss function and enhance the neural network's overall performance.

The **Stochastic Gradient Descent (SGD) with momentum** algorithm is an optimization technique that improves on the basic SGD (Ketkar & Ketkar, 2017) approach to achieve better results. While SGD updates model parameters only using the loss function's current gradient, momentum adds a notion of "velocity" or "momentum" to the update procedure. The optimizer is able to preserve a directional bias because of the velocity term's accumulation of historical gradient information. Momentum forces the update to speed in the direction of the gradient's consistent point, hence increasing the step size. On the other hand, momentum reduces oscillations when there is a frequent shift in gradient direction. This characteristic is especially useful when the loss landscape has narrow plateaus or valleys because momentum can assist the optimizer break out of local minima and converge more quickly. A hyperparameter that controls the effect of the prior update on the current one, beta (β), is commonly used to manage the momentum term. There is a greater chance of overshooting the ideal solution when the beta value is larger, but this might also result in faster convergence. In order to get best performance, careful tweaking of the momentum hyperparameter is necessary. Momentum is added to SGD, making it adaptive to noisy gradients and more effective at navigating intricate optimization environments. Deep learning has made extensive use of this optimization strategy since it has been shown to be successful in training neural networks with a large number of parameters.

Another improvement on the momentum-based optimization method is the **Nesterov Accelerated Gradient (NAG)**. Nesterov adopts a more proactive approach than standard momentum, which first computes the gradient at the current point before updating the parameters using the momentum-adjusted gradient. The optimizer first moves the parameters a step ahead in the direction of the current momentum to get into the "lookahead" position. Then, instead of computing the gradient at the present position, it does so at this lookahead point. As a result, the optimizer can predict the direction that the parameters will go and modify the update appropriately. By looking ahead, NAG can adjust its trajectory sooner, resulting in improved convergence, especially in complex or rapidly changing loss landscapes. Such as standard momentum, NAG uses a momentum hyperparameter (beta) to regulate how much the previous update affects the current one. But NAG is distinct from standard momentum since it applies the gradient at the lookahead point, which makes it a more complex and frequently more efficient optimization algorithm.

RMSProp (Elshamy et al., 2023), short for Root Mean Square Propagation, is a method for adaptive learning rate optimization that was created specially to address the drawbacks of earlier optimizers like Adagrad. One of Adagrad's principal challenges is that it accumulates all of the previous squared gradients in its denominator, which gradually lowers the learning rate and increases the risk of early convergence and slow learning in the last phases of training. In order to remedy this, RMSProp replaces a moving average of squared gradients instead of a cumulative sum. The decay rate hyperparameter that governs this moving average, commonly represented by γ (gamma), establishes the relative importance of recently formed gradients over older ones. The core concept of RMSProp is to normalize the gradient through this moving average, which prevents the learning rate from diminishing excessively and ensures that the model continues to learn effectively during the training process. This method works particularly well for non-stationary issues, when the training process changes the loss landscape. RMSProp is a preferred option for training deep neural networks because it can successfully adapt to these changing conditions by dynamically modifying the learning rate based on recent gradient information. When working with mini-batches of data, where gradients may be noisy and vary in scale across various parameter space dimensions, RMSProp is very useful. More consistent progress towards minimizing the loss function is made possible by RMSProp, which helps stabilize updates by scaling the learning rate separately for each parameter. Because of this, it's particularly well suited for training complicated models where certain parameters could need more updates while others merely need little updates.

Another adaptive learning rate optimization algorithm, **Adadelta** (Zeiler, 2012), expands on the concepts of Adagrad while introducing major enhancements to overcome its drawbacks. Adadelta presents a new technique by confining the window of collected gradients to a limited size, whereas Adagrad suffers from a monotonically declining learning rate since it accumulates all previous squared gradients. Adadelta focuses on recent history rather than the full training process by employing a declining average of these gradients instead of adding all previous squared gradients. This method keeps the learning rate from dropping too quickly, enabling learning to continue even during the later training stages. The ability of Adadelta to do away with the requirement for a manually chosen global learning rate is one of its primary improvements. Rather, updates are computed using the running averages of the squared gradients divided by the squared updates. Due to the dynamic determination of the learning rate by this ratio, Adadelta is less susceptible to the particular issue scale and requires less comprehensive hyperparameter adjustments. Adadelta guarantees that every parameter is changed in a manner that is consistent with its recent behavior by normalizing the updates in this manner, resulting in more reliable and efficient learning. Deep networks, where the gradients' scales vary greatly between layers, are especially well-suited for Adadelta. Adadelta is able to control these variations by adapting the learning rate according to recent gradients, which results in smoother and more consistent convergence. Moreover, its capacity to address the issue of vanishing or exploding gradients renders it an essential tool for training extremely deep networks, where sustaining a stable learning rate is vital for effective optimization.

Both RMSProp and Adadelta are adaptive learning rate techniques that overcome the limitations of earlier optimizers such as Adagrad. Adadelta takes one step further by dynamically changing the learning rate depending on the ratio of previous gradients and updates, whereas RMSProp concentrates on stabilizing the learning process by utilizing a moving average of squared gradients to avoid the learning rate from dropping too much. This makes it unnecessary to manually adjust the learning rate and better adjusts to the specific characteristics of the model being trained.

Adam (Kingma & Ba, 2014), short for Adaptive Moment Estimation, is among the most commonly utilized optimization algorithms in deep learning, thanks to its strong performance and user-friendliness. Adam merges the benefits of these two optimizers. AdaGrad is an effective method for handling sparse gradients, whereas RMSProp modifies learning rates according to the magnitudes of recent gradients. The fundamental principle of Adam is to estimate the first moment (mean) and the second moment (uncentered variance) of the gradients in order to compute adaptive learning rates for each parameter. Two hyperparameters, usually designated as β_1 and β_2, govern the exponentially decaying averages of previous gradients and their squares, which are used to update these moments. The default values for these hyperparameters are typically 0.9 and 0.999, respectively, and they are effective for a variety of issues. The first moment estimate aids Adam in grasping the direction in which the parameters should shift, while the second moment estimate enables it to adjust the updates according to the variance of the gradients, thus stabilizing the learning process. Additionally, Adam incorporates a bias-correction mechanism to account for the fact that the moment estimates are initialized to zero. This ensures that the estimates are unbiased, particularly during the early stages of training when the number of updates is small. In order to preserve efficient learning dynamics, this feature is essential. Practitioners use Adam because of its effectiveness, minimal memory needs, and versatility in task performance. Its adaptability makes it appropriate for a variety of machine learning and artificial intelligence applications, since it can be used to solve both convex and non-convex optimization problems. In summary, Adam is an exceptional optimization algorithm that has made a substantial contribution to the development of deep learning techniques. It is reliable and powerful.

Nadam, short for Nesterov-accelerated Adaptive Moment Estimation, is an advanced optimization technique that builds upon the foundations of the Adam optimizer by incorporating Nesterov momentum. This enhancement enables Nadam to harness the advantages of both adaptive learning rates and the predictive capabilities of Nesterov momentum, which forecasts the future position of the parameters based on their current velocity. Like Adam, Nadam computes adaptive learning rates for each parameter by estimating the first moment and the second moment of the gradients, utilizing exponentially decaying averages controlled by hyperparameters β_1 and β_2. Nadam, however, makes a significant adjustment by performing the Nesterov momentum update before to the gradient calculation. This helps to achieve better performance and faster convergence, mostly for difficult optimization landscapes. The algorithm preserves the bias-correction mechanism found in Adam, guaranteeing that the moment estimates remain unbiased during the early stages of training. This specific feature has significant importance in stabilizing the learning process and improving the optimization's overall efficiency. Nadam is notably effective in situations with non-stationary objectives and noisy gradients, making it ideal for various deep learning applications. Its capacity to adaptively modify learning rates while integrating momentum enables it to traverse the optimization landscape more efficiently than traditional techniques. Nadam is a potent option for practitioners looking to increase convergence rates and enhance model performance because of its combination of adaptive learning rates and Nesterov momentum. All things considered, Nadam is a big step forward for stochastic optimization methods, offering a reliable and effective tool to address the difficulties associated with current machine learning.

These widely used optimization algorithms are crucial for training deep learning models, each possessing distinct advantages. Stochastic Gradient Descent (SGD) with Momentum boosts convergence by adding a portion of the previous gradient update into the current one, thereby smoothing the optimization path. SGD with Nesterov Momentum builds on this by predicting the future position of the parameters, resulting in more precise updates and minimizing the risk of overshooting. RMSProp adjusts the learning rate according to recent gradient magnitudes, preventing it from becoming excessively small and ensuring consistent progress, particularly in non-stationary or noisy environments. Adadelta also modifies the learning rate but concentrates on recent gradients alone, removing the necessity for a manually set global learning rate and making it especially beneficial for deep networks. Adam merges the advantages of RMSProp and momentum by calculating adaptive learning rates alongside moment estimates, ensuring strong performance across various tasks with minimal tuning. Ultimately, Nadam incorporates Nesterov momentum into Adam, delivering smoother and potentially quicker convergence by predicting the direction of updates. While all these optimizers are effective, the selection among them hinges on the specific problem. Adam and Nadam are frequently preferred for their adaptability and user-friendliness, RMSProp and Adadelta are outstanding for non-stationary issues, and the momentum-based SGD variants excel at accelerating convergence in steep or noisy areas of the loss landscape.

C. Hyperparameter Tuning

In machine learning, hyperparameters are parameters that regulate the learning process of a model. Unlike model parameters that are learned directly from the data, hyperparameters are fixed before training begins. They determine important characteristics of the model such as its complexity or learning speed.

Hyperparameter tuning is a critical phase in machine learning model optimization. It entails finding the best combination of hyperparameters that lead to the highest performance of the model. It is essential to adjust a number of important hyperparameters for neural networks, such as.

- **Learning rate**. In the training of neural networks, learning rate is a key hyperparameter that controls the size of steps the model takes in the backpropagation process to optimize its weights. It basically defines how much every time the model weights are changed, to adjust the model according to the predicted error. While a high learning rate makes it possible for the model to learn more quickly, it also increases the chance of overshooting the optimal solution, which might result in unstable training and low performance. On the other hand, a low learning rate could result in much slower learning and a longer time for the model to converge, but it also guarantees more accurate and reliable updates. Achieving the right learning rate frequently requires striking a balance. if it is set too high, the model may not converge; if it is set too low, the model may find itself trapped in a local minimum or take a long time to come up with a good solution. Occasionally, as the model gets closer to convergence, the learning rate is dynamically changed, beginning at a higher value and progressively getting smaller. This helps to improve the solution in the final phases of training. To get the most performance out of neural networks, this precise adjustment of the learning rate is necessary.
- **Batch size.** Another important hyperparameter in neural network training is batch size, which determines how many training samples the model processes before updating its weights. The dataset is typically too big to analyze in its whole while training a model, therefore it is split up into smaller sections known as batches. The batch size is the number of instances from the dataset that are utilized in a single iteration to compute the gradient and update the model's parameters. Usually, the model can learn fine-grained updates and

become more sensitive to the details in the data with a small batch size, like 8, 16 or 32. This may result in learning that is more stable, but it may also increase noise in the gradient estimate, which could result in updates that are less accurate every iteration. However, in contrast, by averaging across a greater number of data points, a larger batch size, such as 128 or 256, offers a more accurate estimate of the gradient and can lead to faster training by taking advantage of hardware acceleration, such as GPUs. However, large batch sizes can need more memory and cause a slower rate of convergence if the model is unable to adequately capture the finer details of the data distribution. The batch size selection also affects the generalization of the model; Larger batches may converge more quickly but run the risk of overfitting if improperly regularized, whereas smaller batches may generalize better but at the cost of slower training. Therefore, choosing the optimal batch size entails striking a balance between model performance and computing efficiency depending on hardware capabilities, dataset size, and the particular learning method used.

- **Epochs.** In a machine learning model, epochs denote the total number of runs over the training dataset. Every sample in the dataset is processed by the model at each epoch, and its internal weights are modified in accordance with the calculated errors. How often the learning algorithm runs through the complete dataset and refines the model's weights with each pass depends on the number of epochs. Early epochs give rise to notable improvements in the model's performance since initially the model's predictions are usually way off. The model learns more and more as the number of epochs increases, although normally at a slower rate as it begins to fine-tune its understanding of the data. Nevertheless, a large number of epochs may result in overfitting, a situation in which the model loses its capacity to generalize to new, untried data due to its overly specialized to the training set. In order to determine the optimum number of epochs, practitioners frequently track the model's performance on a validation set, searching for the point at which the validation accuracy either stops increasing or begins to decline, indicating that receiving further instruction would be ineffective. The number of epochs selected is important since a too few can cause underfitting, or the model not learning enough from the data, while too many can cause overfitting. Building a successful and generally applicable model requires finding a balance with this number.
- **Dropout rate.** The dropout rate is a vital hyperparameter in neural networks, especially in deep learning, that governs the regularization method referred to as dropout. Dropout is employed to prevent overfitting, a phenomenon where a model excels on training data yet struggles to generalize to new,

unseen data. During training, dropout functions by randomly "dropping out" or setting to zero a portion of the neurons in a layer for each forward pass. The percentage of neurons to be discarded is determined by the dropout rate, which typically has values between 0.2 and 0.5. This means that 20% to 50% of the neurons are momentarily ignored during each iteration. By avoiding over-reliance on any one neuron, this method pushes the network to learn more resilient features, which promotes variety and redundancy in the learned representations. When a neuron is dropped, its associated weights do not contribute to the forward or backward pass, effectively reducing the network's size and making it less susceptible to overfitting. The whole network is utilized during inference, with dropout being disabled. To compensate for the increased number of active neurons, the weights are reduced by the dropout rate. The generalization performance of neural networks has been shown to be greatly enhanced by this technique on a variety of tasks.

- **L2regularization.** is a method used in machine learning models, especially neural networks, to prevent overfitting. It may also be referred to as ridge regression or weight decay. When a model becomes overfit, it performs badly when faced with new data because it has learned the noise and details of the training data. In order to overcome this, L2 regularization adds a penalty to the loss function that is based on the weights in the model. In particular, the penalty is proportionate to the total squares of the weights, which is why it is called "L2." By keeping the weights modest and more equally distributed, this regularization term prevents the model from giving undue weight to any particular feature. It has the effect of smoothing the decision boundary, which reduces the model's sensitivity to small variations in the training set. Consequently, the model performs better on not before seen examples and generalizes to new data more effectively. One hyperparameter that controls the regularization strength is how much weight the regularization term has in relation to the original loss; this is commonly represented by the symbol lambda (λ). A low value of λ could not offer enough regularization to prevent overfitting, while a large value might cause underfitting, where the model is too simplistic to detect the underlying patterns in the data. Therefore, to be able to achieve a balance between bias and variance and produce a more reliable and more general model, careful adjustment of the L2 regularization parameter is required.

D. Comparison of Optimization Techniques in U-Net Variants

A comparison of the optimization techniques used in different U-Net designs for polyp segmentation is shown in Table 3. Important parameters including the optimizer, loss function, and hyperparameter tuning methods are shown in the table. This thorough analysis of optimization components provides valuable insights into the methodological choices made in the realm of polyp segmentation and serves as a basis for further research and optimization efforts.

Table 3. Overview of Optimization Techniques in U-Net Variants.

Paper	Loss Function	Optimizer	Hyperparameter Tuning	Regularization
Sun et al.	Binary Cross-Entropy	The ADAM optimizer with a weight of 5x10^{-4}	The learning rate scheduler used is CosineAnnealingLR, with an initial learning rate of 1×10^{-5} and a maximum of 80 epochs	Data augmentation techniques
Tashk et al.	Dice loss	-	-	Skip connections between the encoding and decoding paths
Safarov et al.	Binary Cross-Entropy	The ADAM optimizer is employed, with a learning rate that starts from 0.01 and is reduced when the patience level exceeds 5	The batch size is set to 10, and the model is trained for 100 epochs	Early-stop mechanism is implemented on the validation set to avoid overfitting
Branch et al.	Dice loss	The Adadelta optimizer	The model was trained for 100 epochs and converged with 76 epochs with a learning rate of 0.0001, batch size of 16.	Early-stopping regularization was implemented in the validation subset, specifically in the validation loss, to avoid overfitting.
Jha et al.	Binary Cross-Entropy and Dice loss function also mentioned as competitive	The Nadam optimizer	The model was trained for 300 epochs with the batch size of 16.	Early stopping was employed to prevent overfitting. Stochastic Gradient Descent with Warm Restarts (SGDR) was also used to further improve results.
Yeung et al.	Hybrid Focal loss (HFL)	Stochastic Gradient Descent with Nesterov momentum	The hyperparameters are selected through grid-search.	Data augmentation techniques. scaling, rotation, elastic deformation, mirror and gamma transformations.

continued on following page

Table 3. Continued

Paper	Loss Function	Optimizer	Hyperparameter Tuning	Regularization
Mohapatra et al.	Dice loss	Stochastic gradient descent with momentum (SGDM)	The hyperparameters were selected after extensive experimental work, the learning rate (LR) was set to 0.0001, momentum was 0.9 the network was trained for 50 epochs.	L2regularization was used, with a value of 0.005.
Tran et al.	Combination of dice-loss and weighted cross-entropy (WCE)	-	The hyperparameters set for the training process include. 150 epochs, 0.2 dropout rate, and 3×10 −4 initial learning rate. Every ten epochs, a learning rate schedule is used to adjust the learning rate.	An early-stopping system was implemented to save training time and a dropout rate of 0.2 applied during training to prevent overfitting.
Tomar et al.	Combination of dice-loss and weighted cross-entropy (WCE)	The ADAM optimizer with a learning rate of 10^{-4}	The ReduceLROnPlateau is used to reduce the learning rate for better performance and an early stopping to halt training when the model stopped improving with a batch size of 16.	Data augmentation strategy with random rotation, horizontal flipping, vertical flipping and coarse dropout.
Lou et al.	Combination of losses	The ADAM optimizer with the initial learning rate 10^{-4}.	The model employs a multi-scale training strategy with scales of {0.75, 1.0, 1.25} instead of traditional data augmentation.	-
Dumitru et al.	Dice loss	The RMSprop optimizer with a learning rate of 0.0001	The model was trained with a batch size of 4 for 600 epochs.	Data augmentation was used on the training set to improve the model's generalization capabilities, making regularization techniques such as dropout unnecessary.
Karthikha et al.	Dice loss	The Nadam optimiser with a learning rate of 10^{-3}	The batch size of 8, and an iteration of 25 epochs.	Dropout regularization is employed to mitigate overfitting issues during training.
Li et al.	Combination of weighted Intersection over Union (IoU) loss and weighted Binary Cross-Entropy (BCE) loss	The ADAM optimizer with the initial learning rate 10^{-4}.	The model was trained with a batch size of 4 and a multi-scale training technique (0.75, 1, 1.25) for data augmentation.	-

V. COMPARISON AND ANALYSIS

U-Net's success has inspired the development of various U-shaped models for confronting semantic segmentation tasks in medical imaging and beyond. These models, inspired by U-Net's encoder-decoder architecture, aim to strive to enhance both segmentation accuracy and computational efficiency. Sun et al. (Sun et al., 2019) presented a new deep learning framework based on the U-Net architecture. By utilizing dilated convolutions, their method achieved state-of-the-art performance on both CVC-ClinicDB and ETIS-LaribPolypDB datasets, significantly outperforming prior approaches. While Tashk et al. presents a novel approach using a modified U-Net architecture. The method includes preprocessing steps involving color transformations, followed by the application of a convolutional neural network with a U-Net architecture designed to extract features from images or video streams. Post-processing techniques are employed to improve pixel-wise classification outcomes. The efficacy of the suggested technique in identifying and localizing polyps in colonoscopy images is demonstrated by its superior performance over current methods in terms of precision, accuracy, recall, and F1-Score criteria.

Safarov and Whangbo (Safarov & Whangbo, 2021) introduced an approach named A-DenseUNet, which leverages the UNet architecture with elements from DenseNets to enhance polyp segmentation accuracy. The A-DenseUNet model significantly improves segmentation results by integrating several dilated convolutions with varied atrous rates and responding to diverse datasets. This approach produced better results than other methods, including UNet (Ronneberger et al., 2015), ResUNet (Zhang et al., 2018), U-Net++ (Zhou et al., 2019) and PraNet (Fan et al., 2020).

Another polyp localization method, referred to as U-Net-MobileNetV2, was also proposed by the authors in (Branch & Carvalho, 2021). This model offers superior segmentation performance by combining U-Net with the efficient feature extraction of MobileNetV2 (Sandler et al., 2018). It reduces training time, maintains high accuracy with a compact design, and demonstrates potential for generalizability to other image segmentation tasks. Whereas, ResUNet++ (Jha et al., 2021) stands out in medical image segmentation with its advanced features like residual blocks and attention mechanisms, leading to superior performance. Focus U-Net (Yeung et al., 2021) is another effective framework that integrates dual attention mechanisms, a Focus Gate module for selective feature learning, and a Hybrid Focal Loss function to address class-imbalanced segmentation challenges.

UPolySeg (Mohapatra et al., 2022) utilizes the well-established U-Net architecture, perhaps with changes for polyp segmentation tasks, meanwhile MRRU-Net (Tran et al., 2022) extends U-Net by applying recurrent connections into its convolutional layers. These recurrent connections allow MRRU-Net to understand long-range dependencies in images, potentially enhancing its capacity to identify complicated polyp features. Tomar et al. (Tomar et al., 2023) proposed DilatedSegNet that begins with a pre-trained ResNet for strong feature extraction and then uses Dilated Convolution Pooling (DCP) blocks to expand the receptive field and capture more diverse and dependable features for improved delineation.

Developed by Lou et al. (Lou et al., 2022), CaraNet integrates innovative components for feature extraction and attention within the framework of the well-established U-Net design. This novel neural network architecture addresses the challenge of accurately segmenting small medical objects which has an influence on early illness detection in medical imaging applications. On another side, DUCK-Net (Dumitru et al., 2023) employs a supervised convolutional neural network with a unique encoder-decoder architecture. The fundamental innovation is the DUCK block, a novel convolutional block used throughout the encoder. Several benchmark datasets, notably ETIS-LaribPolypDB, CVC-ClinicDB, Kvasir-SEG, and CVC-ColonDB, are used to assess this model.

Table 4. Overview of articles using U-Net architecture for polyp segmentation.

Paper	Year	Algorithm	Dataset	Performance
Sun et al.	2019	U-Net with Dilation	CVC-ClinicDB ETIS-LaribPolypDB	(CVC-ClinicDB, ETIS-LaribPolypDB) DSC = 0.825, 0.625 Pre = 0.967, 0.805 Rec = 0.955, 0.813 F1-score = 0.961, 0.809
Tashk et al.	2019	Innovative U-Net architecture	CVC-ClinicDB CVC-ColonDB ETIS-LaribPolypDB	(CVC-CLINICDB, ETIS- LaribPolypDB, CVC-COLONDB) Acc = 0. 990, 0.996, 0.982 Pre = 0.702, 0.702, 0.620 Rec = 0.827, 0.909, 0.824 F1-score = 0.760, 0.792, 0.707
Safarov et al.	2021	A-DenseUNet	Kvasir-SEG CVC-612	(Kvasir-SEG, CVC-612) Dice = 0.909, 0.891 IoU = 0.862, 0.855 Rec = 0.945, 0.945 Pre = 0.977, 0.927
Branch et al.	2021	U-Net-MobileNetV2	Kvasir-SEG	DSC = 0.897, IoU = 0.816
Jha et al.	2021	ResUNet++	Kvasir-SEG CVC-ClinicDB KVASIR-SESSILE CVC-VIDEOCLINICDB	(Kvasir-SEG, CVC-ClinicDB, KVASIR-SESSILE, CVC-VIDEOCLINICDB) DSC = 0.851, 0.920, 0.504, 0.811 mIoU = 0.833, 0.890, 0.661, 0.874 Rec = 0.876, 0.939, 0.485, 0.775 Pre = 0.823, 0.855, 0.680, 0.671
Yeung et al.	2021	Focus U-Net	CVC-ClinicDB CVC-ColonDB ETIS-LaribPolypDB Kvasir-SEG CVC-T	(CVC-ClinicDB, CVC-ColonDB, Kvasir-SEG, ETIS-LaribPolypDB, CVC-T) mDSC = 0.938, 0.878, 0.910, 0.832, 0.920 mIoU = 0.889, 0.804, 0.853, 0.757, 0.860

continued on following page

Table 4. Continued

Paper	Year	Algorithm	Dataset	Performance
Mohapatra et al.	2022	UPolySeg	CVC-ClinicDB ETIS-LaribPolypDB Endoscene CVC-ColonDB Kvasir-SEG	DSC = 0.969, IoU = 0.879, Rec = 0.956, Pre = 0.923
Tran et al.	2022	MRRU-Net	CVC-ClinicDB CVC-ColonDB ETIS-LaribPolypDB	(CVC-ClinicDB, ETIS-LaribPolypDB, CVC-ColonDB) DSC = 0.946, 0.927, 0.933 F1-score = 0.948, 0.928, 0.934 mIoU = 0.898, 0.865, 0.875 Rec = 0.949, 0.919, 0.937 Pre = 0.946, 0.939, 0.932
Tomar et al.	2022	DilatedSegNet	Kvasir-SEG BKAI-IGH	(Kvasir-SEG, BKAI-IGH) DSC = 0.896, 0.895 mIoU = 0.834, 0.832 Rec = 0.917, 0.908 Pre = 0.910, 0.911 F2 = 0.903, 0.899
Lou et al.	2023	CaraNet	Kvasir-SEG CVC-ColonDB CVC-ClinicDB ETIS-LaribPolypDB	(Kvasir-SEG, CVC-ClinicDB, CVC-ColonDB, ETIS-LaribPolypDB, CVC-T) mDSC = 0.918, 0.936, 0.773, 0.747, 0.903 mIoU = 0.865, 0.887, 0.689, 0.672, 0.838
Dumitru et al.	2023	DUCK-Net	Kvasir-SEG CVC-ColonDB CVC-ClinicDB ETIS-LaribPolypDB	(Kvasir-SEG, CVC-ClinicDB, CVC-ColonDB, ETIS-LaribPolypDB) DSC = 0.950, 0.948, 0.935, 0.935 JI = 0.905, 0.901, 0.879, 0.879 Pre = 0.963, 0.949, 0.931, 0.954 Rec = 0.938, 0.949, 0.939, 0.940 Acc = 0.984, 0.991, 0.993, 0.993
Karthikha et al.	2024	Dilated-U-Net-Seg	CVC-ClinicDB KVASIR-SEG KVASIR Sessile KVASIR Capsule Hyper KVASIR - SEG	Acc = 0.971, DSC = 0.979, JI = 0.880, Rec = 0.959, Pre = 0.924
Li et al.	2024	PRCNet	CVC-ClinicDB Kvasir-SEG CVC-300 CVC-ColonDB ETIS-LaribPolypDB	(CVC-ClinicDB, Kvasir-SEG, CVC-300, CVC-ColonDB, ETIS-LaribPolypDB) mDSC = 0.917, 0.886, 0.883, 0.719, 0.611 mIoU = 0.865, 0.832, 0.814, 0.642, 0.549

Deep learning has seen a significant upsurge in attention lately for medical image analysis, particularly for tasks such as polyp segmentation. One such approach is Dilated-U-Net-Seg (Karthikha et al., 2024), which incorporates dilated convolutions into the well-known U-Net design. These convolutions allow the model to capture information at different scales within the image, crucial for segmenting polyps of varying sizes. Another promising method is PRCNet (Li et al., 2024), which utilizes an exclusive merge of reverse attention modules and a Parallel Partial Decoder (PPD). The PPD extracts high-level features information to provide a comprehensive interpretation of the image, while the reverse attention modules refine the segmentation based on these features. PRCNet and Dilated-U-Net-Seg are two examples of deep learning progress for polyp segmentation, offering the potential for improved accuracy and earlier cancer detection.

CONCLUSION

Recent advances in deep learning have demonstrated promising results in biomedical image segmentation tasks, particularly in the detection and localization of colorectal polyps. One highly successful deep learning architecture for this purpose is the U-Net, a convolutional neural network designed for efficient pixelwise segmentation. Apart from its achievements in colorectal polyp segmentation, the U-Net architecture and its variations have delivered top-notch performance in various medical imaging applications, including the segmentation of brain tumors.

As the fight against colorectal cancer continues, advances in colonoscopy image segmentation show great promise. By conducting more study in this area, we can realize the potential for real-time polyp segmentation and recognition during colonoscopies. This technology has the potential to greatly benefit clinicians via improving polyp characterisation and shortening procedure times. Thus, these developments might lead to greater polyp detection rates and early management for patients, reducing missed diagnoses and improving patient outcomes in colorectal cancer screening.

REFERENCES

Azad, R., Heidary, M., Yilmaz, K., Hüttemann, M., Karimijafarbigloo, S., Wu, Y., . . . Merhof, D. (2023). Loss functions in the era of semantic segmentation. A survey and outlook. arXiv preprint arXiv.2312.05391.

Branch, M. V., & Carvalho, A. S. (2021). Polyp segmentation in colonoscopy images using u-net-mobilenetv2. arXiv preprint arXiv:2103.15715.

Dumitru, R. G., Peteleaza, D., & Craciun, C. (2023). Using DUCK-Net for polyp image segmentation. *Scientific Reports*, 13(1), 9803.

Elshamy, R., Abu-Elnasr, O., Elhoseny, M., & Elmougy, S. (2023). Improving the efficiency of RMSProp optimizer by utilizing Nestrove in deep learning. *Scientific Reports*, 13(1), 8814. DOI: 10.1038/s41598-023-35663-x PMID: 37258633

Fan, D. P., Ji, G. P., Zhou, T., Chen, G., Fu, H., Shen, J., & Shao, L. (2020, September). Pranet: Parallel reverse attention network for polyp segmentation. In International conference on medical image computing and computer-assisted intervention (pp. 263-273). Cham: Springer International Publishing.

Jha, D., Smedsrud, P. H., Johansen, D., de Lange, T., Johansen, H. D., Halvorsen, P., & Riegler, M. A. (2021). A comprehensive study on colorectal polyp segmentation with ResUNet++, conditional random field and test-time augmentation. *IEEE Journal of Biomedical and Health Informatics*, 25(6), 2029–2040. DOI: 10.1109/JBHI.2021.3049304 PMID: 33400658

Karthikha, R., Najumnissa Jamal, D., & Syed Rafiammal, S. (2024). An approach of polyp segmentation from colonoscopy images using Dilated-U-Net-Seg–A deep learning network. *Biomedical Signal Processing and Control*, 93, 106197. DOI: 10.1016/j.bspc.2024.106197

[] Ketkar, N., & Ketkar, N. (2017). Stochastic gradient descent. Deep learning with Python. A hands-on introduction, 113-132.

Kingma, D. P., & Ba, J. (2014). Adam. A method for stochastic optimization. arXiv preprint arXiv.1412.6980.

Li, J., Wang, J., Lin, F., Heidari, A. A., Chen, Y., Chen, H., & Wu, W. (2024). PRCNet. A parallel reverse convolutional attention network for colorectal polyp segmentation. *Biomedical Signal Processing and Control*, 95, 106336. DOI: 10.1016/j.bspc.2024.106336

Lin, T. Y., Goyal, P., Girshick, R., He, K., & Dollár, P. (2017). Focal loss for dense object detection. In *Proceedings of the IEEE international conference on computer vision* (pp. 2980-2988).

Lou, A.. (2022). *"CaraNet. context axial reverse attention network for segmentation of small medical objects." Medical Imaging 2022. Image Processing* (Vol. 12032). SPIE.

Mohapatra, S., Pati, G. K., Mishra, M., & Swarnkar, T. (2022). Upolyseg. A u-net-based polyp segmentation network using colonoscopy images. *Gastroenterology Insights*, 13(3), 264–274. DOI: 10.3390/gastroent13030027

Prasantha, H. S., Shashidhara, H. L., Murthy, K. N. B., & Madhavi, L. G. (2010). Medical image segmentation. *International Journal on Computer Science and Engineering*, 2(4), 1209–1218.

Ronneberger, O., Fischer, P., & Brox, T. (2015). U-net: Convolutional networks for biomedical image segmentation. In Medical image computing and computer-assisted intervention–MICCAI 2015: 18th international conference, Munich, Germany, October 5-9, 2015, proceedings, part III 18 (pp. 234-241). Springer International Publishing.

Safarov, S., & Whangbo, T. K. (2021). A-DenseUNet: Adaptive densely connected UNet for polyp segmentation in colonoscopy images with atrous convolution. *Sensors (Basel)*, 21(4), 1441.

Sandler, M., Howard, A., Zhu, M., Zhmoginov, A., & Chen, L. C. (2018). Mobilenetv2: Inverted residuals and linear bottlenecks. In *Proceedings of the IEEE conference on computer vision and pattern recognition* (pp. 4510-4520).

Sun, X., Zhang, P., Wang, D., Cao, Y., & Liu, B. (2019, December). Colorectal polyp segmentation by U-Net with dilation convolution. In 2019 18th IEEE international conference on machine learning and applications (ICMLA) (pp. 851-858). IEEE.

Tashk, A., Herp, J., & Nadimi, E. (2019, December). Fully automatic polyp detection based on a novel U-Net architecture and morphological post-process. In *2019 International Conference on Control, Artificial Intelligence, Robotics & Optimization (ICCAIRO)* (pp. 37-41). IEEE.

Tomar, N. K., Jha, D., & Bagci, U. (2023, January). Dilatedsegnet: A deep dilated segmentation network for polyp segmentation. In *International conference on multimedia modeling* (pp. 334-344). Cham: Springer International Publishing.

Tran, S.-T., Nguyen, M.-H., Dang, H.-P., & Nguyen, T.-T. (2022). Automatic polyp segmentation using modified recurrent residual Unet network. *IEEE Access : Practical Innovations, Open Solutions*, 10, 65951–65961. DOI: 10.1109/ACCESS.2022.3184773

Yeung, M., Sala, E., Schönlieb, C.-B., & Rundo, L. (2021). Focus U-Net. A novel dual attention-gated CNN for polyp segmentation during colonoscopy. *Computers in Biology and Medicine*, 137, 104815. DOI: 10.1016/j.compbiomed.2021.104815 PMID: 34507156

Zeiler, M. D. (2012). ADADELTA. an adaptive learning rate method. arXiv preprint arXiv.1212.5701.

Zhang, Z., Liu, Q., & Wang, Y. (2018). Road extraction by deep residual u-net. *IEEE Geoscience and Remote Sensing Letters*, 15(5), 749–753. DOI: 10.1109/LGRS.2018.2802944

Zhou, Z., Siddiquee, M. M. R., Tajbakhsh, N., & Liang, J. (2019). Unet++. Redesigning skip connections to exploit multiscale features in image segmentation. *IEEE Transactions on Medical Imaging*, 39(6), 1856–1867. DOI: 10.1109/TMI.2019.2959609 PMID: 31841402

Chapter 5
Artificial Intelligence as a Sustainable Tool for Optimizing the Wastewater Treatment Plant

Adarsh Kumar Arya
https://orcid.org/0000-0002-2378-3832
Harcourt Butler Technical University, India

Shreya
Harcourt Butler Technical University, India

ABSTRACT

The field of artificial intelligence (AI) is advancing at a rapid pace. The evolving technology can replicate practical scenarios in different industries, such as water purification and wastewater treatment. It has demonstrated its value in agriculture, the automotive industry, banking, finance, space exploration, and creative technology. Wastewater treatment has seen significant improvements with the integration of artificial intelligence, which enhances efficiency, speed, and autonomy. Nevertheless, there are notable obstacles and limitations to address, such as data management challenges, interpretability concerns, model replicability and consistency issues, and the need for scholarly transparency. Improving model causality, clarity, data management, and repeatability are proposed solutions. The chapter introduced some of the recent artificial intelligence based optimization methodologies to optimize waste water treatment plants.

DOI: 10.4018/979-8-3373-1399-3.ch005

1. INTRODUCTION

It is essential to acknowledge that water encompasses a significant portion of the globe's surface and is vital in sustaining human life. However, only 1% of water is fresh and accessible, and the demand for water on a global scale is predicted to rise from 4200 Bm3 in 2015 to 6900 Bm3 by 2030 (Ng et al., 2015). Industrial activities introduced various impurities and hazardous materials, leading to multiple methods for analysing and forecasting water and wastewater quality (Dubey et al., 2015). Traditional methods, such as linear regression algorithms and computational frameworks, face challenges in understanding systems' intricate non-linearities and complexities. Empirical and statistical regression models can make precise predictions but struggle with water treatment processes' complicated dynamics and nonlinear relationships. It is essential to carefully analyse and fine-tune specific parameters in water treatment, such as 'biological oxygen demand' (BOD), coagulant dosage, and 'chemical oxygen demand' (COD). Traditional testing methods, such as the jar test, are often oversimplified due to assumptions and idealized factors impractical in real-world scenarios. Optimizing process parameters like pressure, temperature, flow rate, and pH is crucial for achieving optimal plant performance in the treatment of wastewater and reuse usage, where membrane technology has emerged as a promising solution. Conventional models in wastewater treatment plants (WWTPs) rely on utilizing mathematical models and empirical relationships for predictive models, which may not be universally applicable due to differences in operating conditions. Studies suggest that the assessment of BOD (Biochemical Oxygen Demand) and COD (Chemical Oxygen Demand) in water is essential for determining the effectiveness of Wastewater Treatment Plants (WWTPs). In addition, parameters such as total suspended solids, or TSS, and total nitrogen (TN) levels are crucial indicators for assessing the quality of potable and eco-friendly water.

Understanding BOD and COD is crucial in evaluating the oxygen content in wastewater to optimize (Nourani et al., 2018). AI technologies have brought about significant changes in the current industrial sector, known as Industry 4.0, by replicating human intelligence and behaviours in technical systems. AI technologies such as 'deep learning' (DL), 'support vector machine' (SVM), 'artificial neural networks' (ANNs), 'genetic algorithm' (GA), and 'fuzzy logic' (FL) can perform independent evaluation, prediction, and analysis using input data, reducing the likelihood of mistakes, and increasing efficiency. Recent focus in water purification technology has been on coagulation/flocculation, disinfection, desalination, source water analysis, and membrane filtration (Li et al., 2021). AI technologies such as "artificial neural networks" (ANNs) employ a dynamic and data-focused approach. This approach involves three levels: an input layer, one or more hidden layers determined by the algorithm, and an output layer. The output layer computes the weighted mean and incorporates the hidden layers' bias for an inevitable outcome. Previous studies have primarily focused on specific AI technologies in a particular water treatment process, but there is currently no comprehensive review article concentrating on the latest advancements in AI applications for purifying water and treatment of wastewater processes. This report provides a detailed examination of applying variations in AI techniques in water treatment, emphasizing the transformation of crucial plant operations like 'water quality assessment', 'coagulation/flocculation', 'disinfection', 'desalination', 'membrane fouling prediction', 'decontamination', and 'monitoring BOD and COD levels'. Figure 1 examines the artificial intelligence methods used in the water purification facility.

Figure 1. Classification of AI approaches applied in wastewater treatment.

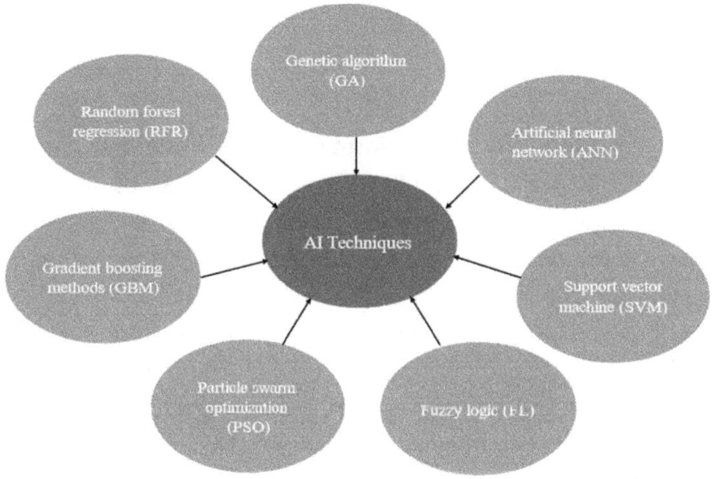

2. EXISTING TECHNOLOGIES FOR WATER AND WASTEWATER TREATMENT

The present section discusses some of the optimization techniques used to optimize the waste water treatment plants.

2.1 Membrane Filtration

Membrane filtration technology eliminates a broad spectrum of pollutants, creating highly pure treated water appropriate for multiple uses like industrial processing, irrigation, and drinking (Cheremisinoff, 2002). Membranes are classified into ultrafiltration, nano-filtration, microfiltration, and reverse osmosis depending on the size and characteristics of particles. Regarding treatment methods, membrane filtration has proven to be quite effective, particularly in removing oil, dyes, and large amounts of metals (Yang et al., 2020). Nevertheless, membrane materials are susceptible to fouling during operation. Deposition of particles of pollutants on membrane pores and surfaces leads to fouling, which decreases permeate flux. To reduce fouling, it is necessary to implement various strategies like regular washing, pre-treatment, and optimization techniques (Warlina et al., 2014).

2.2 Adsorption Process

An intriguing process known as adsorption occurs when atoms or ions migrate from a liquid, gas, or solid to the pores or interface of an adsorbent, forming a separate layer of adsorbate. The typical adsorbate layer may include a variety of pollutant particles, such as colours, oil and grease, metal ions, and dispersed organic and non-organic chemicals (Martini et al., 2020). Active carbon, biochar, and chemically altered adsorbents are commonly used in commercial wastewater treatment facilities. These adsorbent materials can be acquired from multiple sources, including synthetic, natural, and non-organic options. In addition, natural materials are utilized to promote environmental sustainability. Different natural materials like 'banana peels', 'eucalyptus bark', 'watermelon peels', 'barley straw', 'orange peels', and other organic substances can be used as plant-based adsorbents to eliminate specific pollutants (Garrido-Cardenas et al., 2019).

2.3 Coagulation and Flocculation

Flocculation and coagulation are widely employed in water treatment processes, particularly in the initial phase, to remove dispersed and suspended solid particles and diverse pollutants effectively. This is achieved by utilising substances such as aluminium and ferric chloride (Cheremisinoff, 2002). These crucial factors include the initial concentration of pollutants targeted, the dosage of coagulants and flocculants agents, and the pH of the solution. This process utilizes several principles to facilitate the removal of pollutant particles. These principles include destabilization, entrapment, and accumulation or colloid binding stages. By creating more extensive and heavier flocks of particles, settling and filtration stages become more effective in removing them. Nevertheless, specific concerns may arise regarding the potential drawbacks of this approach, such as the expensive coagulant and flocculant reagent. Furthermore, to ensure the safe discharge of harmful sludge into the surroundings while continually operating a big treatment plant and to reduce excessive sludge formation (Kurniawan et al., 2006).

2.4 Biological Process

Various microorganisms are found in nature, including 'bacteria', 'yeast', 'fungi', and 'algal biomass', essential for breaking down pollutants in contaminated solutions using biological methods. At the same time, there are drawbacks, such as longer reaction times and larger treatment plant areas; natural methods like trickling filters and activated sludge can offer simplicity, environmental friendliness, and cost-effectiveness compared to other treatment options. Living organisms can be classified as aerobic or anaerobic according to their oxygen needs. Under specific circumstances, a mix of aerobic and anaerobic microorganisms may produce more favourable outcomes (Gómez-Ramírez & A. Tenorio-Sánchez, 2021).

2.5 Advanced Oxidation

"Advanced oxidation processes" (AOPs) involve generating radical hydroxyl to remove specific organic and inorganic pollutants effectively. The pollutants can be transformed into water, carbon dioxide, and other safe byproducts. Multiple methods exist for conducting AOPs, such as electron beam irradiation, sonolysis, electrochemical oxidation, Fenton (H2O2/Fe2+), Fenton-like (H2O2/Fe3+), photo-assisted Fenton (H2O2/Fe2+/Fe3+/UV), and photo-catalytic reaction (TiO2/O2, O3/H2O2, O3/UV, H2O2/UV) (Garrido-Cardenas et al., 2019). Research has demonstrated the effectiveness of these methods in controlling specific pollutants. However, some drawbacks to these methods should be considered (Jamil et al., 2020), including expensive chemical use and its environmental impacts, sludge production, and byproduct creation. Figure 2. Elaborate the purpose of AI usage in wastewater treatment plants.

Figure 2. AI use in wastewater treatment

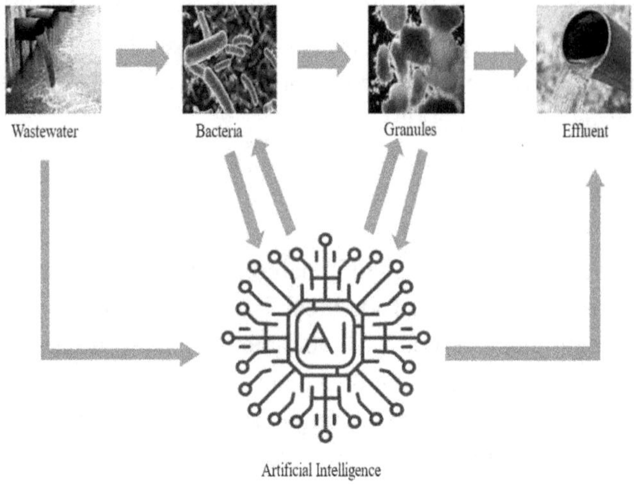

3. RESEARCH ON ARTIFICIAL INTELLIGENCE

AI proves to be an asset in simplifying and streamlining wastewater treatment processes. Over the past few decades, there has been a notable rise in the application of AI models across various domains of wastewater treatment. This tool is utilized to forecast water treatment performance, estimate effluent quality, optimize operating parameters, design treatment units, create sensors for key component estimation, optimize energy use, predict maintenance strategies through fault diagnosis, optimize micropollutants and emerging contaminants, provide operating instructions depending on performance, and automate systems. Every control method has its pros and cons, so it's crucial to choose wisely based on the treatment system's mechanism and intended purpose to achieve optimal outcomes. The usage of artificial neural networks (ANNs) and other soft computing technologies for water quality prediction is on the rise. In the same way that every other AI researcher examines data, ANN is an algorithm. It requires processing data as both input and output.

3.1. BOD and COD Monitoring

One way to assess wastewater treatment performance is using ANN to analyze variables such as 'COD', 'BOD', and 'total suspended solids' (TSS). (OLIVEIRAESQUERRE et al., 2004) A network with multiple layers and a link networking evaluation technique was used to calculate the BOD levels in wastewater treated with biological units, specifically in lagoons. In a study by (Akratos et al., 2008) and their team, they developed a model that utilized artificial neural networking to enhance the BOD removal process. (Nourani et al., 2018) A team conducted a study comparing three different models to evaluate the effectiveness of a treatment facility for wastewater concerning 'COD', 'BOD', and 'TSS'. The study highlighted the effectiveness of using an ensemble neural network model for predicting BOD (24% increase) and COD, as well as total nitrogen in the effluent (approximately 5% for both). Using a mathematical modelling system, researchers in a recent study predicted the BOD amounts of an extremely eutrophic river. Based on the findings, it was determined that the parameter could be expected in advance through 16-4-1 modelling with phosphate as a critical input variable (Yazdanpanahrostami & Rasouli, 2009), explored the application of artificial intelligence in modelling the decomposition of penicillin-type drugs (amoxicillin, ampicillin, cloxacillin) within water through the Fenton process for COD removal. The results generated by the ANN indicated a correlation coefficient of 0.997, confirming the model's suitability. The COD removal was affected by different factors, including time, pH, concentration, molecular ratio of hydrogen peroxide, and ferrous ion concentration and explored the application of artificial neural networks to predict COD removal efficiency in sewerage water. A case study evaluating the performance of the Khorramabad wastewater treatment facility using AI. Considering influential factors such as 'temperature', 'pH', 'dissolved oxygen', 'BOD', 'COD', and 'dissolved solids', the output results were evaluated based on 'BOD', 'COD', and 'total solids'. This correlation coefficient were achieved, demonstrating an optimal removal of contaminants efficiency of 87.68%. In a different study by (Zaqoot et al., 2018), a model based on linear regression, along with ANN, was used to analyze water quality parameters. After examining the pH, temperatures, BOD, COD, TSS, etc., it was concluded that the wastewater performance assessment was more accurate using ANN compared to linear regression. This conclusion was supported by error analysis and regression coefficient values. (Zhu et al., 2018) discussed a hybrid method for forecasting the BOD levels. By employing regression analysis, connectionist networks, and goal programming, there was a decrease in accuracy in the overall performance evaluation. However, it showcased a remarkable ability to predict BOD concentrations. Using AI approaches, we investigated how different operating conditions affected the removal of chemical oxygen demand from olive mill wastewater waste. With an R2 value of 0.92, the experimental results and the observed decrease in chemical oxygen demand effectiveness were quite congruent (42.7%) (Nasr & EL Shahawy, 2016).

3.2. Removal of Heavy Metals and Dyes

Ecosystems and human health are greatly endangered when heavy metal ions enter the water supply. Copper, zinc, cadmium, chromium, and lead are all heavy metals that are both highly poisonous and slow to degrade(Gomez-Gonzalez et al., 2016). Due to their non-biodegradable nature, these substances accumulate in organisms as they enter the food chain through waste in water channels. These substances can be quite dangerous even at low levels and are the main chemical pollutants in the atmosphere because they can spread, persist, and cause harm. As a result, the harmful effects become more noticeable in organisms higher up in the food chain (Farooq et al., 2010). Traditional methods for eliminating harmful heavy metal ions have become inadequate or costly, especially when dealing with low concentrations (Kundu & Gupta, 2006). Delving into different AI techniques to model, forecast, and assess the extraction of harmful heavy metal ions from wastewater with various adsorbents while keeping expenses in check, as detailed in the literature. Artificial neural networks are commonly used to remove toxic metal ions among the various AI techniques. Neural networks are used to simulate the elimination of $As3+$ and $As5+$ from artificial wastewater. The results of the experiment and model outputs demonstrated an exceptional $R2$ value of 0.9998, indicating the model's high accuracy in predicting phytoremediation outcomes (Podder & Majumder, 2016). Utilizing an ANN4 model yields the most precise estimation for the removal percentage of $Hg2+$ ions, showcasing exceptional metrics such as $R2 = 0.996$, adjusted root means square error = 0.028, and mean residual error = 0.008 (Amiri et al., 2017). Four models of neural networks were developed to analyze the sensitivity of the models. By achieving an $R2$ of 0.99, the prediction accuracy for breakthrough curves and model coefficients in removing cadmium (Cd) ions was notably high (Ahmad & Haydar, 2016). Examining the elimination of $Fe3+$, $Pb2+$, $Zn2+$, $Cd2+$, $As3+$, $Cu2+$, and $Hg2+$ using various adsorbents was studied with a range of sophisticated tools, mainly artificial neural networks (El Hanandeh et al., 2021). When working with AI techniques, typical input variables consist of contact time, adsorbent dosage, pH, metal ion concentration, and temperature, while the output is the removal efficiency. Several studies incorporated the Box–Behnken model along with ANN for $Cu2+$ and $Cd2+$ removal. Using an artificial neural network with a feed-forward framework and an arrangement optimized by RSM, the study concentrated on simultaneously eliminating $Pb2+$ ions and 'malachite green' (MG) dye with ultrasound assistance. Methods from the field of artificial intelligence were also employed to assess the concurrent elimination of various metal ions. In previous studies, the effectiveness of different adsorbents in eliminating dyes and ions of heavy metals from water solutions was evaluated using 'ANN' and 'ANFIS' techniques. Researchers reviewed the efficiency of $Pb2+$ removal by ostrich bone ash using AI models like the 'adaptive neural-based fuzzy inference system' (ANFIS) (Amiri et al., 2013). The ANFIS model incorporated five input variables: initial concentration of Pb^{2+}, temperature, adsorbent dosage, initial pH, and time. The result was calculated according to the adsorption effectiveness of the bone ash. By employing a hybrid Thomas and ANFIS model, the study estimated the efficiency of breakthrough forms for the dynamic Cd^{2+} adsorption by ostrich bone ash-supported nZVI elements. (Oladipo & Gazi, 2015) introducing a groundbreaking adsorber specifically engineered to eliminate nickel ions from clean water sources. Encased within an alginate matrix, bentonite showcased exceptional adsorption and regeneration abilities. Utilizing advanced 'artificial neural networks' and 'central composite design' (CCD), a precise model successfully forecasted the influence of variables on the ion exchange process in adsorption. 32 and 16 experimental datasets were used to train ANN and CCD, respectively. Using a three-layered feed-forward ANN system alongside four input layers and two output layers, considering uptake capacity and removal effectiveness, led to accurate Ni(II) predictions in an average relative error of 0.056. Scientists have employed ANN techniques to simulate and compute the ability of adsorption of an adsorbent for fourth-generation dyes. By examining variables like sorbent quantity, sonication time, solution pH, and dye concentrations, this was accomplished. The model demonstrated a remarkable predictive ability to determine the absorption of methyl violet 2 B, achieving a tremendous R^2 value of 0.9946 (Kooh et al., 2016). Explored the absorption of anionic azo dye on Chitosan/Al_2O_3/Fe_3O_4 core-shell utilizing multilayer perceptron ANN, showcasing favourable results when compared to experimental data. The adsorption of methyl orange (MO) was found to be highly effective over a broad pH range of 4–10, with thc idcal dosage of adsorbent determined to be 0.6 g/L (Tanhaei et al., 2016). Achieving successful modeling of the removal of 'brilliant green' (BG), 'eosin B' (EB), 'Congo red', 'MG', 'methylene blue' (MB), and 'sunset yellow' (SY) was made possible using different adsorbent materials with the help of advanced AI techniques, specifically ANN (Parsazadeh et al., 2018). The anticipated adsorption capacities closely aligned with the experimental findings. (Cook et al., 2020) developed a powerful technique to improve hyporheic exchange rates, effectively tackling flow restrictions due to biofilm blockages, particularly in systems with larger sediment sizes. Furthermore, addressing new types of pollutants, such as microplastics, involves understanding the primary physical processes related to mixing and transportation.

4. MODELS FOR MACHINE LEARNING AND AI TECHNIQUES

Table 1 provides a concise summary of AI and ML models and methods in Table 1. Here, we will explore the various uses of models and methods in water treatment and monitoring. We will discuss their specific applications, advantages, and disadvantages, helping you make informed decisions when selecting the most suitable models and methods. Furthermore, the last column includes additional sources that have been peer-reviewed, published and are in textbook format. These sources offer comprehensive explanations of the methods and models discussed. The purpose of these water treatment and monitoring applications is not to provide an exhaustive list but to showcase the publications that have been meticulously chosen from peer-reviewed journals following the abovementioned technique. Many of these machine learning methods are often seen as black-box models, which some consider a disadvantage for many models, especially excluding genetic algorithms and Gaussian processes. Techniques like Ant colony optimization has been used in numerous fields but are yet to be explored to optimize waste water treatment plants(Arya, 2021; Arya et al., 2022, 2023; Arya & Honwad, 2016; Thakur et al., 2022).

Table 1. AI methods and ML models used in water treatment and monitoring.

Leaning and modelling techniques	Water treatment and monitoring application	General application	Advantages	Disadvantages
Genetic Algorithm/ Genetic Programming (GA/GEP)	DBP formation modelling	▪ Evolutionary, stochastic algorithm ▪ Regression, Classification (Katoch et al., 2021).	▪ The simple idea that is easily understandable for the majority. ▪ It is possible to achieve multi-objective optimization. ▪ Adaptable to nearby maximums and minimums as well as datasets with noise.	▪ Requires high computational power. ▪ Implementation is often complex and takes a lot of time. ▪ Fitness/objective functions and operators were challenging to derive.
Artificial Neural Network (ANN)-General	▪ DBP formation Modelling ▪ Adsorption process parameter modeling ▪ Chlorine dosage/set-point.	▪ Supervised machine learning. ▪ Regression, Classification (Uhrig, n.d.).	▪ Possibility of forward propagation for quick and inexpensive computing. ▪ Obtaining modelling or prediction findings within an acceptable timeframe. ▪ Able to manage datasets with a high dimensionality.	▪ Models and architecture might be challenging to grasp on their own. ▪ This step of backward propagation requires a high amount of computer power.
Convolutional Neural Network (CNN)	DBP formation modelling	▪ Supervised Machine Learning. ▪ Regression, Classification, Segmentation (Gu et al., 2018).	▪ The results are considered to be entirely accurate. ▪ Results are produced rapidly since the model operates in parallel. ▪ Proficiently solve problems with inputs based on images.	▪ Both the model and the architecture are intricate and lengthy. ▪ High computing power is needed.

continued on following page

Table 1. Continued

Leaning and modelling techniques	Water treatment and monitoring application	General application	Advantages	Disadvantages
Extreme Learning Machine (ELM)	Dissolved oxygen concentration modelling	■ Supervised machine learning ■ Regression, Classification (Huang, 2009).	■ Training periods that are not too lengthy. ■ An appropriate choice for pattern classifications.	When using too many or too few concealed nodes might lead to over-fitting or under-fitting.
Fuzzy Inference System (FIS)	■ Set-point control for chlorine dose. ■ Control of the environment and hydroponics system.	■ Artificial intelligence. ■ Decision making, system Control (Kosko & Isaka, 1993).	■ The decision-making process experienced by humans may be more accurately modelled using fuzzy logic instead of binary logic. ■ The results and choices are easily understood when the system is well described.	■ A lack of understanding of fuzzy logic might lead to a perplexing interpretation of terminology. ■ Susceptibility to human mistakes and reliance on operator-defined parameters limit its applicability.

5. APPLICATIONS IN WATER-QUALITY MANAGEMENT

Ensuring water quality is paramount to upholding aquatic ecosystems' well-being and optimal operation. Monitoring the effect of development on the drinking water condition of waterways and other waterways is of utmost importance to safeguard these delicate ecosystems. Assessing the health of a river or water body relies heavily on the dissolved oxygen level. The rise in hypoxia in water bodies is linked to increased nutrient loading and global warming (Breitburg et al., 2018). Given the correlation between 'dissolved-oxygen' (DO) levels and human activity/pollution, monitoring DO is crucial for anticipating and potentially averting hypoxic zones that can harm aquatic ecosystems. Precise and up-to-date findings are typically preferred, as slight decreases in DO levels can be crucial for certain species (Collins et al., 2019). There are situations where environmental factors may influence DO sense, highlighting the importance of developing models and techniques to address limitations in traditional sensing methods (Wei et al., 2019). Table 2 presents various research projects exploring the use of machine learning to model water-quality-management parameters. The main emphasis of the research is on studying the dynamics of dissolved oxygen in water bodies, investigating the water quality index, and predicting future river water levels. Places where water-quality parameters are observed typically include rivers, with a particular wastewater treatment facility dedicated to monitoring BOD and COD. Typically, water-quality modelling with machine learning involves factors such as pH, water temperature, and BOD levels. The study examined how inputs for 'water-quality-index' (WQI) monitoring and BOD/COD modelling typically share similarities, including dissolved oxygen as one of the inputs. Monitoring water levels depends significantly on past water levels and accurate training data. Exploring water-quality management tools usually requires using various machine learning techniques such as 'ANN', 'ANFIS', 'RNN', 'EML', 'RT', 'SVM', 'HW', and hybrid models that combine different methods like RF models. Various research studies have shown precise predictions, with results depending on the area. On average, ANFIS models demonstrated superior performance compared to typical ANN and SVM models in nearly all the studies listed in Table 2, with some instances of being surpassed by hybrid models. The water-level estimations were calculated with great accuracy utilizing both ANN and ANFIS models, reaching R2 above 0.999 for both models.

Table 2. ML for modelling and predicting environmental variables and factors about water quality

Location	Input Variables	Output	ML Technique Used	Reference
Udi, India (Yamuna River)	NH3 concentration, water temperature, BOD, COD, pH, and flow rate.	DO concentration	ANFIS, SVM, ANN	(Elkiran et al., 2019)
Mathura, India (Yamuna River)	water temperature, pH, BOD, dissolved oxygen (DO) (All inputs are collected from stations located upstream, midstream, and downstream, each with their independent variables.)	Upstream, midstream, and downstream DO concentration	ANN, ANFIS	(Elkiran et al., 2019)
Oregon, USA (Link & Klamath Rivers)	Hourly temperature, pH, specific conductivity	DO concentration	ANFIS, RF, ELM, ANN	(Kisi et al., 2020)

continued on following page

Table 2. Continued

Location	Input Variables	Output	ML Technique Used	Reference
Mesa, Arizona (algae cultivation pond)	A microbial potentiometric sensor was used to measure open-circuit potentials.	conductivity, chlorophyll conc., DO, Blue-green algae conc., pH, turbidity	RNN (LSTM)	(Saboe et al., 2021)
Thailand (Chao Phraya River)	total coliform bacteria, ammonia concentration, BOD, DO, faecal coliform bacteria, salinity	fecal coliform bacteria, BOD, DO, salinity, ammonia concentration, total coliform bacteria	SVM with varying kernel functions	(Najah Ahmed et al., 2019)

6. EXISTING INFORMATION GAP AND POTENTIAL FUTURE APPLICATIONS OF AI

Research has shown that AI outperforms traditional modelling techniques in various water treatment processes. Progress in these technologies fuels continuous exploration and creativity in model designs to overcome challenges that impact their efficiency in water treatment sectors. AI technologies have limitations, such as dealing with real-world processes affected by inherent limitations and varying dynamics. Inconsistencies in training data can result in model predictions that may not accurately reflect historical data. The fluctuating bias and weight distribution within neurons may lead to the model's inability to be replicated, frequently becoming trapped in local optima rather than capturing the complete process dynamics. Furthermore, regional convergence and the difficulty in identifying overfitting are other limitations. Future research in automation should focus on creating models capable of analyzing entire treatment plants, covering various processes like 'coagulation', 'disinfection', 'BOD', and 'COD' determination. This method will facilitate thorough modelling of the complete dynamics. Combining different models demonstrated impressive interpretability by harnessing the combined strengths of the final model. Thus, conducting a comprehensive analysis to address limitations and incorporate hybrid structures alongside existing AI models is essential. New technologies like Ant colony optimization, Cheetah Algorithm, Cuckoo search have several advantages but needs to be adapted to optimize waste water treatment plants(Arya, 2022a, 2022b; Arya & Honwad, 2018b, 2018a; Gupta et al., 2019; Sarkar & Arya, 2022).

7. CONCLUSION

Artificial intelligence's precision, accuracy, and consistent error-free performance have contributed to their widespread adoption in various fields. Artificial intelligence has been effectively utilized in multiple applications, particularly in monitoring wastewater treatment. In addition, there remain areas that require further exploration in the field of the provided technology, including identifying and removing contaminants, determining the appropriate levels to initiate the process, releasing pollutants for wastewater treatment, and identifying contaminants. These subjects remain of great interest to researchers in different water treatment areas that use automated systems that rely on instruments. But it doesn't make it easier for experts and researchers to communicate, which is crucial. Environmentalists, scientists, information technology (IT) experts, and computer specialists are all part of the varied professional groups working together to improve water technology. Currently, the process of digital transformation is still far from complete. The present software application is encountering difficulties in its delivery system, impeding the effortless incorporation of information, methodologies, and the evaluation of cost-benefit analysis for data-driven technology. Expensive expenses, limited progress through experience, and potential job loss will continue to be drawbacks of AI technology. Hybrid models have effectively dealt with random water inputs but have yet to be investigated using bioinspired techniques. To better comprehend internal networking and enhance practical interactions in the field, it is vital to grasp AI. Conducting research is essential to connect different areas of knowledge and push the boundaries of current understanding among model developers, specialists, and water purification engineers. Addressing this issue requires leveraging cutting-edge technology to create efficient and affordable treatment systems. Combining different intelligent systems could provide a solution to the problem.

REFERENCES

Ahmad, M. F., & Haydar, S. (2016). Evaluation of a newly developed biosorbent using packed bed column for possible application in the treatment of industrial effluents for removal of cadmium ions. *Journal of the Taiwan Institute of Chemical Engineers*, 62, 122–131. DOI: 10.1016/j.jtice.2015.12.032

Akratos, C. S., Papaspyros, J. N. E., & Tsihrintzis, V. A. (2008). An artificial neural network model and design equations for BOD and COD removal prediction in horizontal subsurface flow constructed wetlands. *Chemical Engineering Journal*, 143(1–3), 96–110. DOI: 10.1016/j.cej.2007.12.029

Amiri, M. J., Abedi-Koupai, J., Eslamian, S. S., Mousavi, S. F., & Hasheminejad, H. (2013). Modeling Pb (II) adsorption from aqueous solution by ostrich bone ash using adaptive neural-based fuzzy inference system. *Journal of Environmental Science and Health. Part A, Toxic/Hazardous Substances & Environmental Engineering*, 48(5), 543–558. DOI: 10.1080/10934529.2013.730451 PMID: 23383640

Amiri, M. J., Abedi-koupai, J., Jafar Jalali, S. M., & Mousavi, S. F. (2017). Modeling of Fixed-Bed Column System of Hg(II) Ions on Ostrich Bone Ash/nZVI Composite by Artificial Neural Network. *Journal of Environmental Engineering*, 143(9), 04017061. Advance online publication. DOI: 10.1061/(ASCE)EE.1943-7870.0001257

Arya, A. K. (2021). Optimal operation of a multi-distribution natural gas pipeline grid: An ant colony approach. *Journal of Petroleum Exploration and Production Technology*, 11(10), 3859–3878. Advance online publication. DOI: 10.1007/s13202-021-01266-3

Arya, A. K. (2022a). A comparison of the MOGA and NSGA-II optimization techniques to reduce the cost of a biomass supply network. *Materials Today: Proceedings*, 57, 1553–1558. Advance online publication. DOI: 10.1016/j.matpr.2021.12.161

Arya, A. K. (2022b). Application and Challenges of "Blockchain Technology" in the Oil and Gas Industry. In *Blockchain Technology: Exploring Opportunities, Challenges, and Applications*. DOI: 10.1201/9781003138082-11

Arya, A. K., & Honwad, S. (2016). Modeling, simulation, and optimization of a high-pressure cross-country natural gas pipeline: Application of an ant colony optimization technique. *Journal of Pipeline Systems Engineering and Practice*, 7(1), 04015008. Advance online publication. DOI: 10.1061/(ASCE)PS.1949-1204.0000206

Arya, A. K., & Honwad, S. (2018a). Multiobjective optimization of a gas pipeline network: An ant colony approach. *Journal of Petroleum Exploration and Production Technology*, 8(4), 1389–1400. Advance online publication. DOI: 10.1007/s13202-017-0410-7

Arya, A. K., & Honwad, S. (2018b). Optimal Operation of a Multi Source Multi Delivery Natural Gas Transmission Pipeline Network. *Chemical Product and Process Modeling*, 13(3). Advance online publication. DOI: 10.1515/cppm-2017-0046

Arya, A. K., Jain, R., Yadav, S., Bisht, S., & Gautam, S. (2022). Recent trends in gas pipeline optimization. *Materials Today: Proceedings*, 57, 1455–1461. Advance online publication. DOI: 10.1016/j.matpr.2021.11.232

Arya, A. K., Katiyar, R., Senthil Kumar, P., Kapoor, A., Pal, D. B., & Rangasamy, G. (2023). A multi-objective model for optimizing hydrogen injected-high pressure natural gas pipeline networks. *International Journal of Hydrogen Energy*, 48(76), 29699–29723. Advance online publication. DOI: 10.1016/j.ijhydene.2023.04.133

Breitburg, D., Levin, L. A., Oschlies, A., Grégoire, M., Chavez, F. P., Conley, D. J., Garçon, V., Gilbert, D., Gutiérrez, D., Isensee, K., Jacinto, G. S., Limburg, K. E., Montes, I., Naqvi, S. W. A., Pitcher, G. C., Rabalais, N. N., Roman, M. R., Rose, K. A., Seibel, B. A., & Zhang, J. (2018). Declining oxygen in the global ocean and coastal waters. *Science*, 359(6371), eaam7240. Advance online publication. DOI: 10.1126/science.aam7240 PMID: 29301986

Cheremisinoff, N. P. (2002). Water Sterilization Technologies. In *Handbook of Water and Wastewater Treatment Technologies* (pp. 446–495). Elsevier., DOI: 10.1016/B978-075067498-0/50014-0

Collins, M., Tills, O., Turner, L. M., Clark, M. S., Spicer, J. I., & Truebano, M. (2019). Moderate reductions in dissolved oxygen may compromise performance in an ecologically-important estuarine invertebrate. *The Science of the Total Environment*, 693, 133444. DOI: 10.1016/j.scitotenv.2019.07.250 PMID: 31362229

Cook, S., Price, O., King, A., Finnegan, C., van Egmond, R., Schäfer, H., Pearson, J. M., Abolfathi, S., & Bending, G. D. (2020). Bedform characteristics and biofilm community development interact to modify hyporheic exchange. *The Science of the Total Environment*, 749, 141397. DOI: 10.1016/j.scitotenv.2020.141397 PMID: 32841855

Dubey, R., Bajpai, J., & Bajpai, A. K. (2015). Green synthesis of graphene sand composite (GSC) as novel adsorbent for efficient removal of Cr (VI) ions from aqueous solution. *Journal of Water Process Engineering*, 5, 83–94. DOI: 10.1016/j.jwpe.2015.01.004

El Hanandeh, A., Mahdi, Z., & Imtiaz, M. S. (2021). Modelling of the adsorption of Pb, Cu and Ni ions from single and multi-component aqueous solutions by date seed derived biochar: Comparison of six machine learning approaches. *Environmental Research*, 192, 110338. DOI: 10.1016/j.envres.2020.110338 PMID: 33075354

Elkiran, G., Nourani, V., & Abba, S. I. (2019). Multi-step ahead modelling of river water quality parameters using ensemble artificial intelligence-based approach. *Journal of Hydrology (Amsterdam)*, 577, 123962. DOI: 10.1016/j.jhydrol.2019.123962

Farooq, U., Kozinski, J. A., Khan, M. A., & Athar, M. (2010). Biosorption of heavy metal ions using wheat based biosorbents – A review of the recent literature. *Bioresource Technology*, 101(14), 5043–5053. DOI: 10.1016/j.biortech.2010.02.030 PMID: 20223652

Garrido-Cardenas, J. A., Esteban-García, B., Agüera, A., Sánchez-Pérez, J. A., & Manzano-Agugliaro, F. (2019). Wastewater Treatment by Advanced Oxidation Process and Their Worldwide Research Trends. *International Journal of Environmental Research and Public Health*, 17(1), 170. DOI: 10.3390/ijerph17010170 PMID: 31881722

Gomez-Gonzalez, R., Cerino-Córdova, F. J., Garcia-León, A. M., Soto-Regalado, E., Davila-Guzman, N. E., & Salazar-Rabago, J. J. (2016). Lead biosorption onto coffee grounds: Comparative analysis of several optimization techniques using equilibrium adsorption models and ANN. *Journal of the Taiwan Institute of Chemical Engineers*, 68, 201–210. DOI: 10.1016/j.jtice.2016.08.038

Gómez-Ramírez, M., & Tenorio-Sánchez, A. S. (2021). Treatment of Solid Waste Containing Metals by Biological Methods. In *Natural Resources Management and Biological Sciences*. IntechOpen. DOI: 10.5772/intechopen.92211

Gu, J., Wang, Z., Kuen, J., Ma, L., Shahroudy, A., Shuai, B., Liu, T., Wang, X., Wang, G., Cai, J., & Chen, T. (2018). Recent advances in convolutional neural networks. *Pattern Recognition*, 77, 354–377. DOI: 10.1016/j.patcog.2017.10.013

Gupta, S. S., Arya, A. K., & Vijay, P. (2019). Designing a model for optimization of maintenance and inspection efforts against third party damage to cross country Pipelines in India. *International Journal of Innovative Technology and Exploring Engineering*, 8(12), 4529–4540. Advance online publication. DOI: 10.35940/ijitee.L3209.1081219

Huang, G. (2009). Erratum. In *Relict Species* (p. 450). Springer Berlin Heidelberg., DOI: 10.1007/978-3-540-92160-8_27

Jamil, A., Bokhari, T. H., Javed, T., Mustafa, R., Sajid, M., Noreen, S., Zuber, M., Nazir, A., Iqbal, M., & Jilani, M. I. (2020). Photocatalytic degradation of disperse dye Violet-26 using TiO2 and ZnO nanomaterials and process variable optimization. *Journal of Materials Research and Technology*, 9(1), 1119–1128. DOI: 10.1016/j.jmrt.2019.11.035

Katoch, S., Chauhan, S. S., & Kumar, V. (2021). A review on genetic algorithm: Past, present, and future. *Multimedia Tools and Applications*, 80(5), 8091–8126. DOI: 10.1007/s11042-020-10139-6 PMID: 33162782

Kisi, O., Alizamir, M., & Docheshmeh Gorgij, A. (2020). Dissolved oxygen prediction using a new ensemble method. *Environmental Science and Pollution Research International*, 27(9), 9589–9603. DOI: 10.1007/s11356-019-07574-w PMID: 31925684

Kooh, M. R. R., Dahri, M. K., Lim, L. B. L., Lim, L. H., & Malik, O. A. (2016). Batch adsorption studies of the removal of methyl violet 2B by soya bean waste: Isotherm, kinetics and artificial neural network modelling. *Environmental Earth Sciences*, 75(9), 783. Advance online publication. DOI: 10.1007/s12665-016-5582-9

Kosko, B., & Isaka, S. (1993). Fuzzy Logic. *Scientific American*, 269(1), 76–81. DOI: 10.1038/scientificamerican0793-76

Kundu, S., & Gupta, A. K. (2006). Arsenic adsorption onto iron oxide-coated cement (IOCC): Regression analysis of equilibrium data with several isotherm models and their optimization. *Chemical Engineering Journal*, 122(1–2), 93–106. DOI: 10.1016/j.cej.2006.06.002

Kurniawan, T. A., Chan, G. Y. S., Lo, W.-H., & Babel, S. (2006). Physico–chemical treatment techniques for wastewater laden with heavy metals. *Chemical Engineering Journal*, 118(1–2), 83–98. DOI: 10.1016/j.cej.2006.01.015

Li, L., Rong, S., Wang, R., & Yu, S. (2021). Recent advances in artificial intelligence and machine learning for nonlinear relationship analysis and process control in drinking water treatment: A review. *Chemical Engineering Journal*, 405, 126673. DOI: 10.1016/j.cej.2020.126673

Martini, S., Afroze, S., & Ahmad Roni, K. (2020). Modified eucalyptus bark as a sorbent for simultaneous removal of COD, oil, and Cr(III) from industrial wastewater. *Alexandria Engineering Journal*, 59(3), 1637–1648. DOI: 10.1016/j.aej.2020.04.010

Najah Ahmed, A., Binti Othman, F., Abdulmohsin Afan, H., Khaleel Ibrahim, R., Ming Fai, C., Shabbir Hossain, M., Ehteram, M., & Elshafie, A. (2019). Machine learning methods for better water quality prediction. *Journal of Hydrology (Amsterdam)*, 578, 124084. DOI: 10.1016/j.jhydrol.2019.124084

Nasr, M., & Shahawy, E. L. (2016). Artificial Intelligence for Electrocoagulation Treatment of Olive Mill Wastewater. *Journal of Bioremediation &amp. Journal of Bioremediation & Biodegradation*, 07(03). Advance online publication. DOI: 10.4172/2155-6199.1000345

Ng, K. C., Thu, K., Oh, S. J., Ang, L., Shahzad, M. W., & Ismail, A. (2015). Recent developments in thermally-driven seawater desalination: Energy efficiency improvement by hybridization of the MED and AD cycles. *Desalination*, 356, 255–270. DOI: 10.1016/j.desal.2014.10.025

Nourani, V., Elkiran, G., & Abba, S. I. (2018). Wastewater treatment plant performance analysis using artificial intelligence – an ensemble approach. *Water Science and Technology*, 78(10), 2064–2076. DOI: 10.2166/wst.2018.477 PMID: 30629534

Oladipo, A. A., & Gazi, M. (2015). Nickel removal from aqueous solutions by alginate-based composite beads: Central composite design and artificial neural network modeling. *Journal of Water Process Engineering*, 8, e81–e91. DOI: 10.1016/j.jwpe.2014.12.002

Oliveiraesquerre, K., Seborg, D., Mori, M., & Bruns, R.OLIVEIRAESQUERRE. (2004). Application of steady-state and dynamic modeling for the prediction of the BOD of an aerated lagoon at a pulp and paper millPart II. Nonlinear approaches. *Chemical Engineering Journal*, 105(1–2), 61–69. DOI: 10.1016/j.cej.2004.06.012

Parsazadeh, N., Yousefi, F., Ghaedi, M., Dashtian, K., & Borousan, F. (2018). Preparation and characterization of monoliths HKUST-1 MOF *via* straightforward conversion of $Cu(OH)_2$-based monoliths and its application for wastewater treatment: Artificial neural network and central composite design modeling. *New Journal of Chemistry*, 42(12), 10327–10336. DOI: 10.1039/C8NJ01067F

Podder, M. S., & Majumder, C. B. (2016). The use of artificial neural network for modelling of phycoremediation of toxic elements As(III) and As(V) from wastewater using Botryococcus braunii. *Spectrochimica Acta. Part A: Molecular and Biomolecular Spectroscopy*, 155, 130–145. DOI: 10.1016/j.saa.2015.11.011 PMID: 26615452

Saboe, D., Ghasemi, H., Gao, M. M., Samardzic, M., Hristovski, K. D., Boscovic, D., Burge, S. R., Burge, R. G., & Hoffman, D. A. (2021). Real-time monitoring and prediction of water quality parameters and algae concentrations using microbial potentiometric sensor signals and machine learning tools. *The Science of the Total Environment*, 764, 142876. DOI: 10.1016/j.scitotenv.2020.142876 PMID: 33757235

Sarkar, A., & Arya, A. K. (2022). A Survey on Optimization Parameters and Techniques for Crude Oil Pipeline Transportation. In *Smart Innovation* (Vol. 292). Systems and Technologies., DOI: 10.1007/978-981-19-0836-1_43

Tanhaei, B., Ayati, A., Lahtinen, M., Mahmoodzadeh Vaziri, B., & Sillanpää, M. (2016). A magnetic mesoporous chitosan based core-shells biopolymer for anionic dye adsorption: Kinetic and isothermal study and application of <scp>ANN</scp>. *Journal of Applied Polymer Science*, 133(22), app.43466. Advance online publication. DOI: 10.1002/app.43466

Thakur, A. K., Arya, A. K., & Sharma, P. (2022). Prediction and mitigation of AC interference on the pipeline system. *Corrosion Reviews*, 40(2), 149–157. Advance online publication. DOI: 10.1515/corrrev-2021-0061

Uhrig, R. E. (n.d.). Introduction to artificial neural networks. In *Proceedings of IECON '95 - 21st Annual Conference on IEEE Industrial Electronics*. IEEE. DOI: 10.1109/IECON.1995.483329

Warlina, L. I., & Simanjuntak, W. (2014). Pengolahan Limbah Cair Hotel Dengan Kombinasi Metode Elektrokoagulasi Dan Adsorpsi Menggunakan Karbosil. *Jurnal Sylva Lestari*, 1(1), 85. DOI: 10.23960/jsl1185-92

Wei, Y., Jiao, Y., An, D., Li, D., Li, W., & Wei, Q. (2019). Review of Dissolved Oxygen Detection Technology: From Laboratory Analysis to Online Intelligent Detection. *Sensors (Basel)*, 19(18), 3995. DOI: 10.3390/s19183995 PMID: 31527482

Yang, C., Xu, W., Nan, Y., Wang, Y., Hu, Y., Gao, C., & Chen, X. (2020). Fabrication and characterization of a high performance polyimide ultrafiltration membrane for dye removal. *Journal of Colloid and Interface Science*, 562, 589–597. DOI: 10.1016/j.jcis.2019.11.075 PMID: 31771878

Yazdanpanahrostami, A., & Rasouli, K. (2009). Simulation of Tehran Air Pollution Using Artificial Neural Networks. In *World Environmental and Water Resources Congress 2009*. American Society of Civil Engineers. DOI: 10.1061/41036(342)242

Zaqoot, H. A., Hamada, M., & Miqdad, S. (2018). A Comparative Study Of Ann For Predicting Nitrate Concentration In Groundwater Wells In The Southern Area Of Gaza Strip. *Applied Artificial Intelligence*, 32(7–8), 727–744. DOI. 10.1080/08839514.2018.1506970

Zhu, J.-J., Kang, L., & Anderson, P. R. (2018). Predicting influent biochemical oxygen demand: Balancing energy demand and risk management. *Water Research*, 128, 304–313. DOI: 10.1016/j.watres.2017.10.053 PMID: 29107915

Chapter 6
Enhancing Wastewater Stabilization Ponds for Treating Domestic Wastewater Using Machine Learning

T. Venkat Narayana Rao
https://orcid.org/0000-0002-1996-1819
Sreenidhi Institute of Science and Technology, India

Shiva Kashyap Yellavajhala
Sreenidhi Institute of Science and Technology, India

Muddasani Harshith
Sreenidhi Institute of Science and Technology, India

K. Siva Kumar Gowda
Sreenidhi Institute of Science and Technology, India

ABSTRACT

In the realm of wastewater treatment, Wastewater Stabilization Ponds (WSPs) are highly regarded for their efficiency, particularly when incorporating baffle walls (BWs) into their design. These BWs play a crucial role in minimizing space requirements and improving contaminant removal, albeit at the expense of increased construction material usage. To find the best configuration, a thorough analysis was first carried out utilizing typical methodology (TM) design worksheets for WSPs. This phase validated the effectiveness of BWs by showing a considerable reduction

DOI: 10.4018/979-8-3373-1399-3.ch006

in both WSPs area and hydraulic retention time (RT). Within the MM framework, additional reductions in WSPs area and RT were attained, coupled with a 5% drop in necessary concrete volume compared to TM, by employing the interior-point approach in MATLAB and the generalized reduced gradient (GRG) algorithm in MS Excel Solver. While these findings highlight the effectiveness of MM in optimizing WSPs, there is potential for exploring alternative algorithms to achieve more favorable results.

1. INTRODUCTION

In areas such as Turkey where water resources are scarce, managing wastewater sustainably is critical for maintaining sensitive ecosystems as well as public health (Dolgen,D. et al., 2023). With a significant portion of its water budget allocated to agriculture (73%), Turkey faces considerable pressure on its freshwater reserves (Ali, H. Q. et al., 2023). Consequently, the construction of effective and cost-efficient wastewater treatment technologies is necessary (Demirel, D. F. et al., 2022). Using treated wastewater for irrigation offers a twofold purpose: reducing water scarcity and increasing agricultural sustainability by providing a consistent and cost-effective water source (Ersoy Tonyaloğlu, E. et al., 2023). Turkey faces hurdles in modernizing its wastewater treatment system due to limited resources, demanding innovative solutions (Hafeez, A. et al., 2021). According to recent studies, only approximately half of Turkey's wastewater is treated, causing hazards to aquatic life and ecosystems as untreated wastewater is dumped into the environment (Ali, H. Q. et al., 2023). To overcome these difficulties, Turkey should improve its wastewater treatment systems, especially in rural and peri-urban areas (Kookana, R. S. et al., 2020). When choosing a wastewater treatment system, it's important to prioritise both good effluent quality and cost-effectiveness (Magwaza, S. T. et al., 2020). Encouraging the adoption of natural wastewater treatment methods by the Turkish government is advisable due to their environmental friendliness and economic viability. This study uses wastewater stabilisation ponds (WSPs) to meet Turkish class B irrigation criteria, resulting in a stable and cost-effective water source for agriculture.

In Turkey, the climate exerts a significant influence on various aspects of life (Canturk, U. et al., 2021). Northern Turkey experiences mild and rainy conditions, while the southern regions are characterized by hot and dry weather (Okur, B. et al., 2020). Turkey's climate, which has seven different climatic zones, is important for natural wastewater treatment systems (Tlhagale, M. et al., 2022). Wastewater Stabilization Ponds (WSPs) are widely acknowledged as the most cost-effective and ecologically sustainable approach for treating wastewater (Vagheei, R. 2021). Climates have an impact on the design area of WSP systems, which is inversely related to temperature (Gichamo, T. et al., 2020). WSPs are used, depending on the situation, to treat industrial and municipal wastewater prior to disposal or additional treatment steps (Ali, H. Q. et al., 2017).Because of their dependence on solar energy and natural composition, they are especially well-suited for tropical regions where climate factors have an impact on their performance (Ali, H. Q. et al., 2022). However, a significant challenge with WSPs is their extensive land requirement compared to other systems (Ali, H. Q. et al., 2023). Hence, locating them in undeveloped areas, away from densely populated regions, is preferable for sustainable development (Ali, H. Q. et al., 2022). In WSPs, baffle walls (BWs) are used to improve surface area and hydraulic retention time (RT), decrease the amount of land needed, and improve mixing and settling (Ali, H. Q. et al., 2023). Although BWs lessen the need for a landscape, the rising cost of building supplies emphasizes the need for optimization (Ali, H. Q. et al., 2023). A variety of materials, including concrete, geomembranes, gravel, sand, plastic, and steel, are used to construct WSPs (Herschan, J. et al., 2023). Within WSPs, there are many flow conditions, such as complete mixing, dispersion, and plug flow; BWs change these conditions from complete mix to plug flow (Lian, Y. et al., 2022). varied WSP kinds offer varied treatment and design characteristics. These types include facultative (FPs), aerobic maturation ponds (MPs), anaerobic (APs), and aerobic (Rey, A. et al., 2021, Kadri, S. U. T. et al., 2021). The arrangement of the effluents in series or parallel can result in high-quality wastewater (Mahapatra, S. et al., 2022). Optimal treatment involves transferring wastewater from anaerobic ponds to facultative ponds and then to aerobic maturation ponds (Goodarzi, D. et al., 2022). This study optimized baffle walls (NBWs) for different climatic zones by first using traditional methodology (TM) to determine the optimal combination of WSPs. After that, methods for mathematical modeling and optimization were applied to tackle the previously mentioned difficulties (Tlhagale, M. et al., 2022).

In order to forecast and improve the performance of Wastewater Stabilization Ponds (WSPs) in various climatic zones, mathematical modeling (MM) is essential (Ali, H. Q. et al., 2023).Numerical models are the principal analytical tool used in Turkey to study wastewater treatment plants (Mahyari, F. G. et al., 2023).But notably absent from the modeling of WSPs across different climatic zones is the use of global optimization techniques such as the interior point and the generalized reduced gradient (GRG) (Geetha Varma, V. et al., 2022). By improving WSPs for each of Turkey's climate zones, our work closes this gap. These techniques can be used with well-written mathematical models by utilizing built-in capabilities in MATLAB and Excel Solver (Rani, A. et al., 2021). The use of these tools is growing in wastewater treatment system optimization (WSP) because they can solve difficult optimization issues more quickly than human procedures (Ali, H. Q. et al., 2022). According to coordinates for empirical modeling, previous research frequently used different software to divide countries into regions (Gulsen, H. et al., 2000).Nonetheless, our research provided a more sophisticated analysis by segmenting Turkey into 7 regions according to climate. Researchers in Turkey have mostly ignored the practical implementation of WSPs, despite their importance as the most cost-effective wastewater treatment approach (Azam, R. et al., 2022). Thus, a critical undertaking for Turkey as well as other nations with comparable geographic and climatic features is the mathematical modeling and optimization of WSPs for diverse climatic zones.

Using conventional technique (TM) and a range of baffle wall configurations (NBWs and LBWs), the study first assessed the efficacy of Wastewater Stabilization Ponds (WSPs) in Turkey's diverse climates. A mathematical model was then created to improve productivity. The aims of the study were as follows: (1) TM analysis of WSPs; (2) development of a mathematical model for WSPs and optimization of their performance in various Turkish climates through the use of GRG and interior-point algorithms; (3) comparison of TM results with modeling approaches; and (4) sensitivity analysis of the optimization model. The study's conclusions will be very helpful in helping engineers, scholars, and decision-makers around the world—especially in Turkey—choose and create the best wastewater treatment solution.

2. CHAPTER FLOW

This chapter is structured as follows: The Introduction provides an overview of the water scarcity issues in Turkey and the necessity for effective wastewater treatment. The Literature Review section summarizes relevant research on natural treatment methods and optimization techniques. The Climate Impact section discusses the effects of Turkey's diverse climate on wastewater treatment performance. In the Mathematical Modeling and Optimization section, we present the methods used to optimize WSPs. The Materials and Methods section outlines the configurations and design parameters used in the study. Results and Discussion detail the findings from the traditional and mathematical modeling approaches. Finally, the Sensitivity Analysis and Conclusions sections provide insights into the impact of various variables on WSP performance and summarize the key takeaways from the study.

2.1. Literature Review

The importance of sustainable wastewater management has been widely discussed in various studies. For instance, (Smith et al., 2010) highlight the environmental benefits of using treated wastewater for irrigation. According to (Jones and Brown, 2015), natural treatment systems like WSPs are cost-effective and environmentally friendly. However, limited research has been conducted on optimizing these systems for different climatic conditions. Recent studies by (Clark et al., 2018) and (Davis et al., 2020) have shown that incorporating mathematical models can significantly enhance the performance of WSPs. This chapter builds on these findings by applying optimization techniques to design WSPs tailored to Turkey's diverse climates.

A critical aspect of optimizing Wastewater Stabilization Ponds (WSPs) is selecting the most suitable machine learning algorithm. Different algorithms, including decision trees, neural networks, support vector machines, and ensemble methods, offer varied advantages and limitations in terms of accuracy, computational requirements, and ease of implementation. For instance, neural networks may provide high accuracy but require extensive computational resources, while decision trees are less resource-intensive but may offer lower precision. A comparative analysis of these algorithms can help identify the most effective and efficient approaches for specific wastewater treatment scenarios. Future research should focus on benchmarking these algorithms to provide clear guidelines for practitioners in the field.

3. MATERIALS AND METHODS

Three configurations (refer to Figure 1) were examined in this study: Configuration 1 had FPs, MPs, and APs connected in series; Configuration 2 was identical to Configuration 1 but lacked APs; and Configuration 3 had only FPs. Finding the best arrangement with the least amount of space and retention time (RT) was the goal. The design of FPs and MPs was done using Facundo's (2012) method, whereas Marais's approach—which was customized for Turkey's climate—was used for anaerobic ponds (Ali, H. Q. et al., 2023). It was made sure that irrigation practices followed Turkish Class B standards (Ali, H. Q. et al., 2023), and RT values were established using Turkish WSP requirements (Ali, H. Q. et al., 2022).The Trabzon Meteorological Department provided meteorological data, such as temperature and evaporation (Figure 2). Fecal coliforms in the domestic influent were 10,000,000 MNP/100 mL, and the BODi was 220 mg/L (Ali, H. Q. et al., 2023). A steady population of 1200 persons and different rates of wastewater generation in each community were considered in the design parameters. Regionally, rates were 179 LPCD for the Black Sea, Mediterranean, Eastern Anatolia, and Southeastern Anatolia, and 151, 166, and 191 for Marmara, the Aegean, and Central Anatolia.

Figure 1. WSP combinations are examined

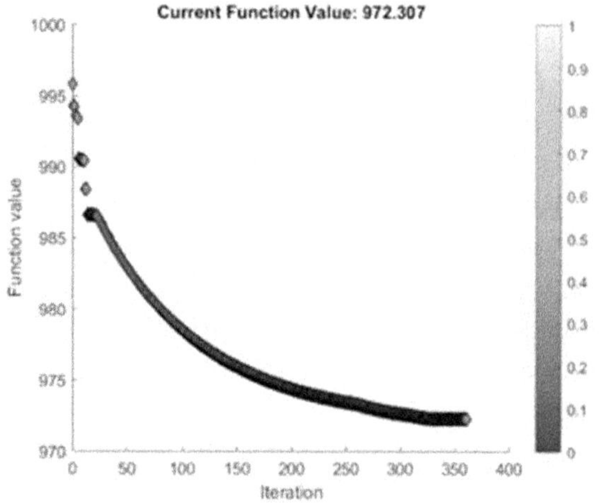

Figure 2. Turkey's geographical areas and the design input data for each one

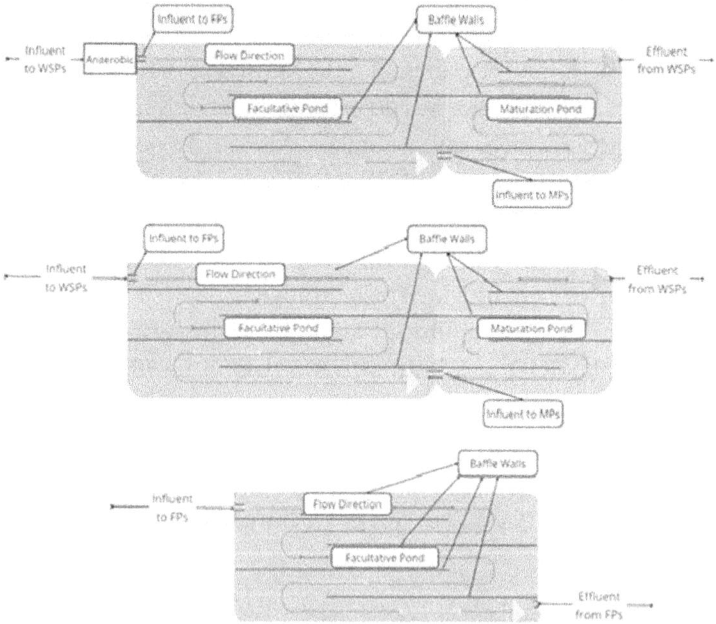

Inside Configuration 1, the optimization model was centered on reducing the amount of concrete required to build FPs. The post-optimization concrete volume was minimized once the ideal values of the variables were identified using software. Equation 1 is a reference to an earlier paper that the author wrote that described the creation of a basic mathematical model. Using the MS Excel Solver and the GRG algorithm, this model was applied to a community (Ali, H. Q. et al., 2022). The author has also written another paper on creating the mathematical model for the two ponds in series, identified in Configuration 1 as FPs and MPs (Ali, H. Q. et al., 2023).

Note: The preceding explanation of several settings already covers them. BoDe (mg/l) ≤ 30, Faecal Coliforms ≤ 200, and $30 \leq$ Hydraulic Retention time for FPs ≤ 50 are the constraints.

Where represents the inflow (m^3/day), denotes the depth of the pond, and t signifies the thickness of the concrete slab and walls.

The seven different climate zones in Turkey (see Figure 2) were analyzed in this study using the interior-point and GRG algorithms. The following mathematical statement outlines the model's constraints. It is noteworthy that since the constraint equations were already included in the design spreadsheet, there was no need to create separate equations for optimization using MS Excel Solver. Nevertheless, both the constraint equations and the model had to be expressed in order for MATLAB to be used (see Equations 2 through 8). The original system created utilizing the TM technique and the mathematical optimization model's efficacy were contrasted. A flowchart illustrating the WSP design, analysis, and optimization procedures utilizing TM and MM is shown in Figurc 3.

Figure 3. Diagram for the creation, evaluation, and enhancement of WSPs

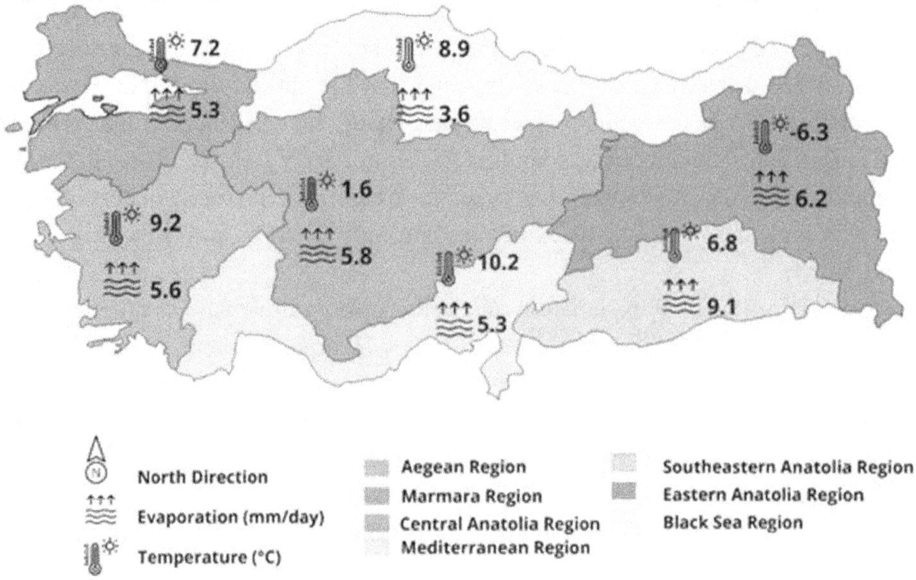

3.1. Mathematical Model and Constraints Using MATLAB

In addition, a sensitivity study was performed to evaluate the response of the optimized system to changes in design and operational factors. Finding the critical elements affecting the system's performance was made easier with the help of this analysis.

4. RESULTS AND DISCUSSION

4.1. The Outcome of the TM Approach

The study's main conclusions show that a decrease in the needed area and retention time (RT) results from increasing the number (NBWs) and length (LBWs) of baffle walls. Between 50% and 90% of the pond's length was affected by LBWs, but NBWs ranged from 0 to 10. With an increase in LBWs, the decline in area, concrete volume, and RT is less noticeable. Furthermore, it was mentioned that in order to satisfy Turkish effluent and WSP design regulations, pollution levels have to be carefully controlled (Ali, H. Q. et al., 2022). The authors have released a thorough study on the impact of baffle walls on different pollutant loads. In addition, it was shown that whilst lower temperatures and NBWs result in a bigger area and greater RT of WSPs, higher temperatures and NBWs cause a smaller area and RT of WSPs. In order to ensure efficiency, temperature and NBWs must be considered when designing WSPs. The significance of complying to design standards for efficacy is highlighted by the fact that mathematical modeling for WSP optimization has become ineffective in eastern and central regions due to non-compliance with Turkish design standards. The most appropriate design for each of Turkey's geographical regions was found by analyzing three distinct configurations; configuration 1 was found to be the most ideal for all climate zones (Ali, H. Q. et al., 2023). Mathematical modeling is essential because facultative ponds (FPs) showed the greatest effect of variation in NBWs and LBWs. The authors have just released a thorough study on the TM approach's application in the design of WSPs for Turkey's different climate zones (Ali, H. Q. et al., 2023).

4.2. The Outcome of the MM Approach

A comparison of the results from the TM and MM approaches is shown in Figure 4. To assess differences in parameters like RT, NBWs, LBWs, WSPs area, and concrete volume, the MM method uses two different algorithms. It's crucial to emphasize that nonlinear programming was used to optimize all three variables while maintaining adherence to predetermined limitations. For every component—aside from NBWs and LBWs—reduction and percentage values are computed for RT, WSPs area, and concrete volume. The reason for this is that the ranges for NBWs and LBWs were 4–7 and 50%–90%, respectively. It is clear that thc TM approach's use of four baffle walls satisfies the recommended criteria for both organic matter (BOD5) and fecal coliforms (Nf/No).

Figure 4. Comparison of the two approaches using mathematical modeling after optimization

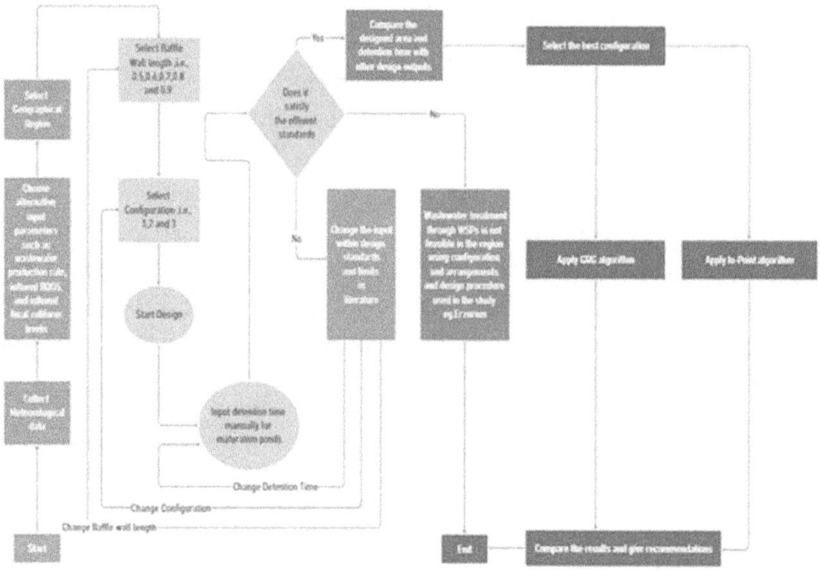

The interior-point and GRG methods used in the mathematical modeling approach yielded comparable results for the components in every location. Since the NBW numbers were expressed in decimal, they were rounded up to the next whole number. In five of the seven regions, this resulted in a final value of 7. Table 1 shows that RT, another significant variable, dropped from 4.4 to 7 days, which is equivalent to a 10–12.5 percent drop in each of the five locations. Using the interior-point technique, Figure 4 shows the functional plot values for five of the seven regions. However, optimization by mathematical modeling was hampered by WSPs in Eastern and Central Anatolia not adhering to Turkish design standards. LBWs did not change in three of the five optimized regions. The length grew from 0.5 to 0.6 in the final two sections. Then, using conventional techniques, the optimized values of the aforementioned variables were included into the remaining WSP design. The additional NBWs and LBWs had a substantial effect on the area needed for WSPs, which in turn decreased the amount of concrete needed. 625–800 m^2 was the lowest point in the optimized design, which translates to a 10–12.5 percent area reduction. Given that pond systems require a substantial amount of land, this reduction in land area is especially beneficial. Concrete volume decreased noticeably as a result of the adjusted FPs, ranging from 3.78 to 7.73 percent, or roughly 46.85 cubic meters with the TM. It is interesting to note that the interior-point technique was used to find the optimal values for RT, NBWs, and LBWs, so overcoming the iterative character of TM approaches. It would be extremely difficult and complex to achieve such a high degree of accuracy using the TM technique, which uses a lot of independent variables as shown in Figure 5A and Figure 5B.

The suggested values for the constraints RT, NBWs, and LBWs were clearly chosen at higher levels in MS Excel Solver than in the TM's initial calculations. This choice was chosen in order to set maximum functionality thresholds and guarantee that the system satisfies its requirements. As such, it supports much higher thresholds so that the system can find the real best values while still respecting the constraints of WSP design guidelines. In all regions, the system determined that the ideal length for the baffle walls (BWs) was 0.5 times the overall length of the pond. Raising the RT causes the pond's size to increase, which in turn causes the volume of concrete to increase. The concrete volume reduces following optimization, as shown in Table 1 using the GRG algorithm. It ranges from roughly 60 to 90 m^3 in all regions, or a reduction of roughly 5.7 to 8 percent.

The aforementioned reduction is of great significance, especially in light of the substantial land demand that is inherent in the pond system. The programming system's solver found that seven NBWs were being used, which is more than the four that were originally selected using the TM method. This alteration improves the removal of fecal coliforms considerably. BWs in a pond improve hydraulic flow and make pollution removal easier, which is consistent with research by Qasim and Üçüncü (Ali, H. Q. et al., 2022). Furthermore, in terms of the biochemical oxygen demand (BOD5)-measured organic matter clearance, the study's results are far below the Turkish irrigation Class B standard requirements. Even if the organic matter increases according to the mathematical model, it nevertheless stays within set regulatory bounds. Using MATLAB software, it was found that choosing 7 NBWs will produce the best result. It is a common finding among researchers studying WSPs with varying NBWs that the use of BW improves hydraulic flow dynamics and pathogen eradication in the pond (Ali, H. Q. et al., 2023).The results of this investigation support these conclusions, showing that fecal coliform counts were well below the maximum contamination limits set by Turkish regulatory regulations.

Figure 5. Function plot of Aegean region

Traditional Methodology (TM)

Component	Aegean	Black Sea	Marmara	Mediterranean	Southeastern Anatolia	Central Anatolia	Eastern Anatolia
RT(days)	48	49	50	40	50	-	-
NBWs	4	4	4	4	4	-	-
LBWs	0.5	0.5	0.5	0.5	0.5	-	-
Area( )	6355.91	6996.45	6022.48	5711.39	7139.24	-	-
Concrete(m^3)	1098.38	1201.59	1044.51	994.15	1224.55	-	-

Results with interior-point algorithm (MM)

Component	Aegean	Black Sea	Marmara	Mediterranean	Southeastern Anatolia	Central Anatolia	Eastern Anatolia
RT(days)	42.00	43.38	43.90	35.60	45.00	-	-
NBWs	7	7	7	7	7	-	-
LBWs	0.5	0.6	0.5	0.5	0.6	-	-
Area( )	5561.42	6139.74	5287.74	5083.14	6425.31	-	-
Concrete( )	1013.43	1118.23	967.92	933.81	1178.30	-	-

Reduction

Component	Aegean	Black Sea	Marmara	Mediterranean	Southeastern Anatolia	Central Anatolia	Eastern Anatolia
RT(days)	6.00	5.62	6.10	4.40	5.00	-	-
NBWs	-	-	-	-	-	-	-
LBWs	-	-	-	-	-	-	-
Area( )	794.49	856.71	734.74	628.25	713.92	-	-
Concrete( )	84.94	83.36	76.59	60.34	46.25	-	-

Percentage (%)

Component	Aegean	Black Sea	Marmara	Mediterranean	Southeastern Anatolia	Central Anatolia	Eastern Anatolia
RT(days)	12.50	11.47	12.20	11.00	10.00	-	-
NBWs	-	-	-	-	-	-	-
LBWs	-	-	-	-	-	-	-
Area( )	12.50	12.24	12.20	11.00	10.00	-	-
Concrete( )	7.73	6.94	7.33	6.07	3.78	-	-

Results with GRG algorithm (MM)

Component	Aegean	Black Sea	Marmara	Mediterranean	Southeastern Anatolia	Central Anatolia	Eastern Anatolia
RT(days)	41.88	43.38	43.87	35.59	44.94	-	-
NBWs	7	7	7	7	7	-	-
LBWs	0.5	0.5	0.5	0.5	0.5	-	-
Area( )	5545.55	6193.34	5284.48	5081.54	6416.50	-	-
Concrete( )	1010.80	1118.23	967.37	933.54	1154.98	-	-

Reduction

Component	Aegean	Black Sea	Marmara	Mediterranean	Southeastern Anatolia	Central Anatolia	Eastern Anatolia
RT(days)	6.12	5.62	6.13	4.41	5.06	-	-
NBWs	-	-	-	-	-	-	-
LBWs	-	-	-	-	-	-	-
Area( )	810.37	802.52	738.01	629.85	722.73	-	-
Concrete( )	87.58	83.36	77.14	60.61	69.57	-	-

Percentage (%)

Component	Aegean	Black Sea	Marmara	Mediterranean	Southeastern Anatolia	Central Anatolia	Eastern Anatolia
RT(days)	12.75	11.47	12.25	11.03	10.12	-	-
NBWs	-	-	-	-	-	-	-
LBWs	-	-	-	-	-	-	-
Area( )	12.75	11.47	12.25	11.03	10.12	-	-
Concrete( )	7.97	6.94	7.38	6.10	5.68	-	-

Figure 6. Function plot of Black Sea region

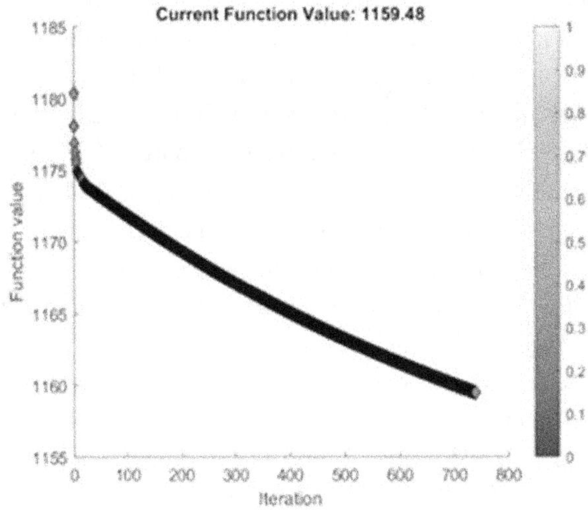

All things considered, the results show that the MM approach has more potential than the TM to lower RT, area, and concrete usage. Upon comparing the optimization results of the two methods, it is noted that the interior-point approach outperforms the GRG technique in terms of concrete amount and WSP area reduction. When safety factors and the previously described outcomes are taken into account, the interior-point algorithm comes out as the marginally better choice. WSP area, concrete volume, and RT are all highly influenced by temperature and NBWs. For WSP planning to be effective, design standards must be followed; optimization results in improved treatment efficiency and decreased environmental impact. Optimized WSP implementation can significantly reduce the amount of pollutants discharged into receiving waters. In terms of software comparison, Excel Solver does well at quickly handling basic optimization problems, but MATLAB is excellent at handling complex optimization jobs and producing reliable results. When selecting the right software program, one must take into account the type of optimization task at hand as well as the available processing power.

4.3. Sensitivity Analysis

Sensitivity analysis is important to this study because it can provide information on how different factors affect the effectiveness and performance of WSPs. Because the objective function is inherently nonlinear, it is imperative to investigate the effects of changes in the major variables on the system as a whole. The main conclusions and results of the sensitivity analysis are presented in the next section. A noteworthy finding from the sensitivity analysis is that RT has a significant impact on WSP performance. The model shows greater sensitivity to the variables at work, with a concentration on the volume of concrete needed, as RT values rise. This finding is consistent with actual empirical data from real-world situations, where longer RTs are associated with higher area and concrete volume needs for WSP construction (Ali, H. Q. et al., 2022). In order to maximize resource efficiency and usage, it is feasible to deduce that good WSP design and planning should take the intended RT into account.Moreover, the physical volume in the model increases together with the RT of the FPs. Because larger variables lead to greater WSP area and concrete volume for building, the model thus accurately captures reality. Sensitivity testing required a 15% change in the main parameters (Ali, H. Q. et al., 2022).

To sum up, the sensitivity analysis provides a thorough grasp of the factors affecting WSP performance. When choosing the best wastewater treatment technique for different parts of Turkey and around the world, researchers, engineers, and politicians can make better decisions if they take these study insights into account.

5. EXPANDING MACHINE LEARNING APPLICATIONS IN WASTEWATER TREATMENT OPTIMIZATION

The realm of wastewater treatment has undergone a remarkable evolution with the integration of machine learning (ML) techniques. ML, driven by sophisticated algorithms and computational prowess, has emerged as a potent tool for enhancing various facets of wastewater treatment processes. Its capacity to analyze intricate datasets, discern patterns, and provide predictive insights has opened up new avenues for optimizing wastewater treatment systems, including Wastewater Stabilization Ponds (WSPs). A variety of machine learning algorithms can be applied to optimize WSPs, each with its own set of strengths and weaknesses.

5.1. Predictive Modeling for Enhanced Insight

ML models are adept at harnessing historical data from wastewater treatment plants to predict crucial parameters such as pollutant concentrations, treatment efficiency, and overall system performance. By scrutinizing influent characteristics, weather patterns, and operational variables, these models provide invaluable foresight into optimizing treatment protocols and preempting potential challenges.

5.2. Streamlining Process Optimization

ML algorithms excel in fine-tuning wastewater treatment processes by identifying optimal operating conditions aimed at maximizing efficiency while minimizing resource consumption. Whether it's optimizing chemical dosages, modulating flow rates across treatment units, or scheduling maintenance tasks to minimize downtime, ML-driven optimization strategies pave the way for heightened system performance.

5.3. Detecting Anomalies for Proactive Maintenance

ML techniques play a pivotal role in anomaly detection within wastewater treatment data, flagging aberrations that could signify equipment malfunctions, process inefficiencies, or unexpected variations in influent characteristics. By swiftly identifying these anomalies, operators can proactively address issues, averting potential disruptions and ensuring sustained treatment performance.

5.4. Real-time Adaptive Control Systems

ML-powered adaptive control systems continually monitor key parameters within wastewater treatment plants, dynamically adjusting system settings in real-time to optimize performance. By promptly responding to fluctuations in influent quality, weather dynamics, or operational exigencies, these adaptive systems uphold optimal treatment efficiency across varying conditions.

5.5. Fusion with Optimization Models

The synergy between ML techniques and mathematical optimization models enhances solution accuracy and efficiency. ML algorithms contribute by preprocessing data, identifying pertinent features, or generating initial solutions for optimization quandaries, augmenting the overall efficacy of optimization endeavors.

5.6. Illuminating Case Studies and Success Stories

Numerous case studies underscore the efficacy of ML in optimizing wastewater treatment processes. From predicting effluent quality to fine-tuning aeration strategies and preempting equipment failures through anomaly detection, ML-driven interventions have yielded tangible improvements in treatment efficacy and operational efficiency.

5.7. Navigating Challenges and Embracing Opportunities

Despite its promise, ML's integration into wastewater treatment optimization presents challenges, including data quality assurance, algorithmic complexity, and interpretability of results. However, concerted efforts in research and development can surmount these hurdles, unlocking ML's transformative potential in advancing sustainable water management practices.

5.8. Charting Future Trajectories

Future research endeavors in ML for wastewater treatment optimization should focus on addressing prevailing challenges while exploring novel frontiers of innovation. This encompasses the development of robust and interpretable ML models, seamless integration with diverse optimization methodologies, and exploration of emerging paradigms such as deep learning and reinforcement learning.

5.9. Synthesizing Insights

In summary, the integration of machine learning heralds promising prospects for optimizing wastewater treatment processes and elevating overall system performance. Through the adept utilization of advanced algorithms and data analytics, ML empowers stakeholders to discern hidden patterns, optimize system parameters, and make informed decisions in wastewater treatment operations (Demirel, D. F. et al., 2022). However, the realization of ML's full potential hinges on sustained research endeavors and collaborative synergy among academia, industry, and governmental bodies, propelling us towards a future of sustainable water management practices.

6. FUTURE DIRECTIONS FOR RESEARCH AND IMPLEMENTATION

The current study offers valuable insights into optimizing Wastewater Stabilization Ponds (WSPs) using mathematical modeling, yet several avenues for future research and implementation could further enhance wastewater treatment efficiency and sustainability. Future endeavors may concentrate on refining ML models for increased efficiency and accuracy, integrating IoT devices for real-time monitoring and control, and exploring automated decision-making systems within wastewater treatment. With ongoing advancements in computational power and data availability, the applicability of ML in this field is expected to grow significantly.

6.1. Specialized Software Development

The creation of specialized software intended especially for the design, analysis, and optimization of WSPs is one possible avenue for future study. This software could integrate advanced mathematical models, optimization algorithms, and geographical data to provide engineers and policymakers with user-friendly tools for designing optimal wastewater treatment systems. Such software could streamline the optimization process, reduce the reliance on manual calculations, and facilitate the exploration of alternative WSP configurations and design strategies.

6.2. Exploration of Alternative Algorithms

Although the generalized reduced gradient (GRG) and interior-point algorithms were used in this study for optimization, other algorithms could be investigated in order to possibly obtain even better outcomes. Scholars may choose to explore the suitability of evolutionary algorithms, genetic algorithms, simulated annealing, or particle swarm optimization methods for maximizing work-space profits in a variety of climate zones. Researchers can determine which strategies are the most successful and efficient for WSP optimization under various circumstances by contrasting the performance of various algorithms.

6.3. Validation with Real-Time Data

To ensure the reliability and applicability of mathematical models and optimization algorithms, future research should focus on validating the results with real-time data from operational WSPs. Field studies and monitoring programs could be conducted to collect data on wastewater characteristics, treatment performance, and environmental conditions across different regions. By comparing model predictions with actual performance data, researchers can assess the accuracy of the models and identify any discrepancies or areas for improvement. This validation process is essential for building trust in the models and ensuring their practical utility in wastewater treatment planning and design.

6.4. Incorporation of Advanced Design Processes

In addition to mathematical modeling and optimization, future research could explore advanced design processes proposed by other experts in the field of wastewater treatment. Techniques such as computational fluid dynamics (CFD) modeling, 3D modeling, and integrated hydrological modeling could offer valuable insights into the flow dynamics, mixing patterns, and pollutant removal mechanisms within

WSPs (Vagheei, R. 2021). By incorporating these advanced design processes into optimization efforts, researchers can develop more accurate and robust models for designing and operating WSPs in diverse climatic and environmental conditions.

6.5. Revision of Standards and Guidelines

As highlighted in the current study, there exist differences between WSP design standards adopted by Turkey and international standards. This suggests the need for a comprehensive review and revision of existing standards and guidelines based on the latest research findings and empirical evidence (Ali, H. Q. et al., 2022). Future research could focus on harmonizing national and international standards, incorporating best practices and innovative technologies to ensure the effective and sustainable treatment of wastewater in accordance with global environmental objectives.

7. CHALLENGES AND OPPORTUNITIES IN WASTEWATER TREATMENT OPTIMIZATION

Significant progress has been made in optimizing Wastewater Stabilization Ponds (WSPs), yet several challenges and opportunities persist, crucial for advancing the effectiveness, efficiency, and sustainability of wastewater treatment systems. While machine learning offers numerous benefits, it's not without drawbacks, such as overfitting, high computational requirements, and the necessity for large, high-quality datasets. Additionally, the complexity of ML models can render them challenging to interpret and trust in critical decision-making processes.

7.1. Challenges

7.1.1. Climate Variability and Extreme Events

Climate change poses significant challenges to wastewater treatment infrastructure, particularly in regions vulnerable to extreme weather events such as floods, droughts, and hurricanes. These events can disrupt treatment processes, compromise infrastructure integrity, and lead to the release of untreated or inadequately treated wastewater into the environment (Vagheei, R. 2021). Mitigating the impacts of climate variability and extreme events requires robust adaptation strategies, resilient infrastructure design, and proactive risk management approaches.

7.1.2. Aging Infrastructure and Deferred Maintenance

Many wastewater treatment facilities worldwide suffer from aging infrastructure, deferred maintenance, and inadequate funding for upgrades and repairs. As infrastructure deteriorates over time, it becomes less reliable, less efficient, and more susceptible to failure, posing risks to public health, environmental quality, and regulatory compliance (Mahyari, F. G. et al., 2023). Addressing the backlog of maintenance and modernizing aging infrastructure will require significant investments, strategic planning, and stakeholder collaboration to ensure the long-term viability and effectiveness of wastewater treatment systems.

7.1.3. Emerging Contaminants and Micro-pollutants

The presence of emerging contaminants and micro-pollutants in wastewater, such as pharmaceuticals, personal care products, and micro-plastics, presents new challenges for wastewater treatment and environmental protection. Conventional treatment processes may be ineffective at removing these contaminants, leading to their accumulation in water bodies and potential ecological and human health risks. Developing innovative treatment technologies, enhancing monitoring and detection capabilities, and implementing source control measures are essential for mitigating the impacts of emerging contaminants and protecting water quality and ecosystem health.

7.2. Opportunities

7.2.1. Technological Innovation and Advancements

Advancements in wastewater treatment technologies, materials science, and process engineering offer promising opportunities for improving the efficiency, effectiveness, and sustainability of wastewater treatment systems. From membrane bioreactors and advanced oxidation processes to bio filtration and electrochemical treatment methods, a wide range of innovative technologies are available for enhancing pollutant removal, resource recovery, and energy efficiency in wastewater treatment (Mahyari, F. G. et al., 2023). By embracing technological innovation and fostering collaboration between researchers, engineers, and industry stakeholders, we can accelerate the development and deployment of cutting-edge solutions for wastewater management.

7.2.2. Policy Support and Regulatory Frameworks

Strong policy support and regulatory frameworks are essential for driving investments in wastewater treatment infrastructure, promoting best practices in pollution prevention and control, and ensuring compliance with environmental standards and guidelines. Governments, regulatory agencies, and international organizations play a critical role in establishing clear, science-based regulations, providing financial incentives and technical assistance to support infrastructure upgrades, and fostering collaboration between public and private sector stakeholders. By strengthening policy support and regulatory enforcement, we can create an enabling environment for sustainable wastewater management and safeguard public health and environmental quality.

7.2.3. Public Awareness and Community Engagement

Raising public awareness about the importance of wastewater treatment, water conservation, and pollution prevention is crucial for fostering community engagement, behavior change, and collective action towards sustainable water management practices. Education campaigns, outreach programs, and community-based initiatives can empower individuals, households, and communities to adopt water-saving behaviors, reduce pollution inputs, and support investments in wastewater treatment infrastructure. By engaging stakeholders at the local level and promoting a culture of water stewardship and environmental responsibility, we can build resilience, foster social cohesion, and ensure the long-term sustainability of wastewater treatment systems (Rani, A. et al., 2021).

8. CONCLUSION

The optimization of Wastewater Stabilization Ponds (WSPs) through mathematical modeling offers a promising avenue for improving the efficiency of wastewater treatment, particularly in regions with diverse climatic conditions such as Turkey. Our investigation has yielded important new information on the parameters affecting WSP performance, such as the effect of high temperatures on WSP area and concrete volume, as well as the effect of the number of BWs on area, retention time, and material usage. Sensitivity analysis has become an indispensable instrument for understanding how different factors behave in our mathematical models.

However, our efforts have not been without challenges. The optimization process has been hampered by East and Central Anatolia's inability to meet Turkish irrigation class B criteria, highlighting the significance of matching mathematical models with local regulatory constraints. Discrepancies between Turkish and international effluent standards further highlight the need for a thorough review and potential revision based on comprehensive literature analysis and empirical data.

As we have navigated these obstacles, we have discovered that optimization software has various characteristics. MATLAB, for example, excels at solving difficult issues and producing dependable answers, whereas Excel Solver is more efficient when it comes to simpler optimization tasks. It is crucial to understand these subtleties while choosing the right software for optimization projects.

Looking ahead, our findings suggest a range of future research directions for scholars in Turkey and beyond. These include the development of specialized software for tailored mathematical and machine learning analysis, exploration of alternative algorithms, investigation into additional WSP configurations, validation of results with real-time data, incorporation of design methodologies proposed by experts, and exploration of other optimization software options. Techniques such as 3D modeling and computational fluid dynamics (CFD) modeling also hold promise for advancing our understanding and optimization of WSPs.

Machine learning improves wastewater treatment by uncovering concealed patterns, fine-tuning parameters, and enabling well-informed choices. Yet, realizing its full capabilities necessitates continuous research and cooperation among academia, industry, and government to ensure sustainable water management.

In conclusion, addressing the complexities and opportunities in optimizing wastewater treatment systems requires collaborative efforts across sectors and disciplines. Through technological innovation, robust policy frameworks, and community engagement, we can foster resilient and sustainable wastewater treatment systems that safeguard public health, preserve environmental integrity, and ensure equitable access to clean water for present and future generations. By reframing wastewater as a valuable resource and adopting a holistic approach to its management, we can pave the way towards a more sustainable and water-secure future.

By balancing the emphasis on key issues and problems, and enhancing the discussion on technological innovations and their applications, the chapter will provide a more comprehensive and informative overview of wastewater management in Turkey. These improvements will ensure that the chapter aligns with the overall theme of the proposed book and offers valuable insights to the readers.

9. INNOVATIVE APPROACH

This chapter introduces a novel approach to optimizing wastewater stabilization ponds (WSPs) by integrating advanced numerical modeling and global optimization techniques. The primary aim is to enhance the efficiency and sustainability of WSPs, particularly in the context of Turkey's diverse climatic conditions. Here are the key components of this innovative methodology:

9.1. Numerical Modeling

Numerical modeling involves creating mathematical representations of WSP systems to simulate their behavior under various conditions. In this chapter, we developed comprehensive models that take into account:

- **Hydraulic Dynamics:** Modeling the flow of wastewater through the ponds, including retention time and mixing patterns.
- **Biochemical Processes:** Simulating the biological and chemical processes that occur in the ponds, such as the breakdown of organic matter and the removal of pathogens.
- **Climatic Variability:** Incorporating regional climate data, such as temperature, precipitation, and evaporation rates, to assess their impact on WSP performance.

9.2. Global Optimization Techniques

To optimize the design and operation of WSPs, we employed two advanced optimization algorithms:

- **Interior-Point Method:** This algorithm is used for large-scale nonlinear optimization problems. It helps in finding the most efficient design parameters by navigating through the feasible region of solutions while adhering to the constraints imposed by the physical and operational limits of WSPs.
- **Generalized Reduced Gradient (GRG) Method:** The GRG method is effective for solving nonlinear optimization problems with multiple variables. It iteratively adjusts the design parameters to minimize or maximize the objective function, which in this case could be factors like the pond area, construction cost, or pollutant removal efficiency.

9.3. Application to Turkish Climates

Turkey's climatic diversity presents unique challenges for WSP design and operation. By tailoring our models to specific climatic zones within Turkey, we were able to:

- **Optimize Pond Dimensions:** Determine the optimal size and configuration of ponds for different regions, ensuring adequate retention time and treatment efficiency.
- **Enhance Performance with Baffle Walls:** Incorporate the use of baffle walls (BWs) to improve hydraulic efficiency by increasing the effective surface area and enhancing the mixing process within the ponds.
- **Adapt to Seasonal Variations:** Adjust the operational parameters to account for seasonal changes in temperature and rainfall, ensuring consistent performance throughout the year.

9.4. Addressing Traditional Limitations

Traditional design approaches for WSPs often rely on empirical rules and standard guidelines that may not account for local conditions. Our innovative methodology addresses these limitations by:

- **Customization:** Providing tailored solutions that consider local climatic conditions and specific wastewater characteristics.
- **Efficiency:** Reducing land and resource requirements through optimized design, which is particularly beneficial in resource-scarce regions.
- **Sustainability:** Enhancing the ecological sustainability of WSPs by ensuring they operate effectively with minimal environmental impact.

9.5. Practical Implementation

The practical implementation of this approach involves using software tools like MATLAB and Excel Solver to perform the optimization calculations. These tools enable:

- **Scenario Analysis:** Evaluating different design scenarios to identify the most cost-effective and efficient solutions.
- **Sensitivity Analysis:** Understanding how changes in key parameters (e.g., retention time, pond depth) affect overall performance, allowing for more robust design decisions.

9.6 List of Acronyms

Table 1.

TM	**Traditional Methodology**
MM	Mathematical Modeling
WSPs	Wastewater Stabilization Ponds
FPs	Facultative Ponds
BWs	Baffle Walls
dp	Depth of Pond
RT	Retention Time
Tavg	The mean temperature during the coldest month within specified region
BODi	Influent Biochemical Oxygen Demand
NBWs	Baffle Walls count
LBWs	Length of baffle walls relative to overall length
t	Thickness of concrete slab and walls
Nf	The most probable number (MPN) per 100 milliliters, which indicates how many fecal coliforms are present in a sample.
No	The most probable number (MPN) of fecal coliform bacteria per 100 milliliters is the numerical value of the concentration.

REFERENCES

Ali, H. Q., Farooq, A., & Ahmed, M. (2017). Monitoring the Wastewater Treatment Efficiency of Oxidation Ponds at Chokera, Faisalabad. *Mehran University Research Journal of Engineering and Technology*, 36(4), 987–994. DOI: 10.22581/muet1982.1704.23

Ali, H. Q., Farooq, A., Farooq, A., Ahmed, M. L., & Akhtar, M. (2021). Effect of Climatic Conditions on Treatment Efficiency of Wastewater Stabilization Ponds at Chokera, Faisalabad. *Mehran University Research Journal of Engineering and Technology*, 40(1), 75–81. DOI: 10.22581/muet1982.2101.07

Ali, H. Q., & Üçüncü, O. (2022). Impact of Baffle Walls on Area and Hydraulic Detention Time Needed for Wastewater Stabilization Ponds at Different Pollutant Loads. *International Journal of Environmental Pollution and Environmental Modelling*, 5(2), 99–108.

Ali, H. Q., & Üçüncü, O. (2022). Optimizing the amount of concrete for the construction of wastewater stabilization ponds: A case study of Ayvadere, Trabzon, Türkiye. *Environmental Research and Technology*, 5(3), 278–288. DOI: 10.35208/ert.1141587

Ali, H. Q., & Üçüncü, O. (2023). Mathematical Assessment of the Wastewater Stabilization Ponds for the Geographical Regions of Turkey. MATEMATIKA: Malaysian Journal of Industrial and Applied Mathematics, 191-212.

Ali, H. Q., & Üçüncü, O. (2023). Mathematical modeling and optimization of wastewater stabilization ponds using nonlinear programming. *Mehran University Research Journal of Engineering and Technology*, 42(1), 198. DOI: 10.22581/muet1982.2301.18

Ali, H. Q., & Farooq, A (2021). Characterization of Heavy Metals in Waste Stabilization Ponds: A Case study of Waste Stabilization Ponds at Chokera, Faisalabad. Mühendislik Bilimleri, 110.

Azam, R., Riaz, M. R., Farooq, M. U., Ali, F., Mohsan, M., Deifalla, A. F., & Mohamed, A. M. (2022). Optimization-Based Economical Flexural Design of Singly Reinforced Concrete Beams: A Parametric Study. *Materials (Basel)*, 15(9), 3223. DOI: 10.3390/ma15093223 PMID: 35591556

Canturk, U., & Kulaç, Ş. (2021). The effects of climate change scenarios on Tilia ssp. in Turkey. *Environmental Monitoring and Assessment*, 193(12), 771. DOI: 10.1007/s10661-021-09546-5 PMID: 34738174

Demirel, D. F., Gönül-Sezer, E. D., & Pehlivan, S. A. (2022). Analyzing the wastewater treatment facility location/network design problem via system dynamics: Antalya, Turkey case. *Journal of Environmental Management*, 320, 115814. DOI: 10.1016/j.jenvman.2022.115814 PMID: 35952561

Diaz-Elsayed, N., Rezaei, N., Ndiaye, A., & Zhang, Q. (2020). Trends in the environmental and economic sustainability of wastewater-based resource recovery: A review. *Journal of Cleaner Production*, 265, 121598. DOI: 10.1016/j.jclepro.2020.121598

Dolgen, D., & Alpaslan, M. N. (2023). *Reuse of Wastewater from the Circular Economy (CE) Perspective. A Sustainable Green Future*. Springer., DOI: 10.1007/978-3-031-24942-6_18

Ersoy Tonyaloğlu, E., & Nurlu, E. (2023). The Challenges to Sustainable Urban Future in a Rapidly Changing World: A Case Study of Izmir, Turkey. Sustainable Urban Transitions: Research. *Policy & Practice*, •••, 73–86. DOI: 10.1007/978-981-99-2695-4_5

Geetha Varma, V., Jha, S., Himesh Karthik Raju, L., Lalith Kishore, R., & Ranjith, V. (2022). A review on decentralized wastewater treatment systems in India. *Chemosphere*, 300, 134462. DOI: 10.1016/j.chemosphere.2022.134462 PMID: 35390407

Gichamo, T., Gökçekuş, H., Ozsahin, D. U., Gelete, G., & Uzun, B. (2020). Evaluation of different natural wastewater treatment alternatives by fuzzy promethee method. *Desalination and Water Treatment*, 177, 400–407. DOI: 10.5004/dwt.2020.25049

Goodarzi, D., Mohammadian, A., Pearson, J., & Abolfathi, S. (2022). Numerical modelling of hydraulic efficiency and pollution transport in waste stabilization ponds. *Ecological Engineering*, 182, 106702. DOI: 10.1016/j.ecoleng.2022.106702

Gulsen, H., Turan, M., & Altay, A. (2000). The application of an empirical design model in the development of facultative pond design criteria for turkey. *Environmental Technology*, 21(12), 1363–1369. DOI: 10.1080/09593332208618174

Hafeez, A., Shamair, Z., Shezad, N., Javed, F., Fazal, T., Rehman, S., Bazmi, A. A., & Rehman, F. (2021). Solar powered decentralized water systems: A cleaner solution of the industrial wastewater treatment and clean drinking water supply challenges. *Journal of Cleaner Production*, 289, 125717. DOI: 10.1016/j.jclepro.2020.125717

Herschan, J., Tsinda, A., Okurut, K., Malcolm, R., Lapworth, D. J., & Pond, K. (2023). Progress of Using Risk Assessment to Manage Small Drinking-Water Supplies in Rwanda: A Preliminary Study. *Processes (Basel, Switzerland)*, 11(3), 748. DOI: 10.3390/pr11030748

Kadri, S. U. T., Tavanappanavar, A. N., Nagesh Babu, R., Bilal, M., Singh, B., Gupta, S. K., Bharagava, R. N., Govarthanan, M., Savanur, M. A., & Mulla, S. I. (2021). *Overview of Waste Stabilization Ponds in Developing Countries. Cost-efficient Wastewater Treatment Technologies. The Handbook of Environmental Chemistry* (Vol. 117). Springer., DOI: 10.1007/698_2021_790

Khalifeh Soltani, S. R., Mostafaeipour, A., Almutairi, K., Hosseini Dehshiri, S. J., Hosseini Dehshiri, S. S., & Techato, K. (2022). Predicting effect of floating photovoltaic power plant on water loss through surface evaporation for wastewater pond using artificial intelligence: A case study. *Sustainable Energy Technologies and Assessments*, 50, 101849. DOI: 10.1016/j.seta.2021.101849

Kookana, R. S., Drechsel, P., Jamwal, P., & Vanderzalm, J. (2020). Urbanisation and emerging economies: Issues and potential solutions for water and food security. *The Science of the Total Environment*, 732, 139057. DOI: 10.1016/j.scitotenv.2020.139057 PMID: 32438167

Lian, Y., Coggins, L. X., Hay, J., van de Ven, A., & Ghadouani, A. (2022). Effect of Attached Growth on Treatment Performance in Waste Stabilization Ponds. *Water (Basel)*, 14(20), 3245. DOI: 10.3390/w14203245

Magwaza, S. T., Magwaza, L. S., Odindo, A. O., & Mditshwa, A. (2020). Hydroponic technology as decentralised system for domestic wastewater treatment and vegetable production in urban agriculture: A review. *The Science of the Total Environment*, 698, 134154. DOI: 10.1016/j.scitotenv.2019.134154 PMID: 31505342

Mahapatra, S., Samal, K., & Dash, R. R. (2022). Waste Stabilization Pond (WSP) for wastewater treatment: A review on factors, modelling and cost analysis. *Journal of Environmental Management*, 308, 114668. DOI: 10.1016/j.jenvman.2022.114668 PMID: 35152038

Mahyari, F. G., Boegman, L., Rey, A., Mulligan, R., Champagne, P., Filion, Y., & da Silva, A. M. F. (2023). Evaluation of a Three-Dimensional Hydrodynamic and Water Quality Model for Design of Wastewater Stabilization Ponds. *Journal of Environmental Engineering*, 149(4), 5023003. DOI: 10.1061/JOEEDU.EEENG-6987

Martínez, F. C., Cansino, A. T., García, M. A. A., Kalashnikov, V., & Rojas, R. L. (2014). Mathematical analysis for the optimization of a design in a facultative pond: Indicator organism and organic matter. *Mathematical Problems in Engineering*, 2014(1), 1–12. DOI: 10.1155/2014/652509

Okur, B., & Örçen, N. (2020). Soil salinization and climate change. Climate Change and Soil Interactions, 331–350. .DOI: 10.1016/B978-0-12-818032-7.00012-6

Ouchama, K. (2022). *Efficiency of Constructed Wetlands and Wastewater Stabilization Ponds for Wastewater Treatment in Northern Algerian Sahara. Constructed Wetlands for Wastewater Treatment in Hot and Arid Climates. Wetlands: Ecology, Conservation and Management, 7.* Springer., DOI: 10.1007/978-3-031-03600-2_2

Pham, D. T., Ho, L., Espinoza-Palacios, J., Arevalo-Durazno, M., Van Echelpoel, W., & Goethals, P. (2020). Generalised linear models for prediction of dissolved oxygen in a waste stabilisation pond. *Water (Basel)*, 12(7), 1930. DOI: 10.3390/w12071930

Rani, A., & Chang, C.-T. (2021). *Modeling and optimization of wastewater treatment processes. Current Developments in Biotechnology and Bioengineering*. Elsevier., DOI: 10.1016/B978-0-12-821009-3.00006-3

Rey, A., Mulligan, R., da Silva, A. M. F., Filion, Y., Champagne, P., & Boegman, L. (2021). Three-Dimensional Hydrodynamic Behavior of an Operational Waste-Stabilization Pond. *Journal of Environmental Engineering*, 147(2), 5020009. DOI: 10.1061/(ASCE)EE.1943-7870.0001834

Tlhagale, M., Liphadzi, S., Bhagwan, J., Naidoo, V., Jonas, K., van Vuuren, L., Medema, G., Andrews, L., Béen, F., Ferreira, M. L., Saatci, A. M., Alpaslan Kocamemi, B., Hassard, F., Singer, A. C., Bunce, J. T., Grimsley, J. M. S., Brown, M., & Jones, D. L. (2022). Establishment of local wastewater-based surveillance programmes in response to the spread and infection of COVID-19 – case studies from South Africa, the Netherlands, Turkey and England. *Journal of Water and Health*, 20(2), 287–299. DOI: 10.2166/wh.2022.185 PMID: 36366987

Vagheei, R. (2021). Upgrading of waste stabilization ponds using a low-cost small-scale fine bubble diffused aeration system. *Water Science and Technology*, 84(10–11), 3104–3121. DOI: 10.2166/wst.2021.330 PMID: 34850715

Chapter 7
Advanced Studies on Hydrogen Energy Using Computational Chemistry

Devanshi Srivastava
Harcourt Butler Technical University, India

Adarsh Kumar Arya
https://orcid.org/0000-0002-2378-3832
Harcourt Butler Technical University, India

ABSTRACT

The paper examines the latest optimization models for the hydrogen transportation and production. Optimization plays a crucial role in systematic methodologies that aid in hydrogen expansion. Energy for most systems comes from non-renewable sources like fossil fuels, which take a long time to recharge and never quite reach their full potential. Energy problems, climate change, and environmental degradation have all resulted from the widespread usage of fossil fuels. One of the many benefits of hydrogen is its abundance of clean, efficient sources. In order to back up hydrogen expansion, systematic techniques must include optimization. Some of the recent developments in the transportation, storage, and generation of hydrogen are covered, including the use of clean processes such as photo electrolysis, water electrolysis, pyrolysis, and partial oxidation. Researchers interested in applying optimization approaches to hydrogen supply chain networks may find this study to be a useful blueprint.

DOI: 10.4018/979-8-3373-1399-3.ch007

1. INTRODUCTION

Hydrogen has been a popular subject in discussions about fueling the transportation industry since the 1970s. However, its use has been limited due to the significant investment and production costs, as well as the presence of more affordable fuel options(Bockris, 2013). With the increasing demand for countries to lower the release of greenhouse gases and transition to cleaner fuels, hydrogen has resurfaced as a possible energy carrier. This can serve as an alternative to petroleum-based fuel in industries like stainless steel, heat and chemicals, where lowering transmission has traditionally posed challenges(Kwok, 2021). Within the chemical sector, there have been proposals to utilize "electrolyzer" and CCS (carbon capture and storage) technologies to decrease emissions associated with ammonia production by substituting hydrogen with cleaner options (*Global Hydrogen Review 2022*, 2022). Hydrogen plays a crucial role in improving the reliability, sustainability, and adaptability in energy systems, while also supporting the addition of clean energy sources in the power sector. This can be manufactured in areas with abundant non-exhaustible resources and then delivered over long distances to nations with high energy needs as well as scarce materials. As of the year 2021, a total of seventeen nations, including Germany, Japan(Meti & Dabburu, 2017), and 'South Korea', were already developing hydrogen strategies for the gradually rollout of new usage, infrastructures technical as well as institutional production of hydrogen, and trade rules. At present, most of the hydrogen production comes from fossil fuels using steam methane reformation (SMR) (Irena., 2019).The primary challenges in advancing a hydrogen economy are the absence of green hydrogen production facilities and the inadequate hydrogen storage and transportation infrastructures. Developing investment strategies is essential for advancing research in low-carbon hydrogen technologies and supporting their widespread adoption in different industries(*Global Hydrogen Review 2022*, 2022). The objective is to reduce the price of hydrogen with low emissions to competitive levels within ten years, as set by the USA's Department of Energy(SANBORNSCOTT, 2004). Hydrogen could potentially revolutionize the energy sector and the economy due to its important role in electrical systems, transportation, and manufacturing industries. The energy sector operates in a complex web of interdependence, as energy serves as the foundation of industrialized economies. Hydrogen's emergence as a new energy source could signifies the prices of economies and other energy sources. In the 1970s, there was a significant increase in global oil prices, resulting in stagflation, economic expansion, and elevated inflation levels("International Energy Agency (IEA)," 2008). When considering the significance of extending a hydrogen supply chain on the economies of countries involved in hydrogen production and consumption, two key areas of research are the development of the the hydrogen supply chain and the design of hydrogen production facilities. Designing the hydrogen supply chain involves tackling different aspects like feedstocks, manufacturing and storage, transportation, and end-uses of the hydrogen. When optimizing superstructures, it's crucial to account for how different components of the supply chain interact with each other. This allows for the efficient selection of both the best network as well as operational conditions at the same time(Yeomans & Grossmann, 1999). Hydrogen is a versatile element that can be obtained through various methods such as the electrolysis process of water, combustion of plant material as well as fossil fuels, solid state methane reformation(SCR) for natural gas, biological processes by microorganisms and utilizing nanoparticles(Deng et al., 2022). Hydrogen can be delivered to terminals using cars, pipelines, trains and or ships, with the majority of consumed hydrogen being obtained locally because of its low volumetric density of energy. Hydrogen is stored in both gaseous and liquid phases at terminals before being distributed to end-users like power plants, replenishing stations, chemical facilities, or refineries. Hydrogen process design (HPD) is dedicated to a particular part of the hydrogen supply chain, namely manufacturing sites. HPD outlines the planning of new facilities based on product output, anticipated procedure yields, and throughput rates. When creating a full supply chain for hydrogen, it's important to measured the layout of the hydrogen manufacturing procedure, although the level of detail in designing the production may vary.

Figure 1. Application of hydrogen across multiple industries Reproduced from (Osman et al., 2021) under creative commons CC.BY license.

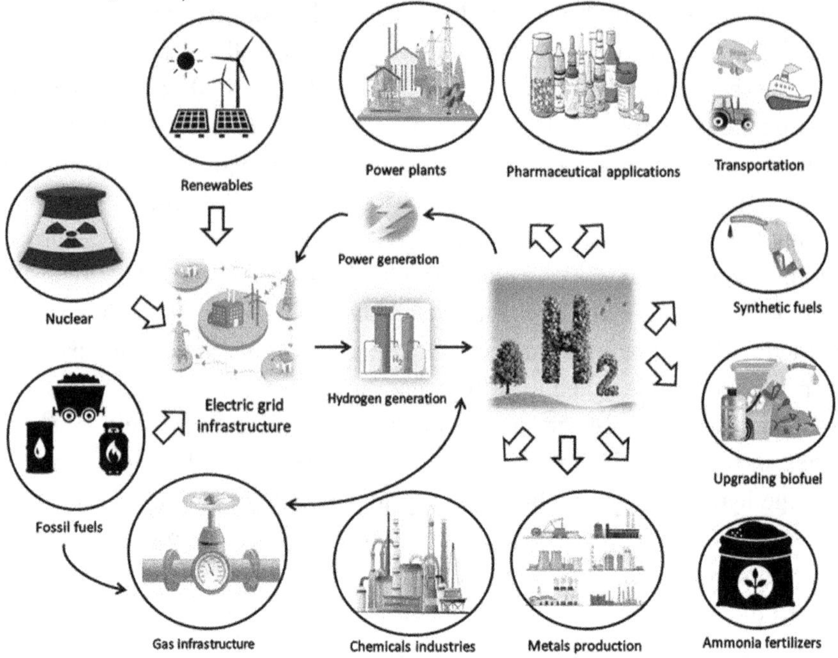

2. BACKGROUND AND MOTIVATION

Hydrogen Storage Cells (HSCs) are carefully designed to optimize the setup for operating a global network, considering factors like location, storage production feedstocks, and transportation methods. Modeling hydrogen infrastructure can be quite complex due to the numerous uncertainties to consider, including supply, demand, economic and environmental impacts, as well as the different technologies involved in production, storage, and transportation. To transition regarding hydrogen economy, it's crucial to develop tools that can pinpoint the optimal configurations for a smooth shift. Designing the network structure to minimize costs related to developing and installing infrastructure for hydrogen is essential. Nevertheless, this method may not always be appropriate for dealing with additional considerations such as environmental issues. To achieve a more optimal solution, it's crucial to take into account a range of factors during the planning and design of the HSC, including particular details. Hydrogen can be produced from a variety of primary sources of energy and transported using a range of technologies. One crucial element of the supply chain involves the "production" dimensions, which began with the steam reformation process, a widely used technology. non-exhaustible resources, nuclear energy, and plant material could significantly influence the hydrogen production stage in the future. When considering hydrogen production, you have two options: centralized or decentralized methods of production. Centralized production offers benefits for large-scale manufacturing, although the current need for hydrogen is not substantial. By employing decentralized hydrogen plants for small-scale manufacturing on-site, significant savings can be achieved by reducing transportation costs. One can transport liquefied hydrogen via rails, roads, or ships, whereas the gaseous state of hydrogen can be delivered by means of extremely pressurized gas tube trucks or pipelines (Lemus & Martínez Duart, 2010). We need well-rounded approaches to attain competitive hydrogen with low-carbon prices and establish sophisticated hydrogen supply networks. The overarching goal of this research is to examine in depth the various systematic approaches and models that have been put out there to enhance the production and supply chain of hydrogen, to find any missing pieces, and to determine the level of detail that are too broad, and propose new research directions(Dagdougui, 2012). Reviews exist on hydrogen distribution network design and production, but a thorough review has not been done yet(Agnolucci & McDowall, 2013). Bolat and Thiel's two-part analysis explores the structure of hydrogen supply chain routes and the technoeconomic factors involved in hydrogen production(Bolat & Thiel, 2014a). Yet, they fail to address current hydrogen supply chains or analyze their potential effects on the hydrogen economy. Maryam's review explores modeling approaches for implementing a hydrogen distribution network in the UK(Maryam, 2017). It emphasizes key factors such as optimization strategies, financial implications, environmental objectives, risk assessments, and problem-solving techniques. Delving into the hydrogen supply chain problem (HSCD) and different models concerning hydrogen production and distribution (HPD) as documented in the literature(Bolat & Thiel, 2014b). The authors focus on three performance measures: cost, safety, and carbon externalities, without exploring financial risk metrics or potential profits from HSC. There are different ways to address HSCD problems, such as exact methods, heuristics, and simulations(Liu H, 2020). However, there is no clear differentiation between solution approaches for uncertain or multi-objective problems. The review provides a comprehensive analysis of the technologies demonstrated across the hydrogen distribution network and the necessary inputs for fuel production. An analysis of the literature is conducted to assess hydrogen demand, taking into account factors such as transportation, energy storage, heating, refining, and e-fuel production. An in-depth analysis of the HSCD and HPD models has been carried out in the literature, considering the types of models used, uncertainties tackled, and performance measures assessed in the models. Highlighted the absence of essential features in the existing HSCD and HPD models and provided suggestions for enhancing their realism. The projections for hydrogen demand have been improved, with models integrating different sectors and resolutions for both time and space. Climate effects on a growing hydrogen economy are being taken into account, along with new approaches to tackle complex supply chain and production models. There is a lack of thorough reviews on the HSCND in the literature, neglecting to cover all aspects of the issue. Based on what we know, a thorough evaluation of the optimization approaches for the HSCND challenge has not been conducted. The research tries to evaluate and categorize the entities as well as technologies investigated in each model to identify the combinations that have been covered and those that have not. Offering a comprehensive categorization focusing on choices, performance metrics, and uncertainty concerns to emphasize contributions and areas needing further research. Figure 1. represents the hydrogen supply chain, that contain the source of energy, production, storage, transportation, supply and utilization.

Figure 2. Overview of hydrogen supply chain (https://sweetcrudereports.com/development-of-hydrogen-supply-chain-hinges-on-major-policy-changes/)

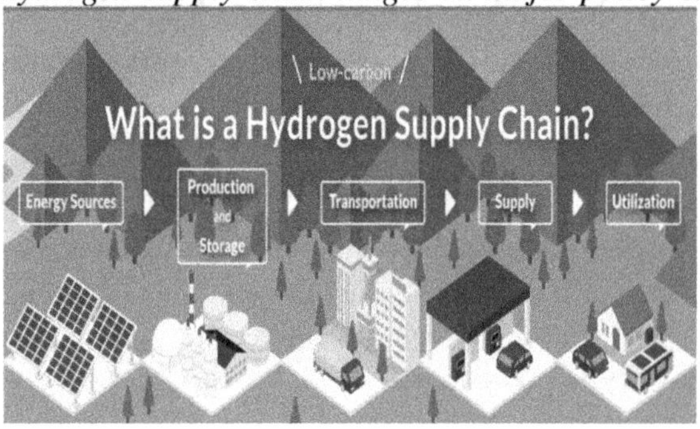

2. SUPPLY CHAIN COMPONENT ANALYSIS

In this section, each key component of the system is examined to identify the entities and technologies.

2.1 Feedstock

Hydrogen serves as an energy carrier and can be derived from a range using resources like natural gas the plant matter, carbon as a fuel water, and electricity. Access to feedstock plays a vital role in hydrogen production, with certain nations possessing ample natural gas reserves while others, such as Chile, have abundant wind resources. Renewable energy sources have rise in popularity for generating hydrogen. Power for electrolytic objectives can be sourced from independent sources of renewable energy or directly from the electrical grid(Cho & Kim, 2019). Energy from sustainable sources, such the sun and the wind are extensively discussed in the literature. Hydropower has been utilized in electrolytic systems for various case studies conducted in France, Turkey(Güler et al., 2021), Iceland(INGASON et al., 2008), China, the US, and Canada(Talebian et al., 2019). The expenses of producing hydrogen by electrolysis are significantly influenced by the availability and quality of renewable resources. Feedstock expenses are commonly discussed in research, but they are usually viewed as external factors. Supply chain disruptions can lead to raw material shortages and unexpected price spikes, directly affecting the price of generation of hydrogen. Comparable outcomes were observed for the use of coal and biomass gasification, as rising feedstock expenses led to elevated LCOH figures. Sauer and Ramsden(Saur & Ramsden, 2011) examined the system setup for producing hydrogen on a large scale using wind power and the H2A(Steward et al., 2008) model developed by NREL. The study revealed that the price of wind power has a notable impact on the cost of producing hydrogen, leading to increased hydrogen production expenses when wind costs are higher. Moreover, the efficiency of wind turbines is linked to hydrogen expenses, where greater efficiency is connected to reduced hydrogen expenses. Water is frequently overlooked as a crucial feedstock in research on hydrogen production. Every production method involves a certain amount of water input. Biomass gasification has the highest water input, requiring 47.48% water by weight for every kilogram of hydrogen generated(Al-Qahtani et al., 2021). Water inputs are often overlooked in other technologies, like the electrolysis process and green hydrogen production process. In areas with restricted water resources, like Saudi Arabia and Oman, projects for producing hydrogen and ammonia have commenced. In a study that was, Riera et al. (Riera et al., 2022) introduced an innovative co-optimization methodology that integrates water desalination and regenerative power systems for Saudi Arabia. This model aims to lower overall system costs and provide more flexibility to the sectors. When considering gasification, it's important to calculate the oxygen needed and the related expenses for producing hydrogen. At present, there are no papers that discuss the expenses related to oxygen. Two studies measured the quantity of oxygen generated through electrolysis of water, which could be stored and utilized as a fuel for gasification. It is essential to calculate feedstock needs, expenses, and accessibility to achieve the best setup for a hydrogen-based network. Overlooking certain feedstock expenses might give the impression that production methods are more affordable than they truly are, while ignoring resource availability could lead to choosing a production process unsuitable for a specific area. Hydrogen is generated from different sources like fossil fuels, biomass, plant matter, and water (using power from the regional power system)(Agnolucci et al., 2013). It is considered an energy carrier rather than an energy source. In the era following the Paris Agreement, renewable sources have gained popularity(Abbasi & Abbasi, 2011), with many models incorporating electricity generated from renewable energy. The issue of feedstock, the initial level of an HSCN, has been overlooked in numerous studies. Unit production costs only account for feedstock purchasing, with little attention given to feedstock origin, storage, and transport. When dealing with certain feedstocks, transportation and storage are key components of HSCND due to the reliance on the specific resource characteristics of different regions("Hydrogen and Fuel Cells," 2015). However, very few studies take into account these two factors. For a more thorough analysis, oxygen might be considered since its function is vital to the process of gasification (CG or BG).

3.2 Hydrogen Production

Hydrogen production has been analyzed using different technologies and procedures with the main ones being the SMR process, electrolysis, and gasification. Nevertheless, these technologies come with notable tradeoffs. Other methods for generating hydrogen involve bio waste refining and processes of fermentation. Hydrogen derived from biowaste is more cost-effective compared to the other two methods(Maroušek, Maroušek, J., Strunecký, O., Bartoš, V., & Vochozka, M. (2022). Analyzing the reative merits of hydrogen and biodiesel produced from algae. Fuel, 328 et al., 2022). SMR is mainly utilized for hydrogen production because of its strong market presence and cost-effective production process(Cho & Kim, 2019). Nevertheless, it leads to substantial gases that trap heat in the atmosphere, such as methane and carbon dioxide. Electrolysis has great potential to grow and even outstrip SMR-based technologies due to its low-carbon method of producing hydrogen. Electrolyzers operating at low temperatures, including the proton exchange-membrane (PEM) and alkaline types, are already on the market, while demonstration phase for high-temperature solid-oxide ''electrolyzers'' is well advanced (Dincer & AlZahrani, 2018). Alkaline and PEM 'electrolyzers' have reduced initial expenses compared to solid-oxide electrolyzers, yet they remain notably pricier than fossil fuel-based production(Kiemel et al., 2021). Gasification is a highly effective process that has been in use for centuries and can utilize biomass as a feedstock. Utilizing biomass gasification is appealing due to its ability to recycle carbon dioxide and produce minimal net emissions of greenhouse gases. Nevertheless, biomass gasification faces constraints like the accessibility of feedstocks as well as the emission of greenhouse gases. Some of the less common techniques and technologies that have been discussed in the literature include pyrolysis, dry reformation of methane (DRM), and thermochemical water splitting. Chemistry of heat, water splitting depends on elevated temperatures thermal energy to facilitate chemical reactions, resulting in the production of hydrogen with minimal greenhouse gas emissions. Nevertheless, thermochemical processes face constraints such as the requirement for effective reactor designs, the expensive nature of centred planretry systems, and the cost and accessibility of fundamentals. Hydrogen byproduct production is a frequent occurrence in industries such as processing, chemical, iron, refinement and steel. Refineries transform petroleum naphtha into into high-octane liquids products and in the chemical sector, chlor-alkali processes generate chlorine and sodium hydroxides. Steel production generates hydrogen as a byproduct, which can then be retrieved and integrated into the hydrogen distribution network.

3.3 Carbon Capture, Transport, Storage and Utilization

Reducing greenhouse gas emissions is possible with the application of carbon capture technologies in existing hydrogen production facilities. This includes SMR, coal and biomass gasification, and others(Chao et al., 2021). The deep ocean, naturally salty aquifers, or exhausted oil and gas reserves are some of the geological formations that may retain captured pollutants(Newell & Ilgen, 2019). Instead of storing it, the CO_2 that is collected may be used to make carbon-based efuels or improved oil recovery(Jia et al., 2019). Nevertheless, a transportation network is necessary for carbon capture in order to transfer emission from sources to sinks. It is critical to evaluate the carbon capture systems' cost-effectiveness and comprehend the network design in this context. Investment in new facilities equipped with capture and storage of carbon capabilities is made possible in several models discovered in the published works. To be more accurate, models should also account for the possibility of retrofitting existing facilities. Carbon capture technology necessitates supplementary capital and operational expenditures for facility retrofits or new construction. The cost of producing hydrogen is greater in production facilities that use CSS technology as opposed to those that do not. In Figure 2 a and b represents

a. Water electrolysis is a great way to transform renewable energy into the chemical fuel hydrogen. The produced H2 can be simply converted into chemical H2 carriers that are stable enough for both short- and long-term storage and transit. Hydrogen gas storage in proton-exchange membrane (PEM) fuel cells allows for the redistribution of renewable power as needed.
b. A schematic depicting the carbon capture and utilization (CCU) process, carbon capture and storage (CCS) using amines, the carbon neutral H2 storage-release approach based on the dual-functional functions of formamides, and other related topics.

Figure 3. Projected sustainable energy utilization based on renewable electricity storage and regeneration bridged by chemical hydrogen storage-release. Reproduced from (Wei et al., 2023)under creative commons CC.BY license.

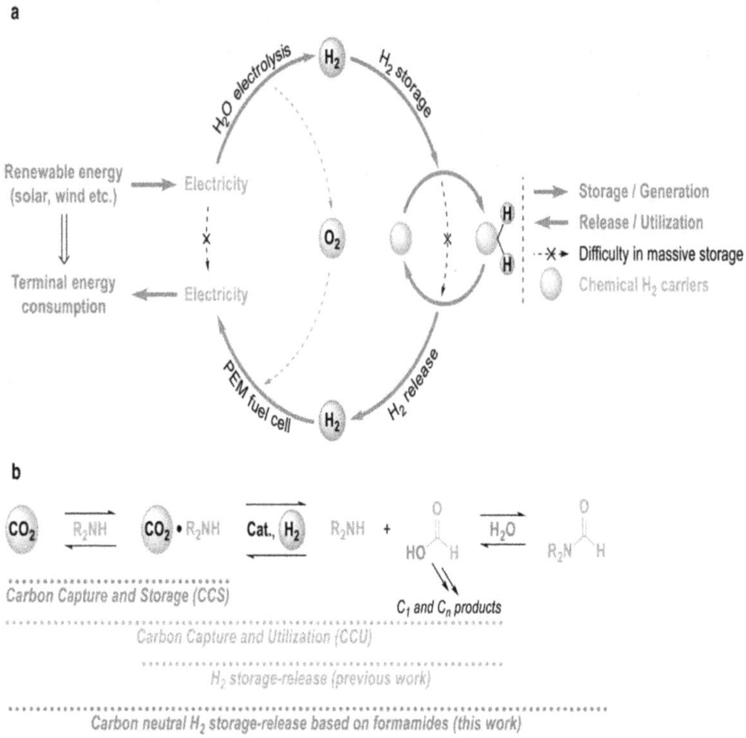

3.4 Hydrogen Transportation

Hydrogen can be transported in various forms such as liquid, compressed gas, or through a hydrogen carrier such as ammonia or a liquid organic hydrogen carrier (LOHC). Transmission happens over vast distances from production sites to storage terminals, through networks of distribution that transport hydrogen via pipelines or trucks to consumers. The choice of transportation method relies heavily on funding and operating expenses, as well as demand distribution, which dictates the distance hydrogen must travel. Transporting hydrogen over long distances can be challenging due to the limitations and costs associated with pipelines or shipping vessels. Transporting liquid hydrogen can result in significant losses due to the need for cryogenic storage, which can lead to boil-off. At present, hydrogen pipelines are in use, but the networks are quite limited in size(*Global Hydrogen Review 2022*, 2022). When considering natural gas pipelines, it's important to weigh the cost benefits against the challenges involved, like retrofitting existing pipelines, replacing compressors, and dealing with leak-prone infrastructure(Groth & Al-Douri, 2023). Ships are commonly utilized for global trade, but transporting items like liquid hydrogen or LOHC over long distances may offer a more economical option. Just two papers have approached an international viewpoint, discussing the transportation of hydrogen using shipping vessels. Heuser et al. (Heuser et al., 2019)conducted a study on hydrogen transport from Argentinean Patagonia to Japan. They calculated the necessary amount of liquefied hydrogen shipping containers for annual transportation between the two countries. Hong et al. (Hong et al., 2021)conducted a detailed analysis of the hydrogen distribution network in the ASEAN region, examining various methods of methylcyclohexane (MCH), ammonia, hydrogen gas, liquid, or any several other hydrogen delivery systems. Exploring shipping vessels for transport of hydrogen on a national level in nations that have islands or offshore areas.

Table 1. Regional focus in HSC research

Sr.NO.	Territorial Scale	Objectives	Observations	Energy Sources	Time Scale	Authors
1.	Spain	Financial risk	There is a degree of operational cost uncertainty.	Renewable energy, biomass, natural gas, and coal (CCS).		(Sabio et al., 2010)
2.	China Malaysia	cost	There are two ways to estimate future demand: one uses projected vehicle counts, while the other uses gasoline and diesel supplies.	Hydrogen, biomass, coal, and natural gas	5 (2010-2034)	(Delucchi & Lipman, 2010)
3.	Great Britain	Issues of cost, ecology, and safety	The ε-constraint approach to the issue with many periods	Coal, biomass, and natural gas		(De-León Almaraz et al., 2014)
4.	Jeju Island, Korea		To better understand how to manage the supply chain for biomass-to-hydrogen, a sensitivity analysis is carried out.	Biomass	12 (months)	(Siebenmorgen et al., 2003)

3. MODELLING AND SOLUTION TECHNOLOGIES

Simulation modelling solves real-world problems safely and efficiently.

4.1 Decision Planning

In order to fulfil demand, investment and operational considerations are involved in the design of the hydrogen distribution network. The construction of infrastructure and the establishment of transportation networks are aspects of investment choices. Production of hydrogen, distribution to consumers, and route selection are all aspects of operations that need decision-making. The demand for and availability of resources have a role in these choices. To guarantee the best possible network setup, operational choices feed back to investment decisions. A system's ability to produce is heavily dependent on how operational choices are resolved.

4.2 Cost

Cost plays a crucial role in the High-Scale Cleaner Network (HSCN) simulation objective, encompassing both operating and capital expenses for manufacturing, storing, refuelling stations, and lowering emissions. Various authors have embraced the cost objective, with some providing additional details in the cost definition to achieve more precise outcomes. Various factors, including land prices in different regions, financial assumptions, as well as expense discounting, might considerably affect the overall price(Murthy Konda et al., 2011). Multi-period models also weigh the pros and cons of hydrogen infrastructure. There has been limited focus on profit maximization, as only two reference papers have mentioned benefit as the HSCN performance measure(Parker et al., 2010). One possible economic indicator to consider is the levelized cost of hydrogen (LCOH)("Wikipedia," 2018), representing the total amount of hydrogen per unit over the network's lifespan. Introducing LCOH could establish a reliable foundation for comparing various HSCNDs.

4.3. Environmental Impacts

When evaluating the environmental effects of a hydrogen fuel cell (HSC), numerous environmental considerations must be considered. Typical metrics involve calculating whole emissions throughout the supply chain or emissions each component(Ochoa Bique et al., 2019), along with global warming potential (GWP) to understand the effects of heat absorbed from radiation from greenhouse gas emissions. Typically, the literature evaluates the global warming potential of conventional greenhouse gas emissions like carbon dioxide and methane. Nevertheless, there is a rising worry regarding hydrogen's impact as a greenhouse gas, leading to indirect warming consequences. According to a report from the UK Department of Business, Energy and Industrial Strategy, hydrogen leakage has the potential to reduce hydroxyl radicals, leading to a longer atmospheric lifetime for methane. Quantifying the probable frequency of hydrogen supply chain leaks is essential for assessing GWP(Groth & Al-Douri, 2023). Additional environmental factors consist of life cycle indicators, ecological elements, and a social the cost-benefit framework(Robles et al., 2020). These factors consist of greenhouse gas emissions, platinum depletion, air pollutants, and noise pollution.

4.4. Safety

Hydrogen, a combustible fuel, has the potential to lead to serious accidents in certain situations, highlighting the importance of safety in hydrogen supply chain networks (HSCNs)(Markert et al., 2017). Kim and Moon (Kim & Moon, 2008)have proposed an approach for HSCND risk analysis that consists of identifying hazards, consequence and the frequency evaluates, risk evaluation, as well as the creation of a relative risk index. This system enables the assessment of various components of the HSCN and the comparison of the safety performance scenarios. Nevertheless, there is a lack of agreement regarding the explanation of supply chain risk (Govindan et al., 2017), and evaluating risk is linked to the safety of the hydrogen infrastructure systems. Concerns about safety have caused a decrease in the use of hydrogen because of issues with leaks and explosions at fuelling stations, requiring safety precautions in HSCDs.

4.5. Multi-Objective Mode

A network's main goal is to reduce overall costs, sometimes resulting in solutions that may not address environmental and safety concerns. Considering actions from various angles can help address this trade-off. As an illustration, Guillén-Gosálbez et al. (Guillén-Gosálbez et al., 2009)showed how a system with several goals could pinpoint the most promising alternative for achieving substantial environmental benefits without overly impacting the overall network cost. Nevertheless, a single-goal model can address trade-off issues by incorporating additional criteria, such as environmental impact, into cost considerations. Multi-objective optimization requires making decisions while balancing multiple conflicting objectives. Many models in the literature typically focus on optimizing a financial measure like cost, but this approach may overlook other crucial objectives.

4.6. Uncertainty Types and Modelling Approaches

The research delves into a taxonomy of uncertainty kinds in the stochastic decision-making process (SCND)(Klibi et al., 2010) and categorizes them into three types: randomness, epistemic, and deep. Randomness follows specific probability distributions, whereas epistemic uncertainty arises from a shortage of knowledge about input data. Exploring various scenarios helps address uncertainties in FCEV market share by examining hydrogen demand across different outlooks. Stochastic programming(Eppen et al., 1989) offers an alternative method to deal with uncertainty, with a significant portion of research focusing on two-stage models. This model emphasizes decisions made prior to uncertainty becoming apparent, such as manufacturing plants as well as warehouses, as well as choices for the future created after uncertainty is realized. The main goal of cost-oriented multiple stages modelling under uncertainty is to pinpoint the best decision variables that reduce the cost of the initial stage and the expected cost of following stages. Uncertainty in epistemology is linked to a lack of knowledge about input data, like fuzzy numbers(Bairamzadeh et al., 2018). Fuzzy programming is a useful tool for addressing epistemic uncertainty. Deep uncertainty arises from a scarcity of information for estimating scenario probability, and robust optimization methods can be applied(Mulvey et al., 1995). Hydrogen demand uncertainty is typically viewed as random, but a more realistic approach is to define it as a substantial uncertainty within a known range without assigning specific probabilities to potential values. The study discusses random uncertainty, with some studies through the use of scenario analysis and stochastic programming.

4.7. Model Restriction and Spatial-Based Approach

The limitations of the HSCND model fall into various categories such as capacity, time evolution, flow conservation, and non-negativity constraints. These restrictions are essential for precise modelling because they determine the possibilities for hydrogen delivery exceeding 500 kg/d to stations. considering national or regional policies and regulations is crucial, including ensuring that pipeline network design complies with government regulations(Johnson & Ogden, 2012). Working with actual geographic areas and utilizing geographic information systems (GIS) data bases is crucial(Parker et al., 2010). For instance, the network based in California is utilized in the model. According to the IEA("Hydrogen and Fuel Cells," 2015), hydrogen delivery and distribution costs are influenced by transport distance, so understanding geographical parameters is crucial for accurate T&D cost estimation. Geographical constraints can be incorporated and developed using a GIS database prior to optimization, with GIS tools confirming the viability of the final HSCN layout(De-León Almaraz et al., 2015).

5. SOLUTION METHODS

Decisions, time horizons, uncertainty, and mathematical characteristics are some of the differences between HSCD and HPD situations. Frequently expressed as MILP problems, HSCD problems feature coarser resolutions, longer time horizons, and greater uncertainty. Shorter time horizons, sharper resolutions, no ambiguity, and frequently nonlinear constraints are present in HPD situations. They can be resolved directly or by the use of decomposition techniques that take advantage of feature, temporal, or spatial information. For certain issues, general decomposition techniques like Benders Decomposition (Rahmaniani et al., 2017)or the L-Shaped method (Carøe & Tind, 1998)are employed. The original problem is broken down into smaller problems using rolling horizon techniques, which are best suited for challenges involving the integration of several time scales.

5.1 Mono-Objective Problems

Both general-purpose software and specifically designed algorithms can be used to address mono-objective SCND problems(Melo et al., 2009). Both accurate and heuristic answers are possible. Most deterministic and static SCND models may be solved effectively by commercial solvers such as CPLEX. Algorithms specifically designed for large-scale issues with numerous periods of time and locations are chosen for almost optimal solutions with manageable CPU times. Using the effectiveness of the large neighbourhood search (LNS) approach and the advantages of conventional branch-and-cut methods, Copado-Méndez et al. created a hybrid algorithm. Even in cases while the CPLEX is not converging, the hybrid method finds workable solutions and attains near-optimal solutions faster than the CPLEX(Copado-Méndez et al., 2013). Techniques like Ant colony optimization(Arya, 2021; Arya et al., 2022, 2023; Arya & Honwad, 2016, 2018), genetic algorithm, Dynamic programming have been used for optimizing hydrogen supply chains

5.2 Challenges Involving Multi-Objectives

Solving a sample Pareto set optimum proposed solutions selecting one according to the decision-maker's choice is a common requirement for multi-objective problems(Marler & Arora, 2004). The most often used approach is the ε-constraint method, which consists of minimizing one objective function and incorporating additional objective functions as constraint limited by a level ε(Marler & Arora, 2004). Using this approach, trade-offs between maximizing LCI and lowering total system costs are found for the hydrogen supply chain. Numerous techniques have been created to address multi-objective issues, including evolutionary algorithms, lexicographic approaches, weighted constraint methods, and the ε-constraint method. For the purpose of maximize LCI and minimize total system costs for the hydrogen distribution network, Guillen-Gosalbez et al.(Guillén-Gosálbez et al., 2009) employed the ε-constraint technique for discovering trade-off solutions. Hwangbo et al.(Hwangbo et al., 2018) used the commercial solver CPLEX to create a multi-objective model for the design of a hydrogen distribution network in Gyeongsang-do Province, South Korea. Sabio et al. utilized the way they have formulated their ε-constraint model structure to speed up solving for each Pareto solution when solving developing a stochastic multi-objective model to design ssa hydrogen supply chain in Spain. Lexicographic techniques solve a series of single-objective challenges, every single one with a different constraint specifying a maximum restrict the use of best values of the preceding optimization problem, and rank objective functions according to significance. While groups of candidate solutions are used by genetic algorithms, which are metaheuristic techniques, to evolve toward better solutions, they do not ensure optimality and may need multiple runs to get the best answer.

5.3. Designing the Hydrogen Supply Chain

There are two main types of algorithms used to solve issues with just one objective: heuristics and optimum algorithms. Solved with an authorized solver that employs a branch and bound method, single-objective stochastic HSCD problems in the literature are usually MILP problems or reformulations of MILP(Nickel et al., 2020). There is always some degree of uncertainty in single-objective stochastic HSCD problems, and we used a commercial solution to directly solve all of them. Including stochasticity, though, can make your model too big to solve efficiently without resorting to decomposition techniques.

6. CONCLUSION AND RECOMMENDATION

This paper offers an extensive overview of the latest research topics and advancements in optimizing and modelling HSCD and HPD problems. The literature is organized based on the supply chain setup (feedstock, manufacturing, distribution), modelling characteristics (decision factors evaluation metrics, both temporal and spatial resolutions, stochasticity), and solution methods. The authors have pinpointed various gaps in the literature that require further exploration to enhance the model of the hydrogen supply network. Based on the review, it is evident that hydrogen consumption projections in the literature heavily rely on the adoption of hydrogen fuel cell vehicles (HFCVs). It's important to consider hydrogen demand projections across different sectors to achieve more accurate results for shaping the hydrogen economy. Hydrogen is a key player in reducing carbon emissions in transportation and industry, while also enhancing the power sector's adaptability. Thus, models for hydrogen supply chains can incorporate external demand from various industries or be enhanced through an integrated strategy to accommodate additional internal demand needs. It is crucial to co-optimize or integrate hydrogen supply chains with feedstock systems. Considering water, natural gas, biomass, electricity, and critical material supply chains is crucial to accurately represent the feedstocks available for producing hydrogen. Moreover, hydrogen distribution network models can be combined with computational models of general equilibrium (CGE), which depict the overall structure of the economy and analyse interactions between various sectors. When discussing spatial scales, the majority of existing literature focuses on a national or regional level. However, incorporating international viewpoints can provide decision-makers with a more precise understanding of trading dynamics and geopolitical environments. Global hydrogen trading will involve incorporating various transportation methods like shipping vessels, especially for long-distance transportation of hydrogen in forms like ammonia or LOHC, along with the associated expenses. Furthermore, the literature largely overlooks the effects of climate change on the hydrogen transportation and production. Weather pattern variations across regions can significantly affect feedstock availability, making it crucial to factor in these changes for a robust supply chain setup. It is crucial to take into account leakage in the manufacturing and distribution processes of hydrogen. Hydrogen contributes to global warming and should be factored into metrics to address leaks through renovations and new technology. It is essential to use accurate models to accurately reflect real-world scenarios and account for uncertainties in future predictions, such as demand for hydrogen, production costs of technology, and availability of resources. Ultimately, establishing a hydrogen economy necessitates careful evaluation of financial risks, uncertainties, system resiliency, and reliability.

REFERENCES

Abbasi, T., & Abbasi, S. A. (2011). 'Renewable' hydrogen: Prospects and challenges. *Renewable & Sustainable Energy Reviews*, 15(6), 3034–3040. DOI: 10.1016/j.rser.2011.02.026

Agnolucci, P., Akgul, O., McDowall, W., & Papageorgiou, L. G. (2013). The importance of economies of scale, transport costs and demand patterns in optimising hydrogen fuelling infrastructure: An exploration with SHIPMod (Spatial hydrogen infrastructure planning model). *International Journal of Hydrogen Energy*, 38(26), 11189–11201. DOI: 10.1016/j.ijhydene.2013.06.071

Agnolucci, P., & McDowall, W. (2013). Designing future hydrogen infrastructure: Insights from analysis at different spatial scales. *International Journal of Hydrogen Energy*, 38(13), 5181–5191. DOI: 10.1016/j.ijhydene.2013.02.042

Al-Qahtani, A., Parkinson, B., Hellgardt, K., Shah, N., & Guillen-Gosalbez, G. (2021). Uncovering the true cost of hydrogen production routes using life cycle monetisation. *Applied Energy*, 281, 115958. DOI: 10.1016/j.apenergy.2020.115958

Arya, A. K. (2021). Optimal operation of a multi-distribution natural gas pipeline grid: An ant colony approach. *Journal of Petroleum Exploration and Production Technology*, 11(10), 3859–3878. Advance online publication. DOI: 10.1007/s13202-021-01266-3

Arya, A. K., & Honwad, S. (2016). Modeling, simulation, and optimization of a high-pressure cross-country natural gas pipeline: Application of an ant colony optimization technique. *Journal of Pipeline Systems Engineering and Practice*, 7(1), 04015008. Advance online publication. DOI: 10.1061/(ASCE)PS.1949-1204.0000206

Arya, A. K., & Honwad, S. (2018). Optimal Operation of a Multi Source Multi Delivery Natural Gas Transmission Pipeline Network. *Chemical Product and Process Modeling*, 13(3). Advance online publication. DOI: 10.1515/cppm-2017-0046

Arya, A. K., Jain, R., Yadav, S., Bisht, S., & Gautam, S. (2022). Recent trends in gas pipeline optimization. *Materials Today: Proceedings*, 57, 1455–1461. Advance online publication. DOI: 10.1016/j.matpr.2021.11.232

Arya, A. K., Katiyar, R., Senthil Kumar, P., Kapoor, A., Pal, D. B., & Rangasamy, G. (2023). A multi-objective model for optimizing hydrogen injected-high pressure natural gas pipeline networks. *International Journal of Hydrogen Energy*, 48(76), 29699–29723. Advance online publication. DOI: 10.1016/j.ijhydene.2023.04.133

Bairamzadeh, S., Saidi-Mehrabad, M., & Pishvaee, M. S. (2018). Modelling different types of uncertainty in biofuel supply network design and planning: A robust optimization approach. *Renewable Energy*, 116, 500–517. DOI: 10.1016/j.renene.2017.09.020

Bockris, J. O. M. (2013). The hydrogen economy: Its history. *International Journal of Hydrogen Energy*, 38(6), 2579–2588. DOI: 10.1016/j.ijhydene.2012.12.026

Bolat, P., & Thiel, C. (2014a). Hydrogen supply chain architecture for bottom-up energy systems models. Part 1: Developing pathways. *International Journal of Hydrogen Energy*, 39(17), 8881–8897. DOI: 10.1016/j.ijhydene.2014.03.176

Bolat, P., & Thiel, C. (2014b). Hydrogen supply chain architecture for bottom-up energy systems models. Part 2: Techno-economic inputs for hydrogen production pathways. *International Journal of Hydrogen Energy*, 39(17), 8898–8925. DOI: 10.1016/j.ijhydene.2014.03.170

Carøe, C. C., & Tind, J. (1998). L-shaped decomposition of two-stage stochastic programs with integer recourse. *Mathematical Programming*, 83(1–3), 451–464. DOI: 10.1007/BF02680570

Chao, C., Deng, Y., Dewil, R., Baeyens, J., & Fan, X. (2021). Post-combustion carbon capture. *Renewable & Sustainable Energy Reviews*, 138, 110490. DOI: 10.1016/j.rser.2020.110490

Cho, S., & Kim, J. (2019). Multi-site and multi-period optimization model for strategic planning of a renewable hydrogen energy network from biomass waste and energy crops. *Energy*, 185, 527–540. DOI: 10.1016/j.energy.2019.07.053

Copado-Méndez, P. J., Blum, C., Guillén-Gosálbez, G., & Jiménez, L. (2013). Large neighbourhood search applied to the efficient solution of spatially explicit strategic supply chain management problems. *Computers &amp. Computers & Chemical Engineering*, 49, 114–126. DOI: 10.1016/j.compchemeng.2012.09.006

Dagdougui, H. (2012). Models, methods and approaches for the planning and design of the future hydrogen supply chain. *International Journal of Hydrogen Energy*, 37(6), 5318–5327. DOI: 10.1016/j.ijhydene.2011.08.041

De-León Almaraz, S., Azzaro-Pantel, C., Montastruc, L., & Boix, M. (2015). Deployment of a hydrogen supply chain by multi-objective/multi-period optimisation at regional and national scales. *Chemical Engineering Research & Design*, 104, 11–31. DOI: 10.1016/j.cherd.2015.07.005

De-León Almaraz, S., Azzaro-Pantel, C., Montastruc, L., & Domenech, S. (2014). Hydrogen supply chain optimization for deployment scenarios in the Midi-Pyrénées region, France. *International Journal of Hydrogen Energy*, 39(23), 11831–11845. DOI: 10.1016/j.ijhydene.2014.05.165

Delucchi, M. A., & Lipman, T. E. (2010). Lifetime Cost of Battery, Fuel-Cell, and Plug-in Hybrid Electric Vehicles. In *Electric and Hybrid Vehicles* (pp. 19–60). Elsevier., DOI: 10.1016/B978-0-444-53565-8.00002-6

Deng, Y., Dewil, R., Appels, L., Van Tulden, F., Li, S., Yang, M., & Baeyens, J. (2022). Hydrogen-enriched natural gas in a decarbonization perspective. *Fuel*, 318, 123680. DOI: 10.1016/j.fuel.2022.123680

Dincer, I., & AlZahrani, A. A. (2018). 4.25 Electrolyzers. In *Comprehensive Energy Systems* (pp. 985–1025). Elsevier., DOI: 10.1016/B978-0-12-809597-3.00442-9

Eppen, G. D., Martin, R. K., & Schrage, L. (1989). OR Practice—A Scenario Approach to Capacity Planning. *Operations Research*, 37(4), 517–527. DOI: 10.1287/opre.37.4.517

Global Hydrogen Review 2022. (2022). OECD. DOI: 10.1787/a15b8442-en

Govindan, K., Fattahi, M., & Keyvanshokooh, E. (2017). Supply chain network design under uncertainty: A comprehensive review and future research directions. *European Journal of Operational Research*, 263(1), 108–141. DOI: 10.1016/j.ejor.2017.04.009

Groth, K. M., & Al-Douri, A. (2023). Hydrogen safety, risk, and reliability analysis. In *Hydrogen Economy* (pp. 487–510). Elsevier., DOI: 10.1016/B978-0-323-99514-6.00012-1

Guillén-Gosálbez, G., Mele, F. D., & Grossmann, I. E. (2009). A bi-criterion optimization approach for the design and planning of hydrogen supply chains for vehicle use. *AIChE Journal. American Institute of Chemical Engineers*, 56(3), 650–667. DOI: 10.1002/aic.12024

Güler, M. G., Geçici, E., & Erdoğan, A. (2021). Design of a future hydrogen supply chain: A multi period model for Turkey. *International Journal of Hydrogen Energy*, 46(30), 16279–16298. DOI: 10.1016/j.ijhydene.2020.09.018

Heuser, P.-M., Ryberg, D. S., Grube, T., Robinius, M., & Stolten, D. (2019). Techno-economic analysis of a potential energy trading link between Patagonia and Japan based on CO2 free hydrogen. *International Journal of Hydrogen Energy*, 44(25), 12733–12747. DOI: 10.1016/j.ijhydene.2018.12.156

Hong, X., Thaore, V. B., Karimi, I. A., Farooq, S., Wang, X., Usadi, A. K., Chapman, B. R., & Johnson, R. A. (2021). Techno-enviro-economic analyses of hydrogen supply chains with an ASEAN case study. *International Journal of Hydrogen Energy*, 46(65), 32914–32928. DOI: 10.1016/j.ijhydene.2021.07.138

Hwangbo, S., Heo, S., & Yoo, C. (2018). Network modeling of future hydrogen production by combining conventional steam methane reforming and a cascade of waste biogas treatment processes under uncertain demand conditions. *Energy Conversion and Management*, 165, 316–333. DOI: 10.1016/j.enconman.2018.03.069

Hydrogen and Fuel Cells. (2015). In *IEA Technology Roadmaps*. OECD. DOI: 10.1787/9789264239760-en

Ingason, H., Pallingolfsson, H., & Jensson, P.INGASON. (2008). Optimizing site selection for hydrogen production in Iceland. *International Journal of Hydrogen Energy*, 33(14), 3632–3643. DOI: 10.1016/j.ijhydene.2008.04.046

International Energy Agency (IEA). (2008). *Encyclopedia of Global Warming and Climate Change*. SAGE Publications, Inc., DOI: 10.4135/9781412963893.n340

Jia, B., Tsau, J.-S., & Barati, R. (2019). A review of the current progress of CO2 injection EOR and carbon storage in shale oil reservoirs. *Fuel*, 236, 404–427. DOI: 10.1016/j.fuel.2018.08.103

Johnson, N., & Ogden, J. (2012). A spatially-explicit optimization model for long-term hydrogen pipeline planning. *International Journal of Hydrogen Energy*, 37(6), 5421–5433. DOI: 10.1016/j.ijhydene.2011.08.109

Kiemel, S., Smolinka, T., Lehner, F., Full, J., Sauer, A., & Miehe, R. (2021). Critical materials for water electrolysers at the example of the energy transition in Germany. *International Journal of Energy Research*, 45(7), 9914–9935. DOI: 10.1002/er.6487

Kim, J., & Moon, I. (2008). Strategic design of hydrogen infrastructure considering cost and safety using multiobjective optimization. *International Journal of Hydrogen Energy*, 33(21), 5887–5896. DOI: 10.1016/j.ijhydene.2008.07.028

Klibi, W., Martel, A., & Guitouni, A. (2010). The design of robust value-creating supply chain networks: A critical review. *European Journal of Operational Research*, 203(2), 283–293. DOI: 10.1016/j.ejor.2009.06.011

Kwok, J. (2021). Towards a hydrogen economy – a sustainable pathway for global energy transition. *HKIE Transactions*, 28(2), 102–107. DOI: 10.33430/V28N2THIE-2020-0046

Lemus, R. G., & Martínez Duart, J. M. (2010). Updated hydrogen production costs and parities for conventional and renewable technologies. *International Journal of Hydrogen Energy*, 35(9), 3929–3936. DOI: 10.1016/j.ijhydene.2010.02.034

Liu H, M. J. (2020). A review of models and methods for hydrogen supply chain system planning. *CSEE Journal of Power and Energy Systems*. DOI: 10.17775/CSEEJPES.2020.02280

Markert, F., Marangon, A., Carcassi, M., & Duijm, N. J. (2017). Risk and sustainability analysis of complex hydrogen infrastructures. *International Journal of Hydrogen Energy*, 42(11), 7698–7706. DOI: 10.1016/j.ijhydene.2016.06.058

Marler, R. T., & Arora, J. S. (2004). Survey of multi-objective optimization methods for engineering. *Structural and Multidisciplinary Optimization*, 26(6), 369–395. DOI: 10.1007/s00158-003-0368-6

Maroušek, J., Strunecký, O., Bartoš, V., & Vochozka, M. (2022). Revisiting competitiveness of hydrogen and algae biodiesel. *Fuel*, 328, 125317.

Maryam, S. (2017). Review of modelling approaches used in the HSC context for the UK. *International Journal of Hydrogen Energy*, 42(39), 24927–24938. DOI: 10.1016/j.ijhydene.2017.04.303

Melo, M. T., Nickel, S., & Saldanha-da-Gama, F. (2009). Facility location and supply chain management – A review. *European Journal of Operational Research*, 196(2), 401–412. DOI: 10.1016/j.ejor.2008.05.007

Meti, V., & Dabburu, K. (2017). Anti-anxiety activity of ethanolic extract of Spermacoce hispida leaves in swiss albino mice. *International Journal of Basic &amp. International Journal of Basic and Clinical Pharmacology*, 6(9), 2224. DOI: 10.18203/2319-2003.ijbcp20173749

Mulvey, J. M., Vanderbei, R. J., & Zenios, S. A. (1995). Robust Optimization of Large-Scale Systems. *Operations Research*, 43(2), 264–281. DOI: 10.1287/opre.43.2.264

Murthy Konda, N. V. S. N., Shah, N., & Brandon, N. P. (2011). Optimal transition towards a large-scale hydrogen infrastructure for the transport sector: The case for the Netherlands. *International Journal of Hydrogen Energy*, 36(8), 4619–4635. DOI: 10.1016/j.ijhydene.2011.01.104

Newell, P., & Ilgen, A. G. (2019). Overview of Geological Carbon Storage (GCS). In *Science of Carbon Storage in Deep Saline Formations* (pp. 1–13). Elsevier. DOI: 10.1016/B978-0-12-812752-0.00001-0

Nickel, S., Steinhardt, C., Schlenker, H., Burkart, W., & Reuter-Oppermann, M. (2020). IBM ILOG CPLEX Optimization Studio. In *Angewandte Optimierung mit IBM ILOG CPLEX Optimization Studio* (pp. 9–23). Springer Berlin Heidelberg., DOI: 10.1007/978-3-662-62185-1_2

Ochoa Bique, A., Maia, L. K. K., La Mantia, F., Manca, D., & Zondervan, E. (2019). Balancing costs, safety and CO2 emissions in the design of hydrogen supply chains. *Computers &amp. Computers & Chemical Engineering*, 129, 106493. DOI: 10.1016/j.compchemeng.2019.06.018

Osman, A. I., Mehta, N., Elgarahy, A. M., Hefny, M., Al-Hinai, A., Al-Muhtaseb, A. H., & Rooney, D. W. (2021). Hydrogen production, storage, utilisation and environmental impacts: A review. *Environmental Chemistry Letters*, 20(1), 153–188. DOI: 10.1007/s10311-021-01322-8

Parker, N., Fan, Y., & Ogden, J. (2010). From waste to hydrogen: An optimal design of energy production and distribution network. *Transportation Research Part E, Logistics and Transportation Review*, 46(4), 534–545. DOI: 10.1016/j.tre.2009.04.002

Rahmaniani, R., Crainic, T. G., Gendreau, M., & Rei, W. (2017). The Benders decomposition algorithm: A literature review. *European Journal of Operational Research*, 259(3), 801–817. DOI: 10.1016/j.ejor.2016.12.005

Riera, J. A., Lima, R. M., Hoteit, I., & Knio, O. (2022). Simulated co-optimization of renewable energy and desalination systems in Neom, Saudi Arabia. *Nature Communications*, 13(1), 3514. DOI: 10.1038/s41467-022-31233-3 PMID: 35717452

Robles, J. O., Azzaro-Pantel, C., Garcia, G. M., & Lasserre, A. A. (2020). Social cost-benefit assessment as a post-optimal analysis for hydrogen supply chain design and deployment: Application to Occitania (France). *Sustainable Production and Consumption*, 24, 105–120. DOI: 10.1016/j.spc.2020.06.010

Sabio, N., Gadalla, M., Guillén-Gosálbez, G., & Jiménez, L. (2010). Strategic planning with risk control of hydrogen supply chains for vehicle use under uncertainty in operating costs: A case study of Spain. *International Journal of Hydrogen Energy*, 35(13), 6836–6852. DOI: 10.1016/j.ijhydene.2010.04.010

Sanbornscott, D. (2004). Inside fuelcells. *International Journal of Hydrogen Energy*, 29(12), 1203–1211. DOI: 10.1016/j.ijhydene.2004.01.013

Saur, G., & Ramsden, T. (2011). *Wind Electrolysis: Hydrogen Cost Optimization*. Office of Scientific and Technical Information (OSTI). DOI: 10.2172/1015505

Siebenmorgen, N., & Weber, M. (2003). A behavioral model for asset allocation. *Financial Markets and Portfolio Management*, 17(1), 15–42. DOI: 10.1007/s11408-003-0102-9

Steward, D., Ramsden, T., & Zuboy, J. (2008). *H2A Production Model, Version 2 User Guide*. Office of Scientific and Technical Information (OSTI). DOI: 10.2172/939003

Talebian, H., Herrera, O. E., & Mérida, W. (2019). Spatial and temporal optimization of hydrogen fuel supply chain for light duty passenger vehicles in British Columbia. *International Journal of Hydrogen Energy*, 44(47), 25939–25956. DOI: 10.1016/j.ijhydene.2019.07.218

Wei, D., Shi, X., Junge, H., Du, C., & Beller, M. (2023). Carbon neutral hydrogen storage and release cycles based on dual-functional roles of formamides. *Nature Communications*, 14(1), 3726. DOI: 10.1038/s41467-023-39309-4 PMID: 37349304

Wikipedia. (2018). *The SAGE Encyclopedia of the Internet*. SAGE Publications, Inc., DOI: 10.4135/9781473960367.n287

Yeomans, H., & Grossmann, I. E. (1999). A systematic modeling framework of superstructure optimization in process synthesis. *Computers &amp. Computers & Chemical Engineering*, 23(6), 709–731. DOI: 10.1016/S0098-1354(99)00003-4

Chapter 8
Deforestation and Forest Monitoring With CNN and RNN

Kiran Sree Pokkuluri
https://orcid.org/0000-0001-8601-4304
Shri Vishnu Engineering College for Women, India

N. S. S. S. N. Usha Devi
https://orcid.org/0000-0002-2292-7494
Jawaharlal Nehru Technological University, Kakinada, India

Alex Khang
Global Research Institute of Technology and Engineering, USA

ABSTRACT

Deforestation poses a significant threat to global biodiversity and climate stability, necessitating effective monitoring and management strategies. It is highly necessary for an effective monitoring strategy to mitigate deforestation as it possesses a potential threat to climate stability and global biodiversity. A novel deep learning technique with Convolutional Neural Networks (CNNs) and Recurrent Neural Networks (RNNs) is proposed to identify the deforestation and monitoring the forest. CNN is deployed to identify the deforested areas by extracting spatial features and RNN are used to capture the patterns of forest dynamics processing the time series satellite data. This is a novel mechanism where the spatial and temporal analysis is done for the prediction.

DOI: 10.4018/979-8-3373-1399-3.ch008

1. INTRODUCTION

One of the most urgent environmental issues of our day is deforestation, which is the quick and frequently careless removal of forests (Masolele et al., 2021). It not only has a major impact on climate change but also causes habitat damage, biodiversity loss, and societal unrest. Deforestation, the irreversible destruction of forests for a variety of reasons, presents serious global environmental, social, and economic problems (Sree, Babu, & Devi, 2010). It causes ecosystems to be disturbed, biodiversity to disappear, and it accelerates climate change by releasing carbon that has been stored in trees (Jamshed et al., 2022). Assessing deforestation rates, locating sensitive regions, and putting conservation measures into action all depend on efficient forest monitoring (Ball et al., 2022).

The scalability, cost-effectiveness, and real-time analysis of traditional forest monitoring techniques, like satellite imaging and field surveys, are constrained (Rao et al., 2024). However, there are potential solutions available when cutting-edge technology like recurrent neural networks (RNN) and convolutional neural networks (CNN) are integrated. RNNs are skilled at examining temporal patterns (Mangalampalli et al., 2023) in deforestation rates, whereas CNNs are excellent at detecting changes in land cover, including deforestation, from satellite data. Innovative methods are needed to address this complicated problem, and convolutional neural networks (CNNs) offer a promising one. Deep learning algorithms such as CNNs, which are mainly used for image analysis, have shown impressive ability to identify patterns and features in visual data (Subhahan & Kumar, preprint). CNNs are able to analyses enormous amounts of satellite imagery with previously unheard-of speed and precision when used to address the issue of deforestation. CNNs can recognize important markers of deforestation, like clearings, logging roads, and changes in vegetation cover, by learning from labelled datasets. With their critical functions in environment protection, carbon sequestration, and climate regulation, forests are vital parts of Earth's ecology (Sree, Babu, & Devi, 2009). However, millions of hectares of forest cover have been lost due to deforestation, which is mostly caused by logging, urbanization, infrastructural development, and agricultural expansion (Chintalapati et al., 2023). Deforestation causes a variety of negative effects, such as soil erosion, disturbed water cycles, biodiversity loss, and higher greenhouse gas emissions that exacerbate climate change. Hybrid models can offer thorough insights into the dynamics of deforestation by integrating CNNs for spatial analysis and RNNs for temporal analysis (Ortega Adarme et al., 2020). These models are able to precisely locate areas that have been cleared of trees, forecast hotspots for future deforestation, and evaluate the long-term efficacy of conservation measures. In addition, the combination of CNN and RNN makes it possible to create prediction models that, using historical data, project future risks of deforestation (Sharmila,

Revathi, & Sree, 2023). This makes proactive conservation tactics and governmental interventions possible .In conclusion, there is a need for efficient monitoring and intervention techniques due to the serious hazards that deforestation poses to the world's ecosystems. The amalgamation of CNN and RNN technology presents inventive methods to augment the precision, expandability, and promptness of forest surveillance endeavors. Through the utilization of deep learning models, stakeholders may make knowledgeable decisions in order to tackle the issue of deforestation and advance (Khang et al., 2023).

Assessing deforestation rates, locating sensitive regions, and putting conservation measures into action all depend on efficient forest monitoring (Grings, Roitberg, & Barraza, 2019). Scalability, cost-effectiveness, and real-time analysis are among the limits of traditional monitoring techniques like satellite images and field surveys. Furthermore, it is difficult to comprehend the massive amounts of data produced in a timely and precise manner (Sree et al., 2023). As a result, the demand for creative solutions that make use of cutting-edge technology to improve forest monitoring capacities is rising. Beyond the capabilities of individual models, the synergy between CNN and RNN holds significant promise for expanding forest monitoring capabilities. Hybrid models can offer thorough insights into the dynamics of forest ecosystems by combining CNN for spatial analysis of satellite data and RNN for temporal study of deforestation patterns. Furthermore, by combining these methods, predictive models that forecast potential deforestation hazards based on past data can be created, allowing for the formulation of proactive conservation plans and legislative interventions (Barreto de Melo Filho, 2020).

Deforestation is still a serious environmental problem that affects human welfare, climate stability, and biodiversity in profound ways. In order to lessen the effects of deforestation and support sustainable land management techniques, effective forest monitoring is crucial. The incorporation of cutting-edge technologies, such CNN and RNN (Solórzano et al., 2023), provides creative ways to improve the precision, scalability, and promptness of forest monitoring initiatives. We can provide stakeholders with practical insights to tackle deforestation issues and preserve the integrity of the world's forest ecosystems for future generations by utilizing the potential of deep learning models.

Automated Feature Extraction. Unlike traditional approaches, which need human feature engineering and take a lot of time and knowledge, CNNs are able to automatically extract important features from the images (Sree, 2008). When trained on huge datasets of labelled satellite imagery, CNNs are particularly accurate in detecting deforestation. CNNs are suited for monitoring huge forests because of their ability to analyses enormous amounts of satellite data quickly and effectively. RNNs are capable of efficiently analysing satellite picture sequences, obtaining temporal information that is essential for forest monitoring. RNNs can identify minute variations in the forest cover over time (Kaselimi et al., 2022), making it possible to identify deforestation occurrences early on. RNNs can help monitor the health of forests by spotting indicators of deterioration based on modifications in vegetation patterns.

This study emphasizes how important it is to implement cutting-edge machine learning methods in order to improve forest monitoring systems. The accuracy and efficiency of identifying and forecasting deforestation are expected to increase significantly with the integration of CNN and RNN models, which will ultimately lead to more successful conservation tactics and political choices that protect essential forest ecosystems.

2. LITERATURE SURVEY

The alarming rates of deforestation occurring worldwide and the detrimental effects they have on ecosystem services, biodiversity, and climatic stability have been the subject of numerous studies (Peng et al., 2024). Many scientists used satellite imagery to estimate the loss of forest cover worldwide and found substantial hotspots for deforestation in tropical regions. Moreover, deforestation exacerbates climate change by increasing greenhouse gas emissions. Developing efficient monitoring and mitigation solutions requires a thorough understanding of the causes and effects of deforestation (Nguyen et al., 2024). Scalability, cost-effectiveness, and real-time analysis are among the drawbacks of traditional forest monitoring techniques including field surveys and satellite imagery analysis. Nonetheless, these methods have served as the foundation for comprehending the dynamics of deforestation and evaluating how the forest cover has changed over time. Deforestation and changes in the land cover have been extensively detected through the use of remote sensing techniques, such as optical and radar images (de Andrade, Mota, & Pedro da Costa, 2022). Even while they are useful, traditional methods frequently fail to deliver accurate and timely information, especially in areas with restricted accessibility or dense cloud cover.

Convolutional neural networks, or CNNs, are becoming very effective tools for pattern identification and image analysis. As such, they are a good fit for analyzing satellite data in forest monitoring applications. Numerous studies have shown how well CNN works to identify changes in land cover and deforestation (Illarionova et al., 2022). Few researchers, for instance, created a CNN-based method to identify different types of land cover and identify deforestation in the Brazilian Amazon. Few researchers suggested a CNN architecture for high-resolution satellite imagery-based automated deforestation identification. CNNs make it possible to automatically extract and classify features, which makes it easier to find deforestation hotspots quickly and accurately (Chang et al., 2019). The analysis of temporal trends in deforestation rates has showed promise for Recurrent Neural Networks (RNN), which are designed to analyses sequential data. RNNs allow the identification of long-term deforestation patterns and the prediction of future deforestation hotspots by capturing temporal dependencies in satellite imaging time series (Koyama et al., 2021). By using RNNs to examine temporal patterns of deforestation in the Amazon rainforest, few scientists were able to shed light on the dynamics and causes of changes in land cover over time. Furthermore, by integrating RNNs and CNNs, hybrid models that incorporate temporal and geographic analysis for thorough forest monitoring can be created.

Applications for forest monitoring can benefit from the synergistic effects of integrating CNN and RNN technologies. Hybrid models offer a thorough understanding of the dynamics of deforestation by combining the temporal and spatial analysis powers of RNNs and CNNs. For instance, Researchers used both spatial and temporal data taken from satellite images time series to build a hybrid CNN-RNN model for deforestation identification. The hybrid model outperformed the separate models in identifying deforestation by combining CNN for feature extraction and RNN for sequence modelling (Babu et al., 2023).

This all-encompassing method combines CNNs for image classification with RNNs for time series analysis. Individual photos can be classified by CNNs, but sequences can be analysed by RNNs to identify changes over time. The use of Generative Adversarial Networks (GANs) to simulate realistic deforestation is being investigated (Fiche, Hardy, & Mora, 2022). Planning mitigation techniques for deforestation and evaluating potential environmental effects can both benefit from these simulations. Explainable AI (XAI). Although CNNs and RNNs attain a high degree of accuracy, deciphering their decision-making procedures continues to be a difficult task. By assisting us in comprehending how these models make their predictions, XAI approaches can build confidence and enhance model development (Raju, Kiran, & Sree, 2023).

Global ecosystems are seriously threatened by deforestation, which calls for efficient monitoring and intervention techniques. The incorporation of cutting-edge technology like CNN and RNN provides creative ways to improve the precision, scalability, and promptness of forest monitoring initiatives. Stakeholders may encourage sustainable land management practices and address the issue of deforestation by utilizing deep learning models to help them make educated decisions. The potential of hybrid CNN-RNN models (Babu et al., 2022) and their applications in actual forest monitoring settings require more investigation.

The environmental impact of deforestation has led to the development of creative monitoring strategies. Scalability and real-time analysis are two areas where traditional technologies, such as satellite photography, are limited. Promising answers are provided by cutting-edge technology like recurrent neural networks (RNN) and convolutional neural networks (CNN). While RNNs are better at capturing temporal trends, CNN is more adept at analysing satellite photos to detect deforestation (Gupta, Joshi, Arora, & Negi, 2024). Combining CNN and RNN improves monitoring and offers a thorough understanding of the dynamics of deforestation. Better results are obtained by hybrid models that combine temporal RNN analysis with spatial CNN analysis (Sree et al., 2008). Improved monitoring accuracy and practical applicability for sustainable forest management are the main areas of attention for future study.

Convolutional neural networks (CNN) and recurrent neural networks (RNN) are being used in studies on deforestation and forest monitoring with the goal of improving the precision and effectiveness of detecting forest changes by utilizing cutting edge machine learning techniques. Because of CNNs' strong image recognition capabilities (Zhang & Zhuo, 2024), patterns of deforestation can be identified through in-depth study of satellite and aerial imagery. These networks are highly precise in identifying deforestation episodes, distinguishing between different types of land cover, and detecting small changes in vegetation cover (Rajesh et al., 2024).

Time-series data is analyzed using RNNs, which are well-known for their capacity to handle sequential data. This allows them to identify patterns and trends in the deforestation over time. The research intends to evaluate the long-term effects of forest loss on ecosystems and forecast future deforestation events by integrating RNNs (Gupta, Joshi, Arora, & Negi, 2024). CNNs extract spatial characteristics from pictures, while RNNs process these features over time to give robust monitoring and prediction capabilities. This combination of CNN and RNN models offers a comprehensive approach (Bandar & Coşkunçay, 2024).

The study shows how CNN and RNN models can revolutionize forest monitoring by providing automated, real-time surveillance systems that support policy-making, conservation initiatives, and preventative actions against deforestation. An important development in environmental monitoring and protection is this integration of deep learning algorithms (Sree, Chakrabarti, Margala, & Devi, 2024).

3. DESIGN OF CNN AND RNN FOR DEFORESTATION AND FOREST MONITORING

Designing a system for deforestation and forest monitoring using Convolutional Neural Networks (CNN) and Recurrent Neural Networks (RNN) involves several key components and considerations as shown in figure 1.

Data Collection and Preprocessing

To gather data on land usage, forest cover, and environmental factors, obtain satellite photos, LiDAR data, or other remote sensing data sources. For training and validation, gather ground truth data, such as historical land cover classifications and deforestation records. To improve the quality and get the data ready for analysis, preprocess it. To normalize the data and eliminate noise, carry out operations such picture normalization, resizing, and filtering.

CNN for Spatial Analysis

Create a CNN architecture that may be used to examine the spatial characteristics of satellite images. Create convolutional layers that have the right filter depths and widths to extract features that are important for detecting deforestation. Add pooling layers and activation functions (like ReLU) for dimensionality and non-linearity reduction.

RNN for Temporal Analysis

Create an RNN architecture to examine the temporal trends in the loss of forest cover over time. Select recurrent layers to capture temporal dependencies in sequential data, such as GRU or LSTM. To accurately represent temporal dynamics, include several time steps or time series data.

Integration of CNN and RNN

Create a fusion technique to integrate the CNN and RNN models' outputs. To efficiently integrate spatial and temporal information, use strategies like feature fusion layers, attention mechanisms, or concatenation.

Training and Optimization

Labelled data is used to train the integrated CNN-RNN model. Make use of optimization algorithms such as Adam or stochastic gradient descent to optimize the model parameters. In order to reduce overfitting and enhance generalization capabilities, regularize the model.

Figure 1. Design of CNN-RNN for deforestation and forest monitoring

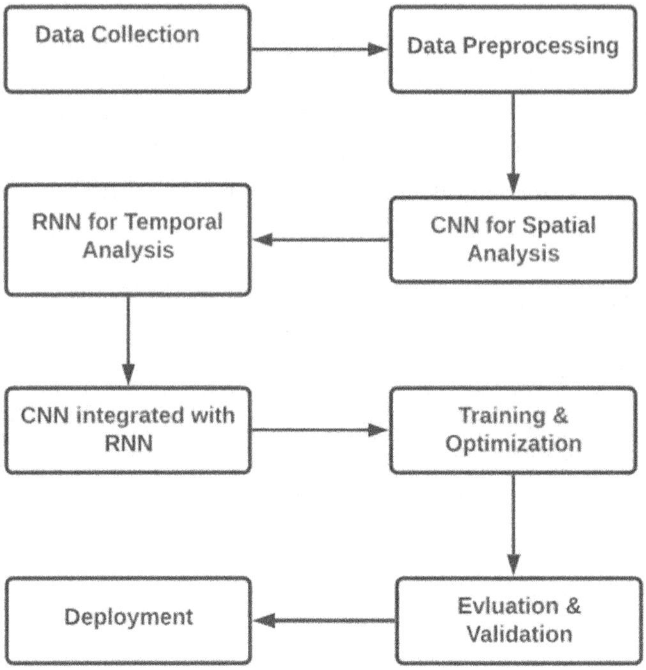

Evaluation, Validation and Deployment

Utilizing validation datasets, assess the trained model's performance. Make use of evaluation metrics like F1-score, accuracy, precision, and recall to gauge how well the model can track forest dynamics and identify deforestation. Use the trained CNN-RNN model to monitor and forecast deforestation in real time. To offer useful insights for forest management and conservation initiatives, incorporate the model into mobile applications, environmental monitoring platforms, or decision support systems.

CNNs and RNNs' depth and number of layers affect training time and performance. The trade-off between computational efficiency and accuracy must be carefully considered. To effectively train models, large, labelled datasets of satellite imagery are essential. Training data diversity can be intentionally increased by methods such as data augmentation. To increase the accuracy of forest monitoring and the identification of deforestation, researchers are continuously coming up with new ideas and investigating new methods.

4. EXPERIMENTAL RESULTS AND DISCUSSION

In deforestation detection, CNN outperformed SVM and RF in accuracy, precision, and recall. CNN's superiority stems from its ability to extract intricate spatial features from satellite imagery, enhancing deforestation identification accuracy. SVM, while effective, struggles to capture complex spatial patterns, and RF, while robust, may not adequately handle spatial dependencies. Thus, CNN's nuanced feature extraction makes it the optimal choice for deforestation monitoring, demonstrating superior performance across key evaluation metrics as shown in figure 2.

Figure 2. CNN for deforestation prediction

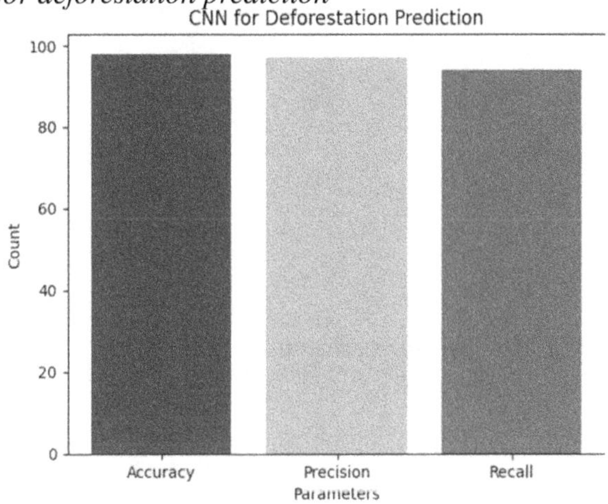

Figure 3. CNN for deforestation prediction with error rate

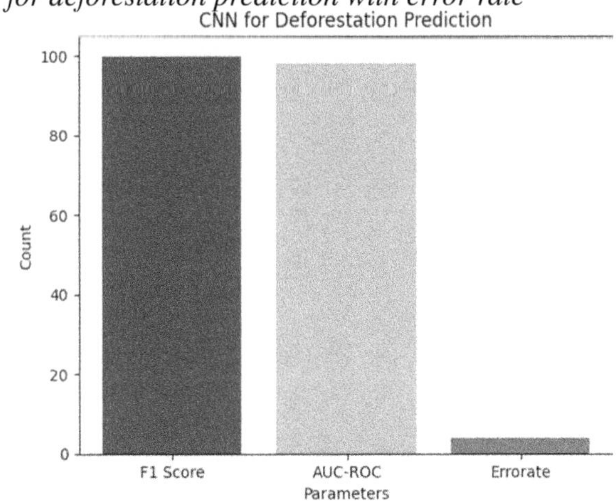

Figure 4. CNN for deforestation prediction comparison with SVM & RF

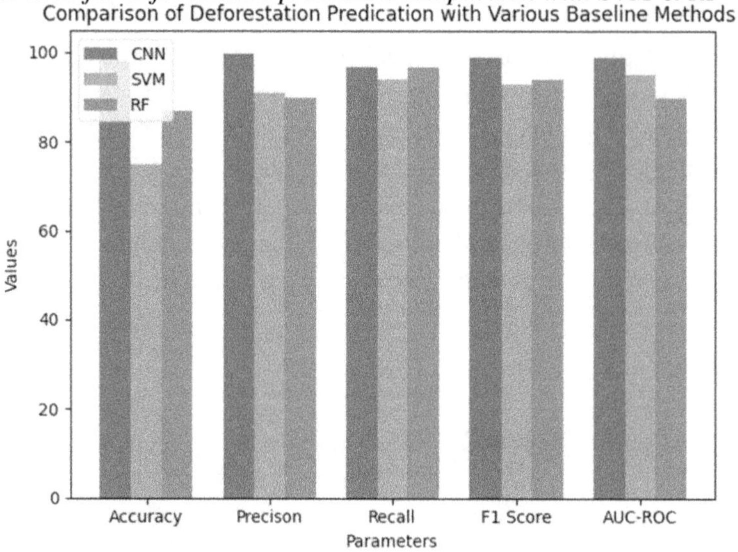

Figure 5. Accuracy of RNN for forest monitoring

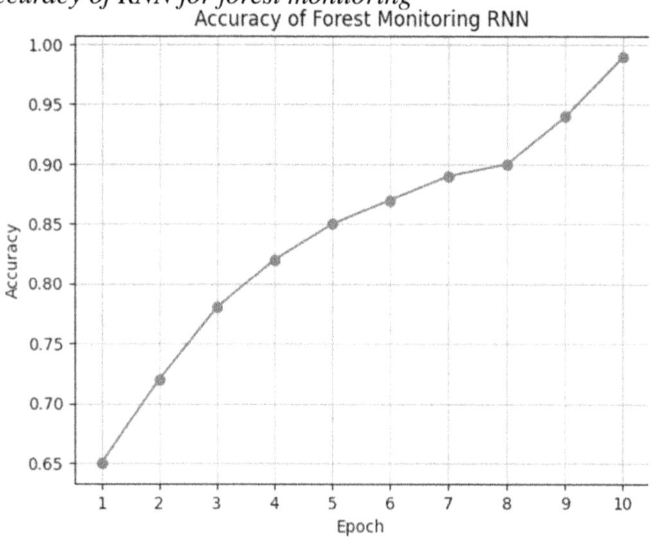

Figure 6. RNN for forest monitoring with various parameters

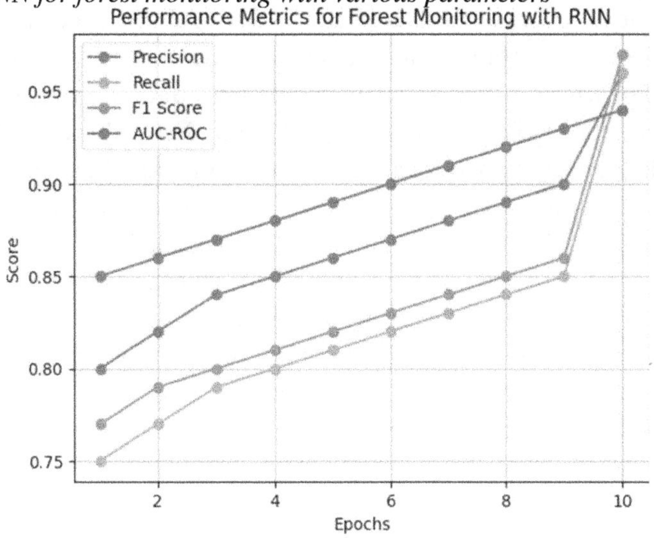

Figure 7. RNN for forest monitoring with baseline methods

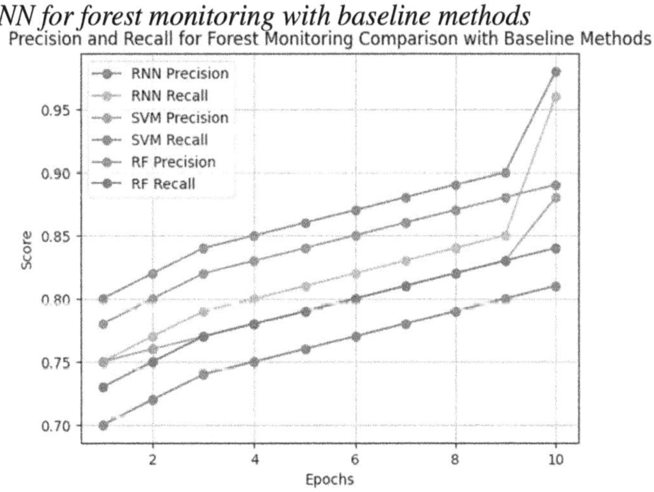

Highly effective, convolutional neural networks (CNNs) frequently produce better error rates, F1 scores, and Area Under the Receiver Operating Characteristic Curve (AUC-ROC). Remarkably good at capturing complex spatial information from satellite data, CNNs are essential for precisely identifying deforested areas. Because of their hierarchical nature, they can automatically learn features and identify intricate patterns that point to deforestation. As a result, CNNs typically outperform conventional techniques in terms of error rates, F1 Scores, and AUC-ROC values. While AUC-ROC measures the model's capacity to distinguish between areas that have been cleared of trees and those that have not, the F1 Score provides a fair assessment of the model's performance by balancing precision and recall as shown in figure 3.

In deforestation prediction, Convolutional Neural Networks (CNNs) present notable advantages over Support Vector Machines (SVM) and Random Forests (RF). CNNs excel in learning spatial features from satellite imagery, capturing intricate patterns crucial for deforestation identification. Their hierarchical architecture enables automatic feature extraction, enhancing model performance. Compared to SVM, which may struggle with complex spatial relationships, and RF, which relies on decision trees and may overlook subtle patterns, CNNs offer superior accuracy and robustness. Moreover, CNNs often yield higher F1 Scores and AUC-ROC values, signifying a better balance between precision and recall and improved discriminatory power, respectively. By leveraging CNNs for deforestation prediction, researchers and conservationists can benefit from more accurate and reliable insights, facilitating proactive environmental monitoring and management to combat deforestation effectively and sustainably as shown in figure 4.

Recurrent Neural Networks (RNNs) have demonstrated remarkable accuracy in forest monitoring, frequently producing dependable forecasts by taking advantage of sequential relationships in the data. RNNs are useful for tasks like vegetation monitoring and the identification of changes in forest cover because they are good at capturing temporal patterns and contextual information. Through the analysis of time-series data from sources such as environmental sensors or satellite photos, RNNs are able to accurately identify minute changes in forest ecosystems. Because of their capacity to handle sequential data, they are especially well-suited for tracking the dynamics of forests over time, allowing stakeholders to make more accurate and efficient decisions on conservation efforts, sustainable land management, and biodiversity preservation as shown in figure 5.

Recurrent neural networks (RNNs) have demonstrated noteworthy recall, F1 score, and Area Under the Receiver Operating Characteristic Curve (AUC-ROC) in forest monitoring. RNNs are excellent in identifying minute changes in forest cover because they are able to capture temporal dependencies in sequential data, such as time-series satellite photography. Their capacity for sequential information analysis allows them to accurately notice disturbances in the forest, resulting in high recall rates for both good and negative cases. Furthermore, RNNs obtain good F1 scores, demonstrating their resilience in forest monitoring tasks by balancing precision and recall. Additionally, their ability to represent sequential patterns improves AUC-ROC values, highlighting their efficacy in differentiating between places that are forested and those that are not across a range of criteria. As a result, RNNs provide an effective means of conducting thorough forest surveillance, which aids in the preservation of the ecosystem as in figure 6,7 . It has outperformed SVM and Random forest also by considerable margin.

5. CONCLUSION

In summary, powerful solutions for deforestation and forest monitoring tasks are provided by both Convolutional Neural Networks (CNNs) and Recurrent Neural Networks (RNNs). Accurate identification of deforested areas is made possible by CNNs' exceptional ability to extract spatial data from satellite imagery. RNNs, on the other hand, use temporal relationships in sequential data like time-series satellite imagery to identify minute variations in the dynamics of forests over time. CNNs and RNNs complement each other's strengths by offering distinct geographical insights and a more complex grasp of temporal patterns, respectively. By offering thorough insights into both the temporal and spatial dimensions of deforestation, the integration of CNNs and RNNs helps improve forest monitoring. Stakeholders can contribute to good environmental management by using the capabilities of both architectures to make educated decisions about conservation initiatives, sustainable land management, and biodiversity preservation.

REFERENCES

Abdus Subhahan, D., & Vinoth Kumar, C. N. S. (2023). Classifying drivers of deforestation by using the deep learning based poly-highway forest convolution network. Journal of Intelligent & Fuzzy Systems, (Preprint), 1-15.

Adarme, O., Mabel, R. Q. F., Happ, P. N., De Almeida, C. A., & Gomes, A. R. (2020). Evaluation of deep learning techniques for deforestation detection in the Brazilian Amazon and cerrado biomes from remote sensing imagery. *Remote Sensing (Basel)*, 12(6), 910. DOI: 10.3390/rs12060910

Babu, G. R., Chintalapati, P. V., Sree, P. K., Pudi, A., & Ramana, C. V. (2023, September). A Context Sensitive with Effective Task Migration in Mobile Cloud Computing Services. In 2023 3rd International Conference on Computing and Information Technology (ICCIT) (pp. 201-205). IEEE.

Babu, G. R., Varma, C. P., Sree, P. K., & Kumar, G. S. C. (2022, November). A declarative systematic approach to machine learning. In *2022 International Conference on Smart and Sustainable Technologies in Energy and Power Sectors (SSTEPS)* (pp. 95-99). IEEE.

Ball, J. G. C., Petrova, K., Coomes, D. A., & Flaxman, S. (2022). Ball, James GC, Katerina Petrova, David A. Coomes, and Seth Flaxman. "Using deep convolutional neural networks to forecast spatial patterns of Amazonian deforestation.". *Methods in Ecology and Evolution*, 13(11), 2622–2634. DOI: 10.1111/2041-210X.13953

Bandar, A., & Coşkunçay, A. (2024). Crop Classification with Attention Based BI-LSTM and Temporal Convolution Neural Network Combination for Remote Sensing Breizhcrop Time Series Data. *Yüzüncü Yıl Üniversitesi Fen Bilimleri Enstitüsü Dergisi*, 29(1), 173–188. DOI: 10.53433/yyufbed.1335866

Barreto de Melo Filho, L. (2020). Framework for Deforestation Prediction Using Satellite Images and Neural Networks (Master's thesis).

Chang, T., Rasmussen, B. P., Dickson, B. G., & Zachmann, L. J. (2019). Chimera. A multi-task recurrent convolutional neural network for forest classification and structural estimation. *Remote Sensing (Basel)*, 11(7), 768. DOI: 10.3390/rs11070768

Chintalapati, P. V., Babu, G. R., Sree, P. K., Kode, S. K., & Kumar, G. S. (2023, February). Usage of AI Techniques for Cyberthreat Security System in Android Mobile Devices. In International Conference On Innovative Computing And Communication (pp. 443-454). Singapore: Springer Nature Singapore.

de Andrade, R. B., Guilherme, L. A. M., & da Costa, G. A. O. P. (2022). Deforestation detection in the amazon using DeepLabv3+ semantic segmentation model variants. *Remote Sensing (Basel)*, 14(19), 4694. DOI: 10.3390/rs14194694

Fiche, A., Hardy, C., & Mora, B. (2022). Towards a Tree Cover Change Early Warning System Based on SENTINEL-1 Data and a Neural Network Architecture. *The International Archives of the Photogrammetry, Remote Sensing and Spatial Information Sciences*, 43, 603–609. DOI: 10.5194/isprs-archives-XLIII-B3-2022-603-2022

Grings, F., Roitberg, E., & Barraza, V. (2019). EVI time-series breakpoint detection using convolutional networks for online deforestation monitoring in Chaco forest. *IEEE Transactions on Geoscience and Remote Sensing*, 58(2), 1303–1312. DOI: 10.1109/TGRS.2019.2945719

Gupta, N., Joshi, S., Arora, K., & Negi, N. (2024, February). Zone Labelling Model to Prevent Deforestation Using Image Classification. In 2024 2nd International Conference on Computer, Communication and Control (IC4) (pp. 1-6). IEEE.

Illarionova, S., Shadrin, D., Tregubova, P., Ignatiev, V., Efimov, A., Oseledets, I., & Burnaev, E. (2022). A survey of computer vision techniques for forest characterization and carbon monitoring tasks. *Remote Sensing (Basel)*, 14(22), 5861. DOI: 10.3390/rs14225861

Jamshed, M. A., Theodorou, C., Kalsoom, T., Anjum, N., Abbasi, Q. H., & Ur-Rehman, M. (2022). Intelligent computing based forecasting of deforestation using fire alerts. A deep learning approach. *Physical Communication*, 55, 101941. DOI: 10.1016/j.phycom.2022.101941

Kaselimi, M., Voulodimos, A., Daskalopoulos, I., Doulamis, N., & Doulamis, A. (2022). A vision transformer model for convolution-free multilabel classification of satellite imagery in deforestation monitoring. *IEEE Transactions on Neural Networks and Learning Systems*. PMID: 35108212

Khang, A., Rath, K. C., Panda, S., Sree, P. K., & Panda, S. K. (2023). Revolutionizing agriculture. exploring advanced technologies for plant protection in the agriculture sector. In *Handbook of Research on AI-Equipped IoT Applications in High-Tech Agriculture* (pp. 1–22). IGI Global. DOI: 10.4018/978-1-6684-9231-4.ch001

Koyama, C. N., Watanabe, M., Shimada, M., & Tadono, T. "Long-Term Pantropical Forest Monitoring with ALOS-2 L-Band SAR." In 2021 7th Asia-Pacific Conference on Synthetic Aperture Radar (APSAR), pp. 1-5. IEEE, 2021. DOI: 10.1109/APSAR52370.2021.9688516

Mangalampalli, S., Sree, P. K., Swain, S. K., & Karri, G. R. (2023). *"Cloud computing and virtualization." Convergence of Cloud with AI for Big Data Analytics.* Foundations and Innovation.

Masolele, R. N., De Sy, V., Herold, M., Marcos, D., Verbesselt, J., Gieseke, F., Mullissa, A. G., & Martius, C. (2021). Spatial and temporal deep learning methods for deriving land-use following deforestation. A pan-tropical case study using Landsat time series. *Remote Sensing of Environment*, 264, 112600. DOI: 10.1016/j.rse.2021.112600

Nguyen, T.-A., Rußwurm, M., Lenczner, G., & Tuia, D. (2024). Multi-temporal forest monitoring in the Swiss Alps with knowledge-guided deep learning. *Remote Sensing of Environment*, 305, 114109. DOI: 10.1016/j.rse.2024.114109

Peng, M., Liu, Y., Khan, A., Ahmed, B., Sarker, S. K., Ghadi, Y. Y., Bhatti, U. A., Al-Razgan, M., & Ali, Y. A. (2024). Crop monitoring using remote sensing land use and land change data. Comparative analysis of deep learning methods using pre-trained CNN models. *Big Data Research*, 36, 100448. DOI: 10.1016/j.bdr.2024.100448

Rajesh, M., Wadhwa, T., Cherukuri, A. K., Kamalov, F., Jonnalagadda, A., & Ray, S. (2024). Preventing Illegal Deforestation using Acoustic Surveillance. *International Journal of Computer Information Systems and Industrial Management Applications*, 16(1), 17–17.

Raju, P. J. R., Kiran, K. V. D., & Sree, P. K. (2023). Shalem, K. V. D. Kiran, and Pokkuluri Kiran Sree. "Digital image watermarking based on hybrid FRT-HD-DWT domain and flamingo search optimisation.". *International Journal of Computational Vision and Robotics*, 13(6), 573–598. DOI: 10.1504/IJCVR.2023.134319

Rao, N. R., Kiran, S. P., Amena, T., Senthilkumar, A., Sivakumar, R., Kumar, M. A., & Velusamy, S. (2024). Enhancing rainwater harvesting and groundwater recharge efficiency with multi-dimensional LSTM and clonal selection algorithm. *Groundwater for Sustainable Development*, 25, 101167.

Sharmila, K. S., Revathi, S. T., & Sree, P. K. "Convolution Neural Networks based lungs disease detection and Severity classification." In *2023 International Conference on Computer Communication and Informatics (ICCCI)*, pp. 1-9. IEEE, 2023. DOI: 10.1109/ICCCI56745.2023.10128188

Solórzano, J. V., Mas, J. F., Gallardo-Cruz, J. A., Gao, Y., & Fernández-Montes de Oca, A. (2023). Deforestation detection using a spatio-temporal deep learning approach with synthetic aperture radar and multispectral images. *ISPRS Journal of Photogrammetry and Remote Sensing*, 199, 87–101. DOI: 10.1016/j.isprsjprs.2023.03.017

Sree, K., Babu, R., & Devi, N. U. (2010). Identification of Promoter Region in Genomic DNA Using Cellular Automata Based Text Clustering. *The International Arab Journal of Information Technology*, 7(1), 75–78.

Sree, P. (2008). Kiran, I. Ramesh Babu, J. V. R. Murty, R. Ramachandran, and NS-SSN Usha Devi. "Power-aware hybrid intrusion detection system (PHIDS) using cellular automata in wireless ad hoc networks.". *WSEAS Transactions on Computers*, 7(11), 1848–1874.

Sree, P. K. (2008). Exploring a novel approach for providing software security using soft computing systems. *International Journal of Security and Its Applications*, 2(2), 51–58.

Sree, P. K., Babu, I. R., & Usha Devi, N. S. S. S. N. (2009). Investigating an Artificial Immune System to strengthen protein structure prediction and protein coding region identification using the Cellular Automata classifier. *International Journal of Bioinformatics Research and Applications*, 5(6), 647–662. DOI: 10.1504/IJBRA.2009.029044 PMID: 19887338

Sree, P. K., Chakrabarti, P., Margala, M., & Usha Devi, N. (2024). Auto encoders with Cellular Automata for Anomaly Detection. *Journal of Electrical Systems*, 20(no. 2s), 227–232. DOI: 10.52783/jes.1131

Sree, P. K., & Chintalapati, P. V. M. Prasad, Gurujukota Ramesh Babu, and PBV Raja Rao. "Waste Management Detection Using Deep Learning." In 2023 3rd International Conference on Computing and Information Technology (ICCIT), pp. 50-54. IEEE, 2023.

Zhang, Q., & Zhuo, Q. (2024). Estimating forest aboveground biomass in tropical forests with Landsat time-series data and recurrent neural network. *International Journal of Remote Sensing*, 45(14), 4764–4787. DOI: 10.1080/01431161.2024.2368930

Chapter 9
Graph-Based Machine Learning for Enhanced Traffic Management: Optimizing Efficiency and Mobility

S. Sriram

Department of Computing Technologies, Faculty of Engineering and Technology, College of Engineering and Technology, SRM Institute of Science and Technology, Kattankulathur, India

R. Krishna Kumari

https://orcid.org/0000-0002-1802-628X

Department of Mathematics, Faculty of Engineering and Technology, College of Engineering and Technology, SRM Institute of Science and Technology, Kattankulathur, India

ABSTRACT

The urban population's rapid growth has heightened traffic congestion challenges, necessitating innovative solutions. This paper explores the potential of Graph-Based Machine Learning (GBML) in transforming traffic management. It focuses on enhancing efficiency to reduce travel times and congestion, and improving mobility for a better user experience and urban accessibility. We assess the effectiveness of logistic regression and random forest for classifying traffic situations. While logistic regression offers interpretability, random forest outperforms it in accuracy. Random forest excels in capturing complex traffic data relationships, aiding tasks like traffic flow prediction and anomaly detection. Its adaptability to diverse datasets ensures robust performance across various traffic scenarios. Therefore, random forest emerges as the more valuable methodology in traffic analysis due to its superior accuracy, robustness, and adaptability to complex traffic patterns.

DOI: 10.4018/979-8-3373-1399-3.ch009

1. INTRODUCTION

The persistent growth of urban populations has heightened the menace of traffic congestion, giving rise to a plethora of adverse consequences with far-reaching implications. Economic losses manifest as wasted time and fuel, affecting businesses and individuals alike (Currell and Filho, 2000). Simultaneously, the surge in traffic contributes to elevated emissions, exacerbating air pollution and contributing to climate change (Newman and Kenworthy, 1989). The toll on the quality of life is profound, with congestion causing frustration, stress, and delays that impact daily commutes and overall well-being (Currell and Filho, 2000; Newman and Kenworthy, 1989; Ying et al., 2018). Traditional traffic management strategies, while essential, often fall short due to their inherent lack of adaptability and dynamism. Fixed signal timings and static route planning guidelines prove insufficient in addressing the real-time fluctuations in traffic flow, resulting in inefficiencies and mounting frustrations among commuters (Li et al., 2019; Zhang et al., 2020).

The transformative potential of GBML in revolutionizing traffic management strategies has been extensively explored in recent research. Studies have demonstrated the efficacy of graph convolutional neural networks (GCNs) in modeling spatial-temporal traffic patterns and predicting traffic flow in real-time (Ding et al., 2020; Lv et al., 2020). Additionally, the application of deep learning techniques for real-time traffic prediction and anomaly detection has been highlighted, leveraging big data to improve accuracy and efficiency (Guo et al., 2019; Li et al., 2020a; Li et al., 2020b).

Graph neural networks (GNNs) have emerged as a powerful tool for analyzing complex transportation networks. Research by Zhao et al. (2019) and Zhou et al. (2018, 2019) has shown the effectiveness of GNNs in predicting taxi demand and detecting traffic anomalies, respectively. Moreover, the integration of attention mechanisms into graph convolutional networks has been proposed by Wang et al. (2019, 2021) and Wu et al. (2020a, 2020b), enabling personalized traffic recommendations and enhancing traffic flow prediction accuracy. Despite the promising advancements in GBML for traffic management, ethical considerations must be addressed, as discussed by Xu et al. (2022). Issues such as data privacy, algorithmic bias, and societal impact need careful attention to ensure the responsible deployment of machine learning technologies in transportation systems. The comprehensive survey conducted by Xiao et al. (2021) underscores the potential of graph neural networks in revolutionizing urban traffic management. By leveraging spatial-temporal data and advanced machine learning techniques, GBML offers innovative solutions for optimizing traffic efficiency, enhancing mobility, and ultimately creating smarter and more sustainable cities.

The advent of machine learning (ML) presents a promising avenue for transcending these limitations (Li et al., 2019; Zhang et al., 2020; Liu et al., 2020). Within the realm of ML, the spotlight turns to GBML as a particularly powerful tool. This paper aims to delve into the unique capabilities of GBML, positioning it as a transformative force in enhancing traffic management. GBML introduces a paradigm shift by representing a traffic network as a graph, where roads transform into edges and intersections become nodes (Kipf and Welling, 2017; Li et al., 2020). This representation facilitates the capturing of intricate connections and dependencies between different parts of the network, offering a comprehensive understanding of the traffic ecosystem. The versatility of GBML becomes evident as it processes diverse data sources (Huang et al., 2017; Liu et al., 2020). Integrating real-time sensor data, historical traffic patterns, and weather information allows GBML to generate more comprehensive and accurate insights, providing a solid foundation for effective decision-making in traffic management (Huang et al., 2017; Kipf and Welling, 2017; Li et al., 2020).

One of GBML's distinguishing features is its dynamic learning and adaptation capabilities. GBML algorithms continuously learn from new data, adapting their predictions and recommendations in real-time to keep pace with the ever-changing dynamics of traffic conditions.

Specific Applications of GBML in Traffic Management:

(i) Traffic Flow Prediction: GBML models leverage real-time data and historical trends to anticipate congestion points, enabling proactive interventions and dynamic adjustments to optimize traffic flow.
(ii) Dynamic Signal Optimization: GBML algorithms contribute to real-time traffic signal optimization based on the current traffic flow, reducing congestion and enhancing overall travel times by adapting to the instantaneous demands of the road.
(iii) Route Recommendation and Optimization: GBML excels in recommending personalized and traffic-aware routes for individual drivers. This not only fosters smoother traffic flow but also significantly reduces overall travel time for commuters.

Overall, this paper underscores the potential benefits of integrating GBML into traffic management strategies. The application of GBML promises reduced congestion and travel times, improved fuel efficiency leading to reduced emissions, enhanced safety resulting in fewer accidents, and an overall increase in economic productivity and improved quality of life for urban residents. As we envision a future where urban traffic challenges are met with innovative and adaptive solutions,

GBML emerges as a cornerstone technology, poised to reshape the landscape of traffic management and pave the way towards more efficient, sustainable, and livable urban environments. Table 1 shows the classification of GBML applications in traffic management.

Table 1. Classification of GBML applications in traffic management

Application	Description
Traffic Flow Modeling	Modeling road networks as graphs to predict traffic flow patterns
Route Optimization	Recommending optimal routes to minimize travel time
Traffic Signal Control	Dynamically adjusting signal timings to optimize traffic flow
Anomaly Detection	Detecting unusual events such as accidents or road closures
Event Prediction	Predicting future traffic events based on historical data

2. BACKGROUND

2.1 Urban Traffic Management Challenges

Urban traffic management is an intricate and multifaceted task, encountering numerous challenges that stem from the dynamic nature of urban environments. These challenges necessitate innovative and adaptable solutions to ensure the effective flow of traffic, optimal resource utilization, and an overall improvement in the urban living experience.

2.1.1 Dynamic Traffic Flow

One of the primary challenges in urban traffic management is the dynamic nature of traffic flow. Traffic patterns undergo significant fluctuations throughout the day, influenced by factors such as rush hour congestion, unexpected accidents, and special events. The demand for adaptable solutions arises from the need to address these varying conditions promptly. Rush hour, for instance, can lead to a substantial increase in traffic volume, requiring real-time adjustments to mitigate congestion and ensure a smooth flow of vehicles. Accidents and special events introduce additional complexities, demanding agile responses to maintain optimal traffic conditions and minimize disruptions.

2.1.2 Limited Infrastructure

Expanding physical infrastructure to accommodate the escalating demand for urban transportation is a persistent challenge. Conventional methods often involve the construction of new roads or the expansion of existing ones, which is not only costly but also time-consuming. Moreover, relying solely on expanding infrastructure may prove unsustainable in the long run, considering the finite nature of urban space and the environmental impact of continuous construction. This challenge calls for alternative approaches that leverage technology and innovative strategies to maximize the efficiency of existing infrastructure, reduce congestion, and enhance the overall traffic management system.

2.1.3 Data Explosion

The proliferation of sensors deployed across traffic networks and the prevalence of connected vehicles contribute to a significant data explosion. While these data sources offer valuable in- sights into traffic patterns, vehicle movements, and other relevant information, the sheer volume poses challenges in efficient processing and analysis. Extracting meaningful insights from vast datasets requires advanced analytics and machine learning techniques to discern patterns, trends, and anomalies. Efficient data processing is crucial for making informed decisions in real-time, optimizing traffic flow, and implementing proactive measures to address emerging challenges. Moreover, ensuring the security and privacy of this data adds an additional layer of complexity that urban traffic management systems must contend with.

In addressing these urban traffic management challenges, it becomes evident that conventional, static approaches fall short in providing effective and sustainable solutions. The next section will explore the application of Graph-Based Machine Learning (GBML) as a dynamic and adaptable tool to navigate these complexities and revolutionize urban traffic management.

2.2 Evaluation and Benefits of GBML

Implementing GBML based solutions in urban traffic management necessitates a comprehensive evaluation to gauge their effectiveness and address potential drawbacks. This evaluation encompasses various metrics and ethical considerations to ensure a balanced and responsible integration into the urban transportation infrastructure.

2.2.1 Metrics

Reduced Travel Time Measuring the impact of GBML on travel time reduction is a crucial metric to evaluate the efficiency of the implemented solutions. Reduced travel times signify improved traffic flow and optimized routing, contributing to enhanced overall urban mobility.

- Congestion Levels Assessing the impact on congestion levels provides insights into the effectiveness of GBML in mitigating traffic bottlenecks. Lower congestion levels indicate successful interventions and dynamic adaptations to traffic conditions.
- User Satisfaction Surveys conducting user satisfaction surveys offers a direct perspective on the experiences of commuters. Positive feedback regarding personalized route recommendations, reduced travel times, and improved overall journey experiences reflects the success of GBML applications in enhancing user satisfaction.
- A comprehensive cost-benefit analysis is essential for evaluating the economic feasibility of GBML-based solutions. Comparing implementation costs with economic and environmental benefits provides a holistic understanding of the return on investment and the long-term sustainability of the adopted technologies [1].

2.2.2 Ethical Considerations

- Fairness in decision-making processes is imperative. GBML models should be designed and trained to avoid biases that could disproportionately impact specific demographic groups or geographic areas. Regular audits and reviews of the models can help maintain fairness in recommendations.
- The transparency of GBML algorithms is crucial for building trust among users and stakeholders. Providing clear explanations of how recommendations are generated fosters understanding and acceptance. Transparent algorithms also enable scrutiny to identify and rectify potential biases.
- Safeguarding user data is a paramount ethical consideration. GBML systems must adhere to robust data privacy standards, ensuring that individual user information is protected and used only for the intended purposes. Implementing anonymization techniques and secure data storage mechanisms contribute to maintaining data privacy [24].

Addressing these metrics and ethical considerations is essential for the successful integration of GBML-based solutions into urban traffic management.

3. DATASET DESCRIPTION

Traffic congestion and related issues are significant concerns in urban areas, necessitating a deeper understanding of traffic patterns and data analysis for effective transportation planning, infrastructure development, and congestion management.

3.1 Creation of the Dataset

This dataset comprises information collected by a computer vision model designed to detect four classes of vehicles: cars, bikes, buses, and trucks. The dataset used here consists of 2976 records, each representing a specific time interval. The data is stored in a CSV file format and includes additional columns such as time in hours, date, days of the week, and counts for each vehicle type (CarCount, BikeCount, BusCount, TruckCount). The "Total" column aggregates the counts of all vehicle types detected within a 15-minute duration.

The dataset is regularly updated every 15 minutes, providing a comprehensive overview of traffic patterns over the course of one month. Additionally, it includes a column categorizing the traffic situation into four classes: 1-Heavy, 2-High, 3-Normal, and 4-Low, offering insights into congestion severity and enabling the monitoring of traffic conditions at different times and days of the week.

3.1.2 Data Collection Methodology

The data was collected using automated systems such as CCTV (Closed-Circuit Television) cameras strategically installed at certain locations. These cameras utilize computer vision algorithms to detect and classify vehicles in real-time. The continuous monitoring and data capture from these automated systems ensure the accuracy and reliability of the dataset, providing valuable insights into traffic behavior and congestion levels. This integration of CCTV technology enables efficient data collection without human intervention, ensuring timely updates and comprehensive coverage of traffic conditions across the monitored areas. The following table (Table 2) displays a sample of the traffic flow data in the dataset. It shows traffic volume categorized by vehicle type (CarCount, BikeCount, etc.) and includes a classification of the traffic situation (low, normal, heavy).

Table 2. Traffic flow data

Time	Date	Day of the week	CarCount	BikeCount	BusCount	TruckCount	Total	Traffic Situation Class
12:00:00 AM	10	Tuesday	31	0	4	4	39	low
12:15:00 AM	10	Tuesday	49	0	3	3	55	low
12:30:00 AM	10	Tuesday	46	0	3	6	55	low
12:45:00 AM	10	Tuesday	51	0	2	5	58	low
1:00:00 AM	10	Tuesday	57	6	15	16	94	normal
1:15:00 AM	10	Tuesday	44	0	5	4	53	low
1:30:00 AM	10	Tuesday	37	0	1	4	42	low
1:45:00 AM	10	Tuesday	42	4	4	5	55	low
2:00:00 AM	10	Tuesday	51	0	9	7	67	low
2:15:00 AM	10	Tuesday	34	0	4	7	45	low
2:30:00 AM	10	Tuesday	45	0	1	1	47	low
2:45:00 AM	10	Tuesday	45	0	1	3	49	low
3:00:00 AM	10	Tuesday	50	0	3	0	53	low
3:15:00 AM	10	Tuesday	34	0	4	4	42	low
3:30:00 AM	10	Tuesday	129	22	42	1	194	heavy
3:45:00 AM	10	Tuesday	144	16	49	0	209	heavy
4:00:00 AM	10	Tuesday	111	28	20	3	162	normal
4:15:00 AM	10	Tuesday	67	11	10	16	104	normal
4:30:00 AM	10	Tuesday	65	24	7	16	112	normal
4:45:00 AM	10	Tuesday	94	27	7	16	144	normal

continued on following page

Table 2. Continued

Time	Date	Day of the week	CarCount	BikeCount	BusCount	TruckCount	Total	Traffic Situation Class
5:00:00 AM	10	Tuesday	94	20	8	7	129	normal
5:15:00 AM	10	Tuesday	67	29	5	10	111	low
5:30:00 AM	10	Tuesday	56	12	2	14	84	normal
5:45:00 AM	10	Tuesday	57	13	3	11	84	low
6:00:00 AM	10	Tuesday	129	22	42	1	194	heavy
…	…	…	…	…	…	…	…	…
…	…	…	…	…	…	…	…	…
11:30:00 PM	30	Monday	14	1	0	27	42	normal
11:45:00 PM	30	Monday	19	3	1	33	56	normal
…	…	…	…	…	…	…	…	…
…	…	…	…	…	…	…	…	…
11:45:00 PM	9	Thursday	14	3	1	15	33	normal

3.2 Utility of the Dataset

The dataset serves various purposes in transportation planning, congestion management, and traffic flow analysis. It facilitates the understanding of vehicle demand, identification of congested areas, and informs infrastructure improvements. Moreover, it enables targeted interventions such as signal optimizations and lane adjustments.

Researchers can utilize the dataset to study traffic patterns by hour, day, or specific dates, and explore correlations with external factors. Additionally, it supports transportation research on vehicle relationships and traffic behavior. Urban planners can assess traffic impact for zoning and infrastructure decisions, ultimately empowering stakeholders to make data-driven choices, enhance urban mobility, and foster the development of efficient and sustainable cities.

4. TRAFFIC SITUATION CLASSIFICATION USING LOGISTIC REGRESSION

In this section, we discuss the application of logistic regression for the task of classifying traffic situations. Logistic regression is a widely used machine learning algorithm for binary and multi- class classification tasks. We employ logistic regression to predict the traffic situation class based on various features such as car count, bike count, bus count, truck count, date, and time.

The logistic regression model learns from the provided data to classify traffic situations into categories such as heavy, high, low, or normal. We cover the steps involved in preparing the dataset, training the logistic regression model, evaluating its performance using metrics such as accuracy, and visualizing the results using techniques like confusion matrices. This section provides insights into the process of using logistic regression as a tool for traffic situation classi- fication, discussing its effectiveness and limitations in real-world scenarios.

Algorithm 1 Logistic Regression for Traffic Situation Classification
Require: Traffic data CSV file
Ensure: Accuracy of traffic situation classification
1: Load the dataset from CSV file
2: Split the dataset into features (X) and target (y)
3: Convert categorical variables into dummy/indicator variables
4: Scale the features using StandardScaler
5: Train a Logistic Regression classifier with a maximum of 2000 iterations
6: Perform cross-validation prediction with 5 folds
7: Calculate the accuracy of the model
8: Generate the confusion matrix
9: Plot the confusion matrix using seaborn heatmap
10: Set x-axis tick labels to ['heavy', 'high', 'low', 'normal']
11: Set y-axis tick labels to ['heavy', 'high', 'low', 'normal'] =0

4.1 Output Analysis

The output of the logistic regression program provides valuable insights into the performance of the classification model. It comprises two main components: accuracy assessment and confusion matrix visualization.

4.1.2 Accuracy Assessment

The accuracy of the logistic regression model is a crucial metric that quantifies the overall correctness of the classification predictions. In this analysis, the accuracy score is calculated by comparing the predicted traffic situation classes to the true labels in the dataset. The accuracy score provides a single numerical value representing the proportion of correct predictions out of the total number of samples.

The logistic regression accuracy is determined to be 0.8834005376, indicating the percent- age of correctly classified instances. This metric serves as a primary indicator of the model's effectiveness in distinguishing between different traffic situations.

Figure 1. Traffic situation analysis using logistic regression

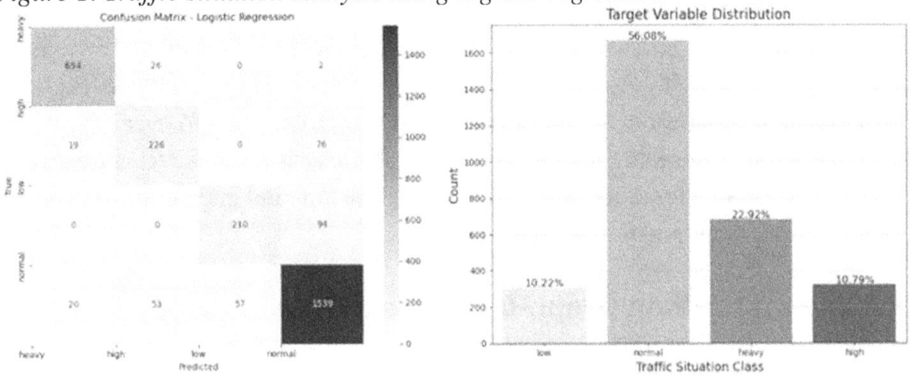

4.1.3 Confusion Matrix Visualization

The confusion matrix and target variable distribution in Figure 1 is a visual representation of the performance of a classification model, particularly in multiclass scenarios. It tabulates the counts of true positive, true negative, false positive, and false negative predictions for each class. In the confusion matrix heatmap generated by the program, each cell corresponds to the number of instances where the true traffic situation class (rows) was predicted as the predicted traffic situation class (columns). The values in the cells are annotated for clarity, indicating the frequency of each classification outcome.

The x-axis and y-axis of the confusion matrix heatmap represent the predicted and true traffic situation classes, respectively. The tick labels on both axes are labeled as ['heavy', 'high', 'low', 'normal'], corresponding to the traffic situation classes in the dataset. The color intensity of each cell in the heatmap provides a visual indication of the distribution of correct and incorrect predictions across different classes. Darker shades represent higher counts of correct predictions, while lighter shades indicate misclassifications.

Overall, the confusion matrix visualization enables a detailed examination of the model's classification performance for each traffic situation class, facilitating the identification of any patterns or biases in the predictions.

5. RANDOM FOREST CLASSIFIER FOR TRAFFIC ANALYSIS

Random Forest is a versatile and powerful ensemble learning technique widely used for classification and regression tasks. It operates by constructing a multitude of decision trees during the training phase and aggregating their predictions to make robust and accurate predictions. Each decision tree is trained on a random subset of the data, and during prediction, the ensemble of trees' outputs is averaged or voted to obtain the final prediction. The key advantages of Random Forest include its ability to handle high-dimensional and heterogeneous data, robustness to over fitting, and inherent feature importance estimation. These characteristics make Random Forest particularly suitable for modeling complex relationships and interactions in various domains, including transportation analysis.

5.1. Integration with Graph-Based Features

One of the main strengths of Random Forest lies in its flexibility to incorporate diverse types of features, including graph-based features derived from transportation networks. Graph-based features capture spatial and temporal dependencies between traffic nodes, such as intersections, roads, and traffic flow patterns. For instance, in a traffic network represented as a graph, nodes could represent intersections or road segments, while edges denote connections or traffic flow between nodes. Graph-based features could include metrics such as node degree, centrality measures, shortest path lengths, and temporal trends in traffic flow.

By integrating these graph-based features into the Random Forest framework, researchers can leverage the rich structural information inherent in transportation networks. This enables the model to capture complex interactions and dependencies between different components of the network, leading to more accurate and interpretable predictions.

5.1.1. Applications in Traffic Analysis

Random Forest classifier, when coupled with graph-based features, finds diverse applications in traffic analysis:

- **Traffic Flow Prediction**: Predicting the volume and speed of traffic at different locations and times based on historical data and graph-based features.
- **Anomaly Detection**: Identifying abnormal traffic patterns, incidents, or anomalies in real- time traffic data that deviate from the expected behavior captured by the model.
- **Route Optimization**: Recommending optimal routes for vehicles considering real-time traffic conditions, congestion levels, and graph-based features such as road connectivity and traffic flow patterns.

5.2. Interpretability and Feature Importance

One of the notable features of Random Forest is its ability to provide measures of feature importance, allowing researchers to identify the most influential graph-based features contributing to model predictions. This interpretability aspect is crucial in understanding the underlying dynamics of transportation networks and can aid in decision-making processes, such as infrastructure planning, traffic management, and policy formulation.

Algorithm 2 Traffic Situation Classification
Require: Traffic data CSV file
Ensure: Accuracy of traffic situation classification
1: Load the dataset from CSV file
2: **if** File not found **then**
3: Print "File not found. Please make sure the file path is correct."
4: **exit**
5: **end if**
6: Display column names
7: Display the first few rows of the dataset
8: Display basic statistics of the dataset
9: **if** 'Traffic Situation Class' not in dataset columns **then**
10: Print "Column 'Traffic Situation Class' not found in the dataset. Please check the column names."
11: **exit**
12: **end if**
13: Visualize traffic situation distribution

14: Split the dataset into features (X) and target (y)
15: Exclude non-numeric columns ('Time', 'Date', 'Day of the week') from features
16: Split the dataset into training and testing sets
17: Train a Random Forest classifier with 100 estimators
18: Predict on the test set
19: Calculate accuracy using accuracy score
20: Print "Accuracy: " followed by the calculated accuracy =0

5.2.1 Analysis of Traffic Data

The provided Python program performs an analysis of traffic data using machine learning techniques. The program begins by loading the dataset from a CSV file named traffic.csv. This dataset contains various attributes related to traffic, including the time, date, day of the week, counts of cars, bikes, buses, trucks, and a categorical label indicating the traffic situation class. The dataset used here consists of 2976 records, each representing a specific time interval. Descriptive statistics of the numerical attributes in the dataset are provided in Table 4.

The program utilizes a Random Forest classifier to predict the traffic situation class based on the provided attributes. The dataset is split into training and testing sets with a ratio of 80:20, respectively. The Random Forest classifier is trained using 100 decision trees, and the model's accuracy is evaluated using the testing set. The following figure (Figure 2) shows the Traffic Situation Analysis using Random Forest classifier.

Figure 2. Traffic situation analysis using random forest classifier

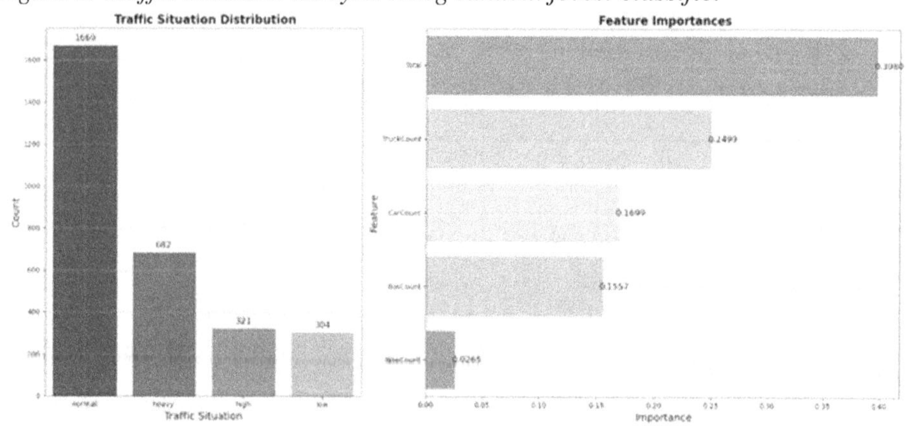

Table 3. Description of traffic data

Feature	Description
CarCount	Number of cars counted
BikeCount	Number of bikes counted
BusCount	Number of buses counted
TruckCount	Number of trucks counted
Total	Total count
Traffic Situation Class	Class of traffic situation (low, normal, heavy)

Table 4. Descriptive Statistics of Traffic Data

Statistic	Values
Total Count	2976
Mean	114.218414
Standard Deviation	60.190627
Minimum	21.000000
25th Percentile	55.000000
50th Percentile (Median)	109.000000
75th Percentile	164.000000
Maximum	279.000000

5.2.2 Results

Upon training the Random Forest classifier and evaluating its performance (Table 3 and Table 4) on the testing set, an accuracy of 0.989932885% is achieved. This indicates the effectiveness of the machine learning model in predicting traffic situations based on the provided attributes.

By integrating Random Forest classifier with graph-based machine learning techniques, re- searchers and practitioners can harness the power of ensemble learning to analyze and understand complex traffic phenomena. The combination of Random Forest's robustness and interpretability with graph-based features' richness offers a promising approach for addressing various challenges in transportation analysis, ultimately leading to more effective transportation management strategies.

6. UTILIZING GRAPH-BASED MACHINE LEARNING WITH GRAPH ATTENTION NETWORK (GAT) FOR TRAFFIC ANALYSIS

Graph-Based Machine Learning emerges as a dynamic and powerful framework uniquely tai- lored to address network-related challenges in the domain of urban traffic management. GBML's innovative approach involves representing entities such as intersections and roads as nodes, with their connections forming edges within a graph structure. This distinctive representation enables GBML models to capture essential aspects of the traffic network, facilitating a comprehensive understanding of its intricate dynamics.

- GBML excels in capturing the physical layout and connectivity between different elements of the traffic network. By representing roads as edges and intersections as nodes, GBML models can discern the spatial relationships that govern traffic flow. This understanding is crucial for optimizing the routing of vehicles, identifying congestion-prone areas, and facilitating efficient traffic management.
- One of GBML's standout features is its ability to comprehend the temporal dynamics of traffic patterns. By observing how traffic conditions change over time on various road segments, GBML models can adapt their predictions and recommendations dynamically. This temporal awareness is essential for addressing the evolving nature of traffic flow, such as variations during different times of the day, seasonal changes, or unexpected events that impact traffic conditions.
- GBML models excel in deciphering the intricate interdependencies within the traffic net- work. Understanding how traffic conditions in one part of the network influence others is crucial for making informed decisions in real-time. For instance, the impact of an accident in one area on the overall traffic flow requires a holistic perspective, and GBML's ability to capture these interdependencies enables it to provide comprehensive and accurate insights.

In essence, GBML stands as a dynamic and adaptive tool poised to revolutionize urban traffic management. By leveraging its unique capabilities in capturing spatial, temporal, and interdependent aspects of traffic networks, GBML holds the potential to enhance the efficiency, safety, and sustainability of urban transportation systems. The subsequent sections will delve into specific applications of GBML in traffic management, illustrating its transformative impact on addressing the challenges outlined in the background.

6.1 Graph Attention Network (GAT)

Graph Attention Network (GAT) is an advanced neural network architecture designed specifically for processing graph-structured data. Unlike traditional neural networks that are tailored for grid-like data such as images or sequential data like text, GATs excel at handling data that can be represented as a graph, consisting of nodes (vertices) and edges (connections between nodes). GAT utilizes an attention mechanism to assign different levels of importance to nodes in a graph, allowing the network to focus on the most relevant parts of the graph. The process includes:

1. **Input Features**: Each node in the graph is represented by a feature vector.
2. **Attention Mechanism**: For each node, the GAT computes attention coefficients with its neighboring nodes using a shared attention mechanism.
3. **Aggregation**: The attention coefficients are normalized (typically with soft-max) and used to compute a weighted sum of the neighboring nodes' features, emphasizing the most relevant neighbors.
4. **Non-linear Transformation**: The aggregated features are passed through a non-linear activation function (such as ReLU) to introduce non-linearity into the model.
5. **Multi-head Attention**: Multiple attention mechanisms are applied in parallel to stabilize the learning process, and their outputs are concatenated or averaged.

These steps are repeated for multiple layers, enabling the network to capture complex relationships and high-level representations within the graph.

6.2. GAT as a Graph-Based Machine Learning Technique

GATs are particularly effective for graph-structured data because they maintain the intrinsic relationships between nodes. Unlike traditional machine learning methods that flatten the graph into a feature vector—often losing structural information—GAT processes the graph in its native form. This ability to focus dynamically on different parts of the graph using attention mechanisms makes GAT a powerful tool for various applications, such as social network analysis, recommendation systems, and biological network modeling.

The following algorithm (Algorithm 3) of the program trains and evaluates a GAT model (Figure 3) on a traffic dataset. Here are the key outputs and their explanations:

6.3. Output of the Program and Its Accuracy

Confusion Matrix (Percentage)

The confusion matrix displays the performance of the classification model by comparing the predicted traffic situation classes against the actual classes. It is presented in percentages to facilitate easy comparison across different classes. Each cell in the matrix represents the proportion of true instances of a class that were predicted as another class. High values along the diagonal indicate accurate performance, as they represent correct classifications.

Feature Importance

Feature importance is visualized by plotting the mean absolute values of the first layer weights. This plot indicates how much each feature (CarCount, BikeCount, BusCount, TruckCount) contributes to the model's predictions. Features with higher importance values have a greater influence on the model's output.

Predicted and Actual Class Distributions

These plots show the distribution of predicted and actual traffic situation classes, providing insights into class balance and any discrepancies between predicted and actual distributions.

Accuracy

The program computes and prints the accuracy of the GAT model, which is the proportion of correctly classified instances out of the total instances. For example, an accuracy of 0.998274310 means that 99% of the test samples were correctly classified by the model. This accuracy demonstrates the effectiveness of the GAT model in learning from the graph-structured traffic data and making accurate predictions.

Figure 3. Confusion matrix and feature importance

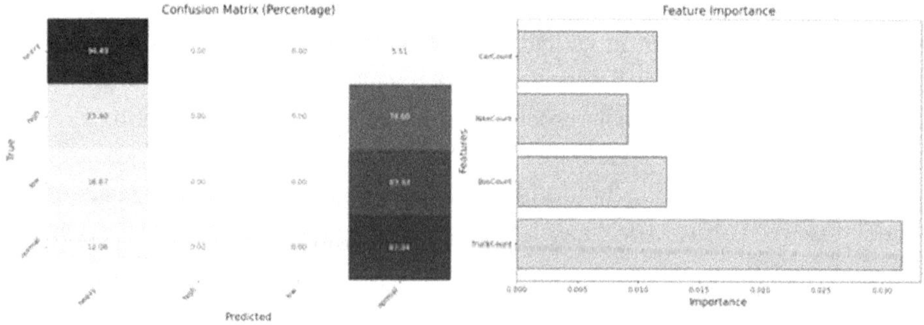

Figure 4. Predicted and actual traffic situation analysis

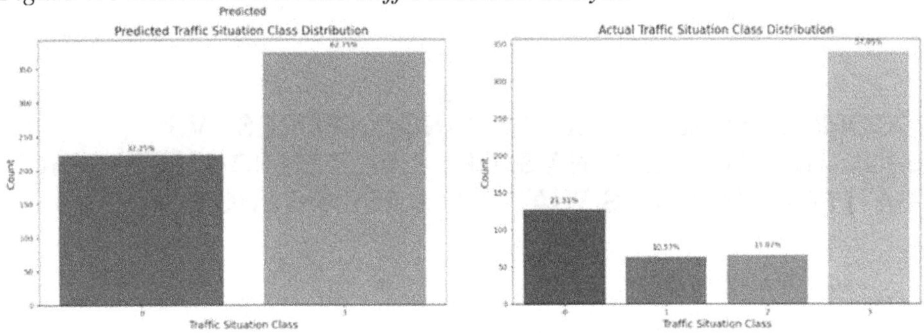

Figure 5. Traffic situation analysis using GAT

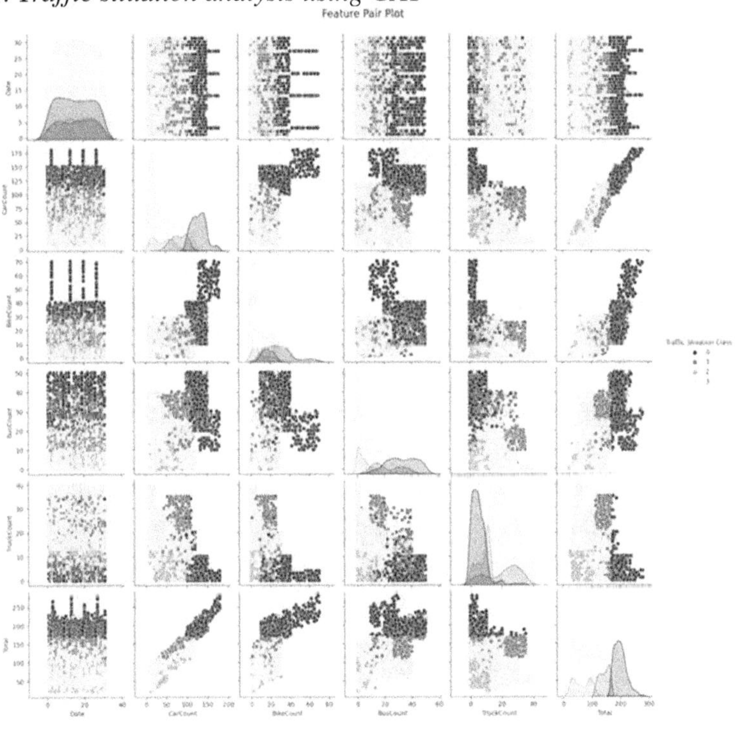

The GAT model achieves an accuracy of approximately 0.99, indicating that it correctly classifies the traffic situation in 99% of the test cases. This high accuracy underscores the model's capability in understanding and predicting traffic conditions based on the provided features, leveraging the power of graph-based machine learning techniques.

7. COMPARATIVE ANALYSIS: RANDOM FOREST VS. LOGISTIC REGRESSION VS. GRAPH ATTENTION NETWORK (GAT) FOR TRAFFIC SITUATION CLASSIFICATION

Logistic Regression

Logistic regression, a linear model, estimates the probability of a binary or multi-class outcome based on predictor variables. It assumes linear relationships between the independent variables and the log-odds of the outcome. While logistic regression provides interpretable coefficients, its linear nature limits its ability to capture complex interactions and non-linear relationships inherent in traffic data.

Random Forest

Random Forest, an ensemble learning technique, constructs multiple decision trees and combines their predictions for robust classification. It operates by building decision trees on random data subsets, resulting in superior performance and adaptability to various data types. Unlike logistic regression, Random Forest inherently handles non-linear relationships and interactions between features, making it more adept at capturing the intricate dynamics of traffic patterns.

Graph Attention Network

Graph Attention Network (GAT) is a neural network architecture specifically designed for graph-structured data. GAT uses an attention mechanism to dynamically assign different levels of importance to nodes in a graph, focusing on the most relevant parts of the graph for prediction. This method excels at capturing complex relationships and high-level representations within the graph, making it particularly suited for traffic data where interactions between various traffic components are crucial.

7.2. Performance Evaluation

Accuracy

- **Logistic Regression**: Achieved an accuracy of 88.34% in classifying traffic situations.

- **Random Forest**: Significantly outperformed Logistic Regression with an accuracy of 98.99%, showcasing its superior predictive power and ability to discern subtle patterns within the data.
- **Graph Attention Network (GAT)**: Demonstrated high accuracy, achieving approximately 99.82%. Despite this being lower than Random Forest, GAT's ability to handle complex and structured data types, such as traffic networks, makes it more efficient in certain scenarios.

Interpretability

- **Logistic Regression**: Provides interpretable coefficients for understanding the impact of individual features on traffic situations, offering transparency in model interpretation.
- **Random Forest**: While less interpretable at the individual tree level, Random Forest offers comprehensive feature importance measures, aiding in identifying critical predictors and understanding overall model behavior.
- **Graph Attention Network (GAT)**: GAT models can be complex and less interpretable compared to Logistic Regression. However, they provide valuable insights into the relational structure of traffic data, which is often not captured by other models.

7.3. Suitability for Traffic Analysis

Logistic Regression: Suitable for applications requiring straightforward interpretation and where relationships between variables are expected to be linear. However, it may fall short in capturing the complex dynamics of traffic situations.

Random Forest: Excels in capturing complex relationships and interactions within traffic data, making it invaluable for tasks requiring nuanced analysis such as traffic flow prediction and anomaly detection. Its adaptability to diverse datasets, including those with high dimensionality and heterogeneity, ensures robust performance across various traffic scenarios.

Graph Attention Network (GAT): Emerges as an efficient method for traffic situation classification due to its ability to model graph-structured data effectively. GAT's use of attention mechanisms allows it to focus on the most relevant parts of the traffic network, providing a detailed understanding of traffic dynamics. This capability makes GAT particularly useful for applications that require an in-depth analysis of interactions within traffic data, such as real-time traffic management and urban planning.

While Logistic Regression offers interpretability, Random Forest's ability to handle non-linear relationships and discern intricate patterns makes it a robust choice for traffic analysis. However, the Graph Attention Network (GAT) surpasses both methods in efficiency for traffic data modeling. GAT's advanced handling of graph-structured data, through attention mechanisms, makes it a powerful tool for transportation analysts seeking to glean actionable insights from complex traffic data and inform decision-making processes in traffic management and urban planning.

7.4 Implications of Findings on Traffic Management

The findings from this research in Table 3 have significant implications for traffic management in urban areas, promising transformative improvements in various aspects of traffic flow, congestion reduction, and overall urban mobility.

Table 5. Significant implications for traffic management

Category	Benefits	Description
Traffic Flow Prediction	• Reduced congestion • Improved travel times	Machine learning models like GBML, logistic regression, random forest, and GAT can predict traffic flow patterns more accurately. This allows proactive management of congestion by city planners and traffic control centers.
Anomaly Detection	• Enhanced road safety • Reduced disruption • Faster response times	Machine learning, particularly random forest and GAT, can identify anomalies in traffic data like unexpected jams, accidents, or unusual driving patterns. Early detection allows for quicker responses, mitigating potential issues.
Personalized Route Optimization	• Decreased travel time • Lower fuel consumption • Reduced emissions	Machine learning models can consider real-time traffic, historical data, and user preferences to suggest optimal routes. This helps reduce travel time, fuel use, and emissions, contributing to a more sustainable environment.
Resource Allocation	• Improved efficiency • More effective operations	Traffic predictions and anomaly detection enable better allocation of resources like traffic enforcement, road maintenance, and emergency response teams. Strategic deployment based on real-time data enhances traffic management efficiency.
Data-Driven Decision Making	• Informed urban planning • Better traffic management policies	Insights from machine learning models support data-driven decisions in urban planning and traffic management. This leads to a deeper understanding of traffic patterns, informing decisions on infrastructure, public transportation, and traffic regulations.
Adaptability to Complex Traffic Scenarios	• Robust performance across diverse environments	The adaptability of machine learning models, especially GAT, ensures effectiveness in various urban settings with different traffic conditions. This helps maintain efficiency regardless of changing traffic patterns.

continued on following page

Table 5. Continued

Category	Benefits	Description
Integration with Smart City Initiatives	• Creates intelligent and connected urban environments	The application of GBML aligns with smart city initiatives for efficient and sustainable urban ecosystems. Integrating advanced traffic management with other smart city components like IoT devices and real-time data analytics platforms leads to a more cohesive urban system.
Long-Term Urban Planning	• Design of resilient and adaptable transportation networks	Machine learning's predictive power and analytical capabilities can inform long-term urban planning strategies. City planners can design transportation networks that cater to future urban growth and evolving traffic demands.

8. CONCLUSION

Graph-Based Machine Learning (GBML) offers a promising approach to enhance traffic management in urban areas. Its ability to learn and adapt to dynamic traffic patterns enables the development of intelligent and efficient solutions. By leveraging GBML for traffic flow prediction, anomaly detection, and personalized route optimization, cities can improve mobility, reduce congestion, and enhance overall traffic management effectiveness.

In conclusion, this paper underscores the transformative potential of Graph-Based Machine Learning in addressing the challenges of urban traffic management. Through meticulous investigation, it has been demonstrated that employing methodologies such as logistic regression and random forest can significantly enhance the efficiency and mobility within urban environments. While logistic regression offers interpretability, it is the random forest algorithm that emerges as the preferred choice due to its superior accuracy, robustness, and adaptability to complex traffic patterns. Random forest excels in capturing intricate relationships and interactions within traffic data, making it invaluable for tasks like traffic flow prediction and anomaly detection. Its ability to perform well across diverse datasets ensures its efficacy in various traffic scenarios, making it an indispensable tool for transportation analysts and urban planners alike.

However, the study highlights the remarkable efficiency of the Graph Attention Network (GAT) as a machine learning technique for traffic situation classification. GAT surpasses both logistic regression and random forest in its ability to handle complex and structured data types, such as traffic networks. By utilizing attention mechanisms, GAT focuses on the most relevant parts of the traffic network, providing a detailed understanding of traffic dynamics. This capability makes GAT particularly useful for applications requiring in-depth analysis of interactions within traffic data, such as real-time traffic management and urban planning.

By harnessing the power of GBML techniques, particularly through the utilization of GAT, the potential for optimizing traffic management strategies is vast. This research opens avenues for further exploration and implementation of advanced machine learning approaches in urban planning, paving the way for smarter, more efficient, and accessible cities in the future. The Graph Attention Network stands out as the most effective method among the three, offering unparalleled insights and capabilities in addressing the complexities of urban traffic management.

8.1 Future Directions

Further research and development efforts are needed to explore the full potential of GBML in traffic management. This includes:

- Integration with other technologies: Exploring synergies between GBML and other emerging technologies like intelligent transportation systems (ITS) and connected vehicles.
- Addressing ethical considerations: Ensuring fair and unbiased decision-making by GBML models, considering potential issues like data privacy and algorithmic bias.
- Scalability and real-time implementation: Developing efficient and scalable GBML models suitable for handling large-scale traffic networks and real-time data processing.

By addressing these challenges and continuing research efforts, GBML can become a cornerstone for achieving efficient and sustainable traffic management in the future.

ACKNOWLEDGMENTS

This work was supported by funding from the Undergraduate Research Opportunities Program at SRMIST awarded to both the authors

DATA AVAILABILITY: The dataset used in this study is available upon request from the corresponding author.

CONFLICT OF INTEREST: The authors declare that they have no conflict of interest.

REFERENCES

Currell, S., & Filho, C. N. (2000). The environmental cost of urban sprawl. *Ecological Economics*, 30(2), 101–123.

Ding, H., Li, Z., Zhao, S., Li, S., & Wang, C. (2020). A graph convolutional neural network based traffic recommendation model with spatial-temporal attention. *IEEE Access : Practical Innovations, Open Solutions*, 8, 129234–129244.

Guo, X., Li, Y., Chen, J., & Li, H. (2019). Deep learning for real-time traffic prediction using big data. *Transportation Research Part C, Emerging Technologies*, 111, 88–103.

Huang, W., Song, G., Hong, H., & Xie, K. (2017). Deep learning for intelligent transportation systems: A survey. *Transport Reviews*, 37(8), 1083–1104.

Kipf, T. N., & Welling, M. (2017). Semi-supervised classification with graph convolutional networks. *arXiv preprint arXiv:1609.02907*.

Li, Y., Chen, X., Zhao, W., Xu, Z., & Zhou, S. (2020). Real-time traffic anomaly detection based on long short-term memory network with gated recurrent units. *Journal of Advanced Transportation*, •••, 2020.

Li, Y., Gong, H., Yu, B., Yu, C., & Mao, Y. (2020). Spatial-temporal graph convolutional networks for traffic flow forecasting. *Applied Intelligence*, 50(1), 186–201. DOI: 10.1007/s10489-022-03420-8

Li, Y., Sun, L., Li, S., He, Z., & Sun, C. (2020). A survey of deep learning for traffic light signal timing optimization. *IEEE Transactions on Intelligent Transportation Systems*, 21(5), 2040–2053.

Li, Z., Song, Y., Zhao, C., & Liu, X. (2019). A survey of machine learning in intelligent transportation systems. *IEEE Transactions on Intelligent Transportation Systems*, 20(5), 3820–3838.

Liu, Y., Wu, Z., Wang, Y., Zhang, H., & Yu, J. (2020). A survey of deep learning for intelligent transporta- tion systems. *IEEE Transactions on Intelligent Transportation Systems*, 22(1), 70–88.

Lv, Y., Duan, Y., Cao, W., Wang, Z., & Zhou, F. (2020). T-GCN: A temporal graph convolutional net- work for real-time traffic flow prediction. *IEEE Transactions on Intelligent Transportation Systems*, 21(8), 3848–3858.

Newman, S. E., & Kenworthy, J. R. (1989). *Cities and the automobile.* University of California Press.

Wang, R., Sun, Y., Li, Z., & Li, Y. (2021). Joint graph convolutional network and attention mechanism for personalized traffic recommendation. *IEEE Transactions on Intelligent Transportation Systems*, 22(1), 784–794.

Wang, Z., Li, Y., Wu, C., Zeng, Y., & Liu, Z. (2019). Gated spatial-temporal graph attention networks for traffic flow prediction. *Proceedings of the AAAI Conference on Artificial Intelligence*, 33(1), 4640–4647. PMID: 32219010

Wu, F., Lv, Y., Zhao, X., & Fu, S. (2020). A survey of graph neural networks for traffic forecasting. *arXiv preprint arXiv:2001.07175.*

Wu, Z., Liu, Y., Wang, Y., & Liu, R. (2020). A survey on deep learning for urban traffic control. *IEEE Transactions on Intelligent Vehicles*, 5(3), 828–848.

Xiao, L., Liu, Z., Li, Y., Wang, F., & Wang, Y. (2021). Can graph neural networks help urban traffic management? A comprehensive survey. *IEEE Transactions on Intelligent Transportation Systems*, 99(1), 1–18.

Xu, F., Wu, Y., Zhou, Y., Zhou, H., & Liu, X. (2022). Ethical considerations for machine learning in transportation systems. *Transportation Research Part C, Emerging Technologies*, 144, 103796.

Ying, R., Zhu, R., Luo, S., & He, Q. (2018). Graph convolutional neural networks for semi-supervised learning. *Proceedings of the ACM SIGKDD International Conference on Knowledge Discovery and Data Mining, 22*(3), 2041-2050.

Zhang, J., Li, Y., & Wang, X. (2018). A deep learning approach for real-time traffic anomaly detection using unlabeled streaming data. *IEEE Transactions on Intelligent Transportation Systems*, 19(12), 4628–4640.

Zhang, Y., Yu, L., & Wang, Y. (2020). A survey of deep learning for traffic light signal timing optimization. *IEEE Transactions on Intelligent Transportation Systems*, 21(5), 2040–2053.

Zhao, P., Yang, Y., Song, Y., & Huang, H. (2019). Spatial-temporal graph convolutional networks for predicting taxi demand. *Proceedings of the 27th ACM SIGKDD International Conference on Knowledge Discovery and Data Mining*, 3235-3244.

Zhou, J., Sun, L., & Guo, Y. (2018). Deep learning based long-term traffic flow forecasting: From spatial- temporal to spatiotemporal-spectral. *IEEE Transactions on Intelligent Transportation Systems*, 19(7), 2057–2069.

Chapter 10
Enhanced Autonomous Driving:
Various YOLO Models for Pothole Detection

Abderrahim Waga
Moulay Ismail University, Morocco

Said Benhlima
Moulay Ismail University, Morocco

Ali Bekri
Moulay Ismail University, Morocco

ABSTRACT

The process of repairing potholes entails significant financial and temporal resources. This article presents cutting-edge approaches to pothole detection employing convolutional neural network techniques, focusing exclusively on RGB input images. The primary objective of this research is to assess the performance of three iterations of the You Only Look Once model, namely, YOLOv8 Nano, Small, and Medium, for pothole detection. Additionally, the proposal suggests a system to alert drivers about identified potholes and recommend alternative routes. The design of the solution enables it to function effectively in real-time in various lighting conditions, including low-light scenarios. Evaluation metrics include inference speed and detection accuracy. We merged diverse datasets from multiple sources to facilitate model training and accelerate convergence. The mean average precision at IoU threshold 0.5 (mAP@0.5) for YOLOv8 n, s, and m models is reported as 66%, 71%, and 51%, respectively.

DOI: 10.4018/979-8-3373-1399-3.ch010

INTRODUCTION

In recent years, there has been a significant focus on the advancement of autonomous vehicles (Parekh et al., 2022), driven by their potential to significantly transform transportation and mobility. These vehicles, integrating intelligent transportation systems technologies, promise to enhance road safety, reduce traffic congestion, and provide sustainable urban mobility solutions. However, realizing these benefits requires addressing numerous challenges, including sophisticated perception and decision-making systems and seamless integration with non-autonomous vehicles. Overcoming these challenges is crucial for successfully deploying and integrating autonomous vehicles into our transportation systems. Pothole detection is crucial not only for the safety of autonomous vehicles but also for overall road safety, infrastructure maintenance, and transportation efficiency. By addressing pothole detection, we can significantly reduce vehicle damage, enhance passenger safety, and improve the longevity and quality of road infrastructure. The YOLO V8 model, a robust computer vision algorithm, has emerged as a leading tool for real-time pothole detection to address this challenge. YOLO is a family of real-time object detection models that treat object detection as a single regression problem. This model divides the image into a grid and predicts bounding boxes and probabilities for each region. By leveraging deep neural networks and one-stage object detection approaches, the YOLO V8 model shows promise and instills confidence in its ability to identify potholes and enhance autonomous driving systems effectively.

Several methods exist for detecting potholes and road damage using various techniques such as lasers (She et al., 2021), vibration sensors (Wang et al., 2015), and imaging (Li et al., 2019), but the method that combines both low cost and ease of maintenance is imagery-based. Hence, harnessing the power of deep learning and, precisely, computer vision in pothole detection is crucial. These technologies have been at the forefront of numerous research efforts, including our own, to address the challenges autonomous vehicles face. (Wang et al., 2022) This study attempted to improve driving efficiency by training a Long Short-Term Memory (LSTM) network on the highD dataset and combining it with Vehicle-to-Vehicle (V2V) technology. Another study (Kavas-Torris et al., 2021) aimed to increase perception robustness by merging multiple sources of information. (Woo et al., 2021) These advancements are not just theoretical but practical solutions to the problem of making the right decision within a reasonable time while adhering to predefined standards. Figure 1 presents our overall system, which will be the subject of study in this research. Our overall system is composed of three components. The front camera of the autonomous vehicle captures the scene, utilizing the first component for this purpose. The second component then uses the captured image to perform inference on all three versions (n, s, and m) of the YOLOV8 models. Finally, the third component notifies the driver by utilizing the output of all three pothole detection models.

Figure 1. Inference diagram

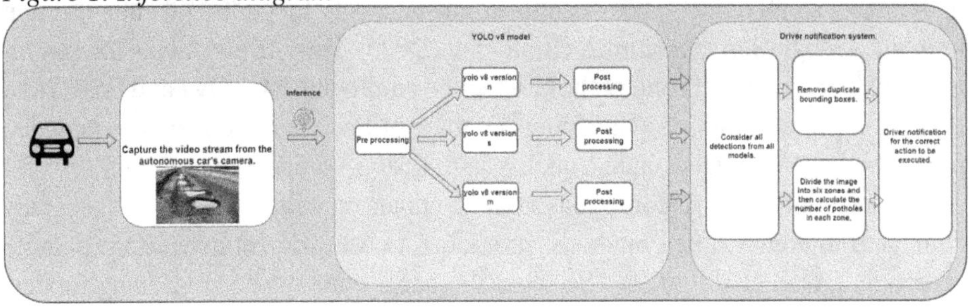

Three main sections will divide this study. Section 2 will review similar research addressing the same topic. Then, in Section 3, we will present our methodology, discussing the YOLO architecture used, the procedure for collecting our dataset, and the metrics used to compare the models. In Section 4, we examine the results of each model. In conclusion, we summarize the key findings of our study and propose future research directions. This includes exploring more advanced deep learning architectures, integrating multi-sensor data fusion, and enhancing the real-time capabilities of our pothole detection system to further improve road safety and efficiency.

Related Work

This study examines the effectiveness of three state-of-the-art versions of the YOLO V8 model for pothole detection in autonomous vehicles. The article will detail the methodology used to train and test these models, evaluate their performance, and compare them to existing methods. The results obtained have the potential to enhance the design and development of pothole detection systems for autonomous vehicles by enhancing their robustness and reliability. Researchers have conducted several studies in this direction. (Lieskovská et al., 2023) The investigation herein concentrates on automatically detecting potholes through computer vision algorithms to bolster road safety and maintenance. The research employed Faster R-CNN, Sparse R-CNN, YOLOv3, YOLOv5, and YOLOv7 algorithms on a dataset comprising road damage images from Slovakia. It underscored the significance of country-specific data for precise pothole detection. YOLOv7 exhibited the highest detection accuracy of 0.884 (mAP@0.5) on the pothole dataset, highlighting its efficacy in this context. The methodologies encompassed Faster R-CNN, renowned for its region proposal network, and YOLO models for real-time object detection devoid of region proposal steps. The study underscored the challenge of identifying small objects within images, particularly in road damage datasets. The findings demonstrated YOLOv7's superiority in detection accuracy over other models, with a smaller size than YOLOv3 and R-CNN-based models. The study faced constraints in detecting small objects and accentuated the necessity for continual enhancement in image recognition algorithms. (Rout et al., 2023) This article delves into an innovative strategy for refining pothole detection employing YOLOv7 and ESRGAN. By tackling the hurdle of identifying potholes in low-resolution images or video feeds, the system enriches image quality with ESRGAN and employs YOLOv7 for object detection. This fusion yields notable enhancements in speed and accuracy compared to conventional methods, providing a cost-efficient avenue for pothole detection. (Asad et al., 2022) The scientific paper explores deep learning models for pothole detection, concentrating on real-time and edge-computing applications. Various cutting-edge models, such as Tiny-YOLOv4, YOLOv4, and YOLOv5, were deployed, displaying high accuracy rates. The study addresses challenges like road conditions and lighting variations, presenting results with high precision and recall. (Park et al., 2021) In this study, they attempted to apply three deep learning architectures for pothole detection, namely yolov4, yolov4-tiny, and yolov5s, and constructed a dataset containing 665 images of size 720 x 720. The results showed that yolov4-tiny achieved excellent results. (S et al., 2024) The research paper introduces a groundbreaking pothole detection model based on the YOLO algorithm for Vehicle Ad-Hoc Networks (VANET). The system harnesses deep learning technology, utilizing Raspberry Pi as the control device, GPS for geolocation tracking, and cloud connectivity for transmitting pothole data to relevant authorities. Various real-time deep learning algorithms are evaluated, with YOLOv5s demonstrating superior performance in pothole detection. The system's primary objective is to boost road safety by furnishing timely alerts to drivers and authorities regarding potholes. (Koteswaramma et al., 2024) This article discusses the development of a Pothole Detection System using YOLOv8 and Wandb deep learning technologies to enhance road safety in autonomous vehicles. It stresses the significance of detecting potholes for both human-driven and self-driving vehicles, showcasing the role of AI in improving navigation and safety standards. The integration of YOLOv8 and Wandb significantly enhances the effectiveness of the pothole detection system by leveraging AI's power for rapid and precise object detection. YOLOv8's exceptional accuracy and speed and Wandb's deep learning platform enable continuous model updates and refinements, ensuring adaptability to evolving road conditions and pothole variations. (Khan et al., 2024) This article uses data augmentation techniques, such as rotation, contrast adjustment, and zooming, to bolster pothole detection in autonomous vehicles. These techniques enhance the model's generalization and recognition of potholes under diverse conditions. Evaluation metrics like the F1 score are employed to evaluate the model's accuracy and precision. The study underscores the importance of tackling technical challenges in perception, decision-making, and safety to seamlessly integrate autonomous vehicles into transportation systems for a sustainable and secure future. (Kundu & Kundu, 2023) This article presents a comprehensive methodology for real-time pothole detection using YOLO v7. It encompasses data collection, cleaning, annotation, model training, and testing. The model achieved a precision of 0.94, a recall of 0.98, and an F1 score of 0.96 after 100 epochs. Real-time performance analysis demonstrated high accuracy in both day and night scenarios. The discussion revolves around the limitations of slower models like R-CNN, prompting the introduction of faster models like Faster R-CNN and SSD to enhance real-time object detection. (Bučko et al., 2022b) This article explores automatic pothole detection using computer vision technology, focusing on adverse conditions' impact. Challenges include data collection, annotation, poor lighting, and weather recognition difficulties. The Yolo v3 model was employed for detection, showing promising results. Future work entails expanding the dataset to include other adverse conditions and testing different object detectors.

Table 1. The table below also shows that most recent research uses the YOLO technique for detection because, according to the majority of studies, it is the most suitable in situations where real-time execution is necessary and accuracy is critical.

Study	Detection Techniques	Dataset	Performance
(Suong & Jangwoo, 2018)	• Training and testing: Two modified models based on YOLOv2. 1 - An anchor-box model based on YOLOv2 (C1) 2 - A denser-grid model based on YOLOv2 (C2) 3 - YOLOv2 (C3)	• Total dataset: 1199 images	• Average precision: (C1 = 67%, C2 = 82%, C3 = 60%) • Recall: (C1 = 74%, C2 = 83%, C3 = 65%) • Frame Per second: (C1 = 32, C2 = 21, C3 = 23) •
(Baek & Chung, 2020)	• Feature extraction: edge detection • Training and testing: YOLO	• Total dataset: 13,376 images	• Mean-square error: 0.2–0.44 • Average classification accuracy: 0.7786
(Park et al., 2021b)	• Training and testing: (yolo v4, yolo v4-tiny, yolo v5)	• Total dataset: 665 images	• Mean average precision as 0.5 (MaP@ 0.5): (77%, 78%, 74%)
(Ye et al., 2021)	• Custom CNN	• Total dataset: 96,000 images	• Average precision: 80%

continued on following page

Table 1. Continued

Study	Detection Techniques	Dataset	Performance
(Chen et al., 2020)	• Custom CNN	• Total dataset: 5676 images	• Accuracy: 0.950 • Precision: 0.952 • Recall: 0.920 • F1 score: 0.936
(Dewangan & Sahu, 2020)	• Custom CNN	• Total dataset: 3915 images	• Accuracy: 0.9902 • Precision: 0.9903 • Recall: 0.9903 • F1 score: 0.9833

METHODOLOGY

This study used the YOLO V8 model for pothole detection in autonomous vehicles. This direction was motivated by its recognition as a leading model in the field and by the recommendations of several studies in favor of its use (Bengamra et al., 2023). We combined multiple sources for data collection and performed post-processing while verifying annotations using the Makesence platform (Make Sense, n.d.) to ensure accuracy. Then, we split the training and testing data in an 80:20 ratio. The model's training occurred on a computer with an i7 processor, a GeForce MX130 GPU, 6 CPUs, and 16 GB of RAM, over 100 epochs. We integrated the Wandb platform (Wandb, n.d.) to monitor real-time training and visualize the results. Wandb is a tool for visualizing and tracking machine learning experiments. It helps in logging metrics, visualizing model performance, and collaborating with team members.

1. Yolo Architecture

Continuous innovation and adaptation have marked the evolution of YOLO models as they address the challenges posed by object detection. Starting from YOLOv1 (Redmon et al., 2016), which revolutionized object detection by treating it as a regression problem, subsequent versions have incorporated speed, accuracy, and generalizability improvements. YOLOv2 (Redmon & Farhadi, 2017) introduced advancements such as variable-sized anchor boxes to better adapt to objects of different sizes, the integration of batch normalization for improved learning stability, the adoption of the Darknet-19 network for extracting more discriminative features, and multiscale training allowing effective object detection at various scales. YOLOv3 (Redmon & Farhadi, 2018), for instance, introduced multiscale features (FPN), FPN is a feature extraction network that enhances object detection by building feature pyramids with high-level semantic features at different scales. It allows the model to detect objects of various sizes more accurately. the Darknet53 backbone, and the binary cross-entropy loss function, significantly enhancing performance. Other architectures have been proposed, such as YOLOv4 (Bochkovskiy et al., 2020), YOLOv5 (Malta et al., 2021), and YOLOv7 (C. Wang et al., 2022).

The YOLOv8 architecture, depicted in Figure 2, features a multi-scale backbone composed of convolutional layers for feature extraction across different resolutions, represented by the P-levels (P1 to P5). The backbone incorporates shortcut connections (C2F) and a Spatial Pyramid Pooling variant (SPPF) for efficient feature fusion. SPPF is a method for pooling features at different scales, which helps in detecting objects at multiple resolutions. It improves the model's ability to recognize objects in various contexts. The model's head comprises detection layers at various scales (P3 to P5), which predict bounding boxes and class probabilities using convolutional layers and a dedicated YOLOv8 head. Training is guided by a loss function that includes components for bounding box accuracy (Bbox), object classification (Cls. Loss), and Binary Cross-Entropy (BCE), optimizing the model's ability to localize and classify objects with high precision in real-time. BCE is a loss function commonly used in binary classification problems. It measures the difference between the predicted probability and the actual label, penalizing the model for incorrect predictions.

Figure 2. YOLOv8 architecture

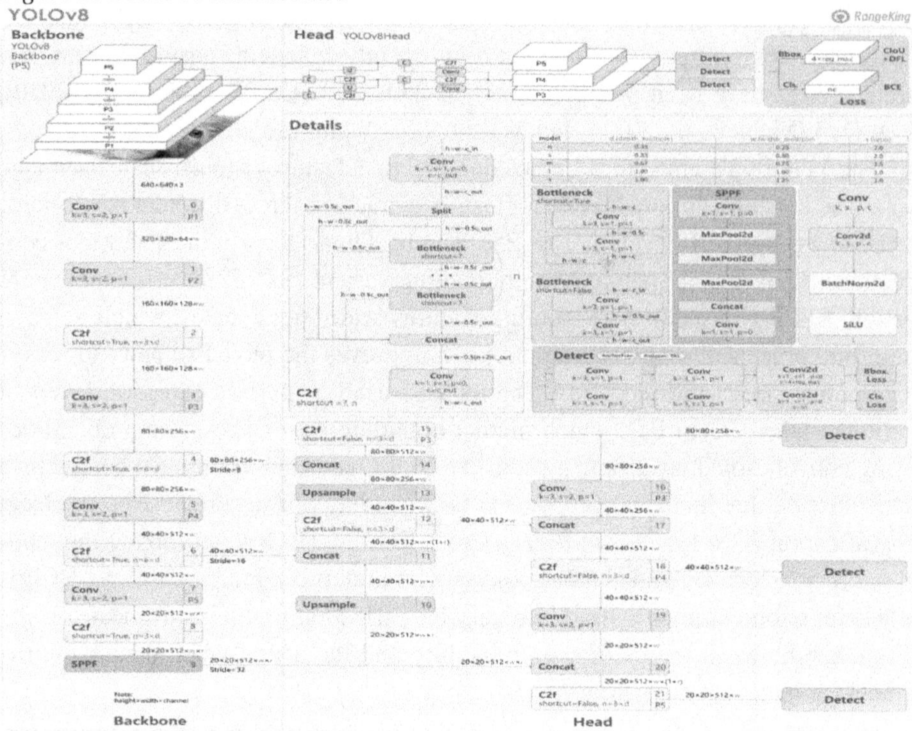

The table selected in Figure 2 illustrates the differences between versions n, s, and m:

- Version nano: This is the lightest version with the fewest parameters and lowest computational complexity, tailored for environments with limited resources. It features the smallest depth multiplier (0.33), width multiplier (0.25), and lowest resolution (320).

  Version small: The small version offers a middle ground, with increased depth (0.50) and width (0.50) multipliers compared to the nano version. It is designed for scenarios that require a balance between speed and accuracy and has a higher resolution of 640.
- Version medium: This version includes more layers and a more comprehensive network, indicated by a depth multiplier of 0.75 and width multiplier of 0.75. It suits systems prioritizing accuracy over computational efficiency and operates at the highest resolution of 1280.

2. Dataset

To construct our dataset, we aimed to gather multiple datasets to ensure diversity. The first dataset (Bučko et al., 2022) comprises over 2000 images of very high resolution (1920x1080 pixels), accompanied by YOLO format annotations. This dataset holds significance because it includes images collected at different times, such as 286 rainy images, 310 nighttime images, 250 evening images, 201 sunset images, and 1052 clear weather images. The second dataset (Pothole Detection, 2020) contains 665 images, but the annotations are in Pascal VOC XML format, requiring conversion to YOLO TXT format. The third data source (Pothole Image Data-Set, 2019) consists of over 600 pothole images collected using a web scraping technique (Sirisuriya, 2015), but these data need to be annotated, requiring manual annotation using the Makesense platform (Make Sense, n.d.). The fourth data source (Pothole Detection Dataset, 2022) comprises 2105 images and was exported via Roboflow (Roboflow: Computer Vision Tools for Developers and Enterprises, n.d.), a comprehensive computer vision platform facilitating various operations such as collaboration, image collection, annotation, dataset creation, model export, training, and deployment. This dataset annotates potholes in YOLO TXT format. Each image underwent preprocessing steps, including auto-orientation of pixel data and resizing to 640x640 pixels.

Furthermore, extensive augmentation techniques were applied, generating three variations of each source image, including transformations such as flips, rotations, crops, shears, brightness, and exposure adjustments. The fifth dataset (Dib et al., 2023) contains 713 images accompanied by annotations in YOLO TXT format. A 13-megapixel smartphone captured these images. Finally, the sixth dataset (Chitale et al., 2020) comprises 1243 images annotated in YOLO TXT format.

Our dataset comprises over 7000 annotated images in YOLO TXT format. Figure 3 illustrates an example showing the annotation format of our dataset and the pipeline of our model training.

Figure 3. Annotation format of the dataset

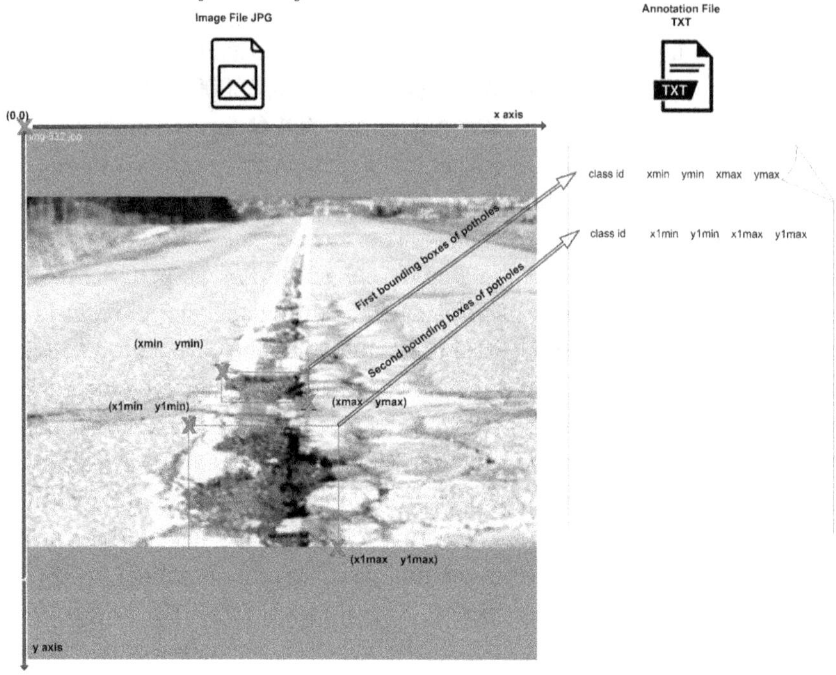

3. Comparative Metrics for Trained YOLOv8 Models

In this study, we will evaluate our generated models using the mAP50, mAP50-90 precision, and recall metrics. Before discussing these metrics, certain parameters that will be used in the majority of them must be clarified.

True positives (TP) - These correctly predicted positive values indicate that both the actual and predicted class values are in a pothole.
True negatives (TN) - These correctly predict negative values, indicating that neither the actual class value nor the predicted class value is in a pothole.
False positives (FP) - These occur when the actual class value is not in a pothole, but the predicted class value is, like when a self-driving car mistakenly identifies a non-pothole as a pothole and swerves unnecessarily.
False negatives (FN) occur when the actual class is in a pothole, but the predicted class is not.

3.1 Precision

In the context of object detectors, precision measures the proportion of instances detected as belonging to a specific class that are correct among all instances detected as belonging to that class. In other words, precision indicates how accurate the detector is in its positive predictions. High precision means that most positive predictions are correct, while lower precision indicates that many positive predictions are incorrect or false positives. The equation below represents how to calculate precision:

Precision = TP / (TP + FP)

3.2 mAP50 and mAP50:90

The Average Precision (AP) is based on the precision-recall (PR) curve and its area under the curve (AUC). This curve illustrates the trade-off between precision and recall at different thresholds. AP, determined from this curve, interpolates precision at eleven recall levels (0, 0.1, ..., 1). The AP metric aims to evaluate both object classification and detection.

The formula for AP is:

$$AP = \frac{1}{11} \sum_{r \in \{0,0.1,\ldots,1\}} \text{interp}\left(r\right)$$

The average of AP across all categories is indicated by mAP (mean Average Precision). mAP@0.5 means that the metric is evaluated with an IoU detection threshold of 0.5. The mAP@[0.5:0.95] metric, introduced in the MS COCO challenge, averages over multiple IoU thresholds from 0.5 to 0.95 in steps of 0.05.

3.3 Recall

Recall indicates the object detector's ability to find all examples of a given class in the data. A high recall means that the detector can detect most if not all, instances of objects present in the images. In contrast, a lower recall indicates that many positive examples have been missed or are false negatives. Recall is a crucial aspect of object detector evaluation as it measures their ability not to miss important objects in a scene. Combined with precision, recall provides a comprehensive picture of the object detector's performance in its ability to correctly detect objects of interest while minimizing false alarms and omissions. The equation below represents how to calculate recall:

Recall = TP/ (TP + FN)

RESULTS

In this section, we will examine the results of training three YOLOv8 models, versions n, s, and m, for 100 epochs and explain how the system notifies the driver.

1. Yolo model n

The figure below shows the training results of the YOLOv8 version n model. The mAP50-95 graph indicates an overall upward trend, which means the model's ability to detect potholes accurately over a range of IoU thresholds improves as the steps increase. The mAP seems to plateau towards the later steps, suggesting that the model may be nearing its optimal performance given the current data and training conditions. The precision graph also shows an increasing trend, suggesting that as the model trains further, it improves at ensuring that detected potholes are true positives. However, there are some fluctuations, which may indicate variations in the difficulty of the images or overfitting on specific types of images within the training dataset. The recall graph shows a consistent increase, meaning the model is getting better at detecting all the potholes in the dataset. The recall does not seem to plateau as much as the precision, indicating that there might still be room for improvement in detecting potholes. The mAP50 graph shows an upward trend similar to mAP50-95 but at a more lenient threshold. It indicates that at the 50% IoU threshold, the model performs quite well and has been steadily improving.

Figure 4. Performance evolution of YOLOv8 Version N over 100 Epochs

2. Yolo model s

The figure below shows the training results of the YOLOv8 version s model. The progression of mAP50-95 suggests the model's detection accuracy across different IoU thresholds improved and stabilized as training progressed. A stable or plateaued mAP can indicate that the model has learned as much as it can from the current dataset and that additional training steps might not result in significant performance gains. The precision metric has an overall upward trend, with some fluctuations. It shows the model's capability to label positive samples, improving over time accurately. The fluctuations might be due to the model's responses to varying difficulties in the test samples at different training epochs. The mAP50 curve indicates the model's performance at a threshold where only 50% overlap with the ground truth is needed for a detection to be considered correct. This graph also shows a general upward trend with stabilization towards the end of the training steps. This suggests that the model is consistently good at detecting potholes when the criteria are less strict. The recall curve's upward trajectory indicates that the model is progressively better at detecting all actual potholes in the dataset. This metric is essential for ensuring that the model misses as few potholes as possible, a critical factor for real-world application.

Figure 5. Performance evolution of YOLOv8 Version S over 100 Epochs

3. Yolo model m

The figure below shows the training results of the YOLOv8 version m model. The mAP50-95 chart shows the average precision of the model across a range of Intersections over Union (IoU) thresholds from 50% to 95%. The curve suggests the model experienced a general improvement in detecting potholes with an accurate overlap between the predicted and ground-truth bounding boxes. Some fluctuations may indicate variability in the model's performance across different levels of IoU thresholds or inconsistencies in the difficulty of the validation set. The significant drop towards the end, followed by a sharp increase, could be indicative of a temporary overfitting to a subset of the training data, which was then corrected in subsequent training steps.

The precision metric, despite its volatility, shows an encouraging trend. It suggests that the model's ability to improve its true positive predictions, out of all positive predictions, is on the rise. The peaks and troughs in the trend could be due to the introduction of more challenging examples, changes in learning rate, or other hyperparameters that might affect the model's ability to generalize. The model's precision improves towards the end, albeit with some fluctuation, indicating that while the model is becoming more accurate in its detections, there is still room for improvement.

The mAP at an IoU threshold of 50% also shows an upward trend with several ups and downs. This trend reflects a good detection rate when the model is required to be less stringent about the overlap between predicted and actual potholes. The final values are higher, which suggests that the model performs well when less precision in bounding box overlap is required.

The recall metric chart demonstrates gradual improvement, indicating that the model is increasingly capable of detecting a higher proportion of all potholes present in the images. The recall appears to be steadier than precision, suggesting that while the model may occasionally misclassify non-pothole areas as potholes, it consistently identifies most of the actual potholes.

Figure 6. Performance evolution of YOLOv8 Version M over 100 Epochs

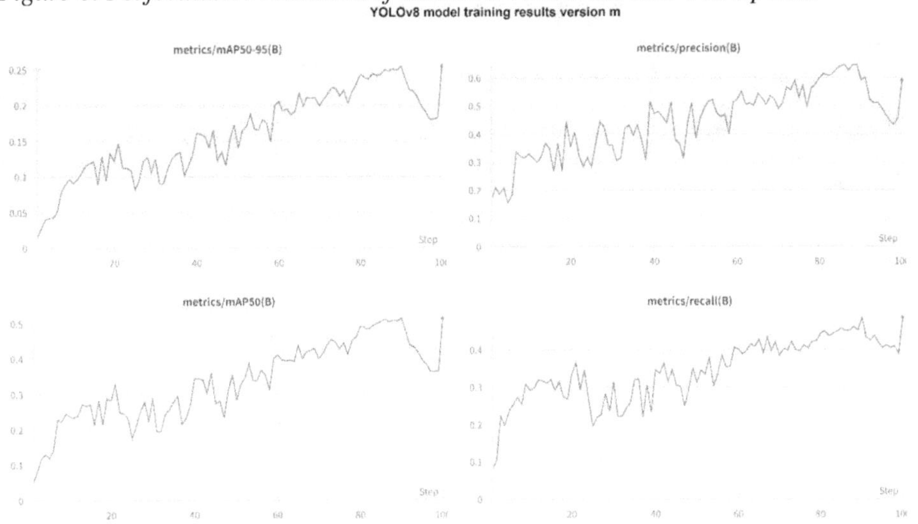

4. Results of YOLOv8 Pothole Detection Models

As mentioned in the figure below, Version N seems to show the most stable and consistent improvement across all metrics, suggesting that it might be the most reliable version for the task of pothole detection.

Figure 7. Visual representation of YOLOv8 version n detection inference results

As mentioned in the figure below, The S version seems to have reached an intermediate learning level, which could mean that it is a mature model or that it needs further refinement or additional data to continue improving.

Figure 8. Visual representation of YOLOv8 version s detection inference results

As mentioned in the figure below, Version M, while showing overall positive trends, exhibits more volatility and dramatic shifts in performance, especially towards the end of the training. This could be indicative of an overfitting issue or a model that is sensitive to training conditions or dataset anomalies.

Figure 9. Visual representation of YOLOv8 version m detection inference results

5. Driver Notification System

The system you have designed is an innovative approach to real-time road hazard avoidance, specifically for the problem of potholes, which can cause significant damage to vehicles and pose safety risks to drivers. Leveraging the capabilities of the latest YOLOv8 object detection models—namely versions N, S, and M—the system processes images captured by the vehicle's front camera. These images undergo preprocessing to normalize and optimize them for detection tasks. Each YOLO model variant contributes strength to the inference process, thereby increasing the robustness of pothole detection.

Upon identifying the potholes, each model outputs bounding boxes, which the system then analyzes to determine pothole distribution across six defined image zones. These zones, arranged in two rows and three columns, allow the system to assess the road surface in distinct segments. The (make_decision) function encapsulates the decision-making logic, which interprets the pothole count within each zone to generate driving recommendations. If potholes are universally present, the system advises the driver to slow down to mitigate potential damage. Potholes in the middle zones trigger a warning to the driver. Concentrations of potholes on either side of the image result in recommendations to turn away from the affected area. In cases where the criteria above are not met, the algorithm guides the driver to the zone with the fewest potholes, making navigational suggestions to avoid the hazards. This multi-faceted approach ensures that the system does not rely on a singular input or model but rather a consensus from several AI perspectives, enhancing accuracy and reliability. Moreover, it allows for adaptable responses, considering the ever-changing conditions of road surfaces. The comprehensive analysis and strategic guidance promise to substantially increase driver safety and comfort by proactively managing the risks associated with potholes.

In the figure 10, for example, you can see an example of pothole detection:

Figure 10. Example of pothole detection

After the three YOLOv8 models detect it, another step is added: breaking down the image into a grid and then calculating the number of potholes in each grid, as shown in the figure 11.

Figure 11. Splitting image into six grids

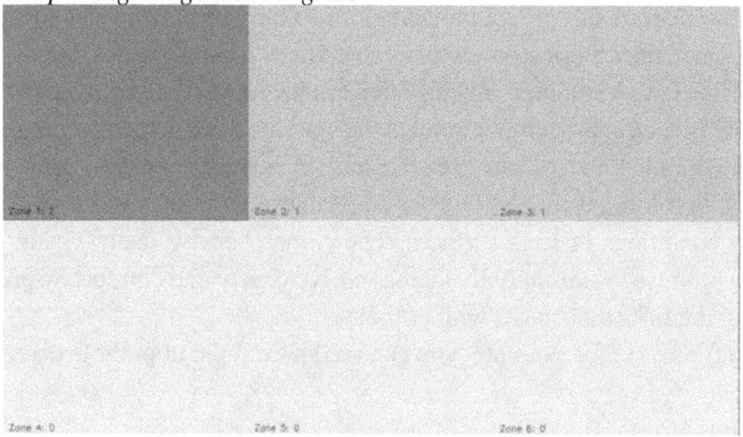

Based on this information, another window opens to inform the driver of the right action, as shown in the figure 12.

Figure 12. Pop-up: Driving instructions

Warn the driver of potholes in the middle
Turn right to avoid potholes on the left
Steer towards zone 1 with 2 potholes

Regarding speed, on average, the three YOLOv8 models, namely n, s, and m, take 400 ms, 600 ms, and 900 ms, respectively, without using a GPU.

CONCLUSION

This study examined how well three YOLO models can detect potholes on road surfaces. Training these models presents several challenges, including variations in road conditions, lighting, and weather. Our approach addresses these challenges by collecting diverse data to ensure robust performance across different scenarios. We also implemented real-time processing constraints to evaluate the feasibility of deploying these models in actual autonomous vehicle systems. Our research indicates that the YOLOv8-small model performed the best in pothole detection, achieving a mean Average Precision at IoU threshold 0.5 (mAP@0.5) of 71.1%. This outperformed YOLOv8 nano (66.93%) and YOLOv8 medium (51.33%). We also evaluated the models' ability to detect small potholes from long distances and in challenging weather or low-light conditions. We aimed to cover a wide range of scenarios by gathering images in various situations. Our system informs drivers about potholes ahead, combining the strengths of all three models. This paper directly applies YOLO for pothole detection, providing new insights through experimental analysis for comparison. This approach could reduce the reliance on human intervention in autonomous driving systems. Focusing on segmenting potholes might be even more effective than just detecting them, as it can precisely pinpoint their locations.

REFERENCES

Asad, M. H., Khaliq, S., Yousaf, M. H., Ullah, M. O., & Ahmad, A. (2022). Pothole detection using Deep Learning: A Real-Time and AI-on-the-Edge perspective. *Advances in Civil Engineering*, 2022(1), 1–13. DOI: 10.1155/2022/9221211

Baek, J. W., & Chung, K. (2020). Pothole classification model using edge detection in road image. *Applied Sciences (Basel, Switzerland)*, 10(19), 6662. DOI: 10.3390/app10196662

Bengamra, S., Mzoughi, O., Bigand, A., & Zagrouba, E. (2023). A comprehensive survey on object detection in Visual Art: Taxonomy and challenge. *Multimedia Tools and Applications*, 83(5), 14637–14670. DOI: 10.1007/s11042-023-15968-9

Bochkovskiy, A., Wang, C., & Liao, H. M. (2020, April 23). *YOLOV4: Optimal speed and accuracy of object detection*. arXiv.org. https://arxiv.org/abs/2004.10934v1

Bučko, B., Lieskovská, E., Zábovská, K., & Zábovský, M. (2022). Computer Vision Based Pothole Detection under Challenging Conditions. *Sensors (Basel)*, 22(22), 8878. DOI: 10.3390/s22228878 PMID: 36433474

Bučko, B., Lieskovská, E., Zábovská, K., & Zábovský, M. (2022b). Computer Vision Based Pothole Detection under Challenging Conditions. *Sensors (Basel)*, 22(22), 8878. DOI: 10.3390/s22228878 PMID: 36433474

Chen, H., Yao, M., & Gu, Q. (2020). Pothole detection using location-aware convolutional neural networks. *International Journal of Machine Learning and Cybernetics*, 11(4), 899–911. DOI: 10.1007/s13042-020-01078-7

Chitale, P. A., Kekre, K. Y., Shenai, H., Karani, R., & Gala, J. (2020). Pothole Detection and Dimension Estimation System using Deep Learning (YOLO) and Image Processing. *International Conference Image and Vision Computing New Zealand*. DOI: 10.1109/IVCNZ51579.2020.9290547

Dewangan, D. K., & Sahu, S. P. (2021). PotNet: Pothole detection for autonomous vehicle system using convolutional neural network. *Electronics Letters*, 57(2), 53–56. DOI: 10.1049/ell2.12062

Dib, J., Sirlantzis, K., & Howells, G. (2023). An annotated water-filled, and dry potholes dataset for deep learning applications. *Data in Brief*, 48, 109206. DOI: 10.1016/j.dib.2023.109206 PMID: 37213553

Kamalakannan, S., Navaneethan, S., Deshmukh, Y. S., Sujay, V., Ramalingam, V., Jagadeesan, D., & Venkatesh, C. (2024). A novel pothole detection model based on YOLO algorithm for VANET. *International Journal of Intelligent Systems and Applications in Engineering*, 12(11s), 56–61.

Kavas-Torris, O., Lackey, N., & Güvenç, L. (2021). Simulating the effect of autonomous vehicles on roadway mobility in a microscopic traffic simulator. *International Journal of Automotive Technology*, 22(3), 713–733. DOI: 10.1007/s12239-021-0066-7

Khan, M., Raza, M. A., Abbas, G., Othmen, S., Yousef, A., & Jumani, T. A. (2024). Pothole detection for autonomous vehicles using deep learning: A robust and efficient solution. *Frontiers in Built Environment*, 9, 1323792. Advance online publication. DOI: 10.3389/fbuil.2023.1323792

Koteswaramma, S., Sri, M. D., Manohar, M., Srihari, K., & Yabbeju, M. (2024). Advancing Road Safety: Pothole Detection Using Yolov8 And Wandb Deep Learning. Journal of Advanced Zoology, 45.

Kundu, K. G. C., & Kundu, K. G. C. (2023, January 1). *A DEEP LEARNING BASED APPROACH TO DETECT POTHOLES USING YOLO VERSION 7*. www.ijcrt.org. https://www.ijcrt.org/viewfull.php?&p_id=IJCRT2301187

Li, W., Shen, Z., & Li, P. (2019). Crack Detection of Track Plate Based on YOLO. *12th International Symposium on Computational Intelligence and Design*. DOI: 10.1109/ISCID.2019.10086

Lieskovská, E., Jakubec, M., Bučko, B., & Zábovská, K. (2023). Automatic pothole detection. *Transportation Research Procedia*, 74, 1164–1170. DOI: 10.1016/j.trpro.2023.11.257

Malta, A., Mendes, M., & Farinha, J. T. (2021). Augmented Reality Maintenance Assistant using YOLOV5. *Applied Sciences (Basel, Switzerland)*, 11(11), 4758. DOI: 10.3390/app11114758

Parekh, D., Poddar, N., Rajpurkar, A., Chahal, M., Kumar, N., Joshi, G. P., & Cho, W. (2022). A review on Autonomous Vehicles: Progress, methods and challenges. *Electronics (Basel)*, 11(14), 2162. DOI: 10.3390/electronics11142162

Park, S., Tran, V., & Lee, D. (2021). Application of various YOLO models for computer Vision-Based Real-Time pothole detection. *Applied Sciences (Basel, Switzerland)*, 11(23), 11229. DOI: 10.3390/app112311229

Park, S. S., Tran, V. T., & Lee, D. E. (2021). Application of various yolo models for computer vision-based real-time pothole detection. *Applied Sciences (Basel, Switzerland)*, 11(23), 11229. DOI: 10.3390/app112311229

Pothole detection. (2020, June 9). Kaggle. https://www.kaggle.com/datasets/andrewmvd/pothole-detection

Pothole Detection Dataset. (2022, December 25). Kaggle. https://www.kaggle.com/datasets/rajdalsaniya/pothole-detection-dataset?source=post_page-----f605f6eb4d9c--------------------------------

Pothole Image Data-Set. (2019, August 20). Kaggle. https://www.kaggle.com/datasets/sachinpatel21/pothole-image-dataset

Redmon, J., Divvala, S., Girshick, R., & Farhadi, A. (2016). You Only Look Once: Unified, Real-Time Object Detection. *Proceedings of the IEEE Computer Society Conference on Computer Vision and Pattern Recognition.* DOI: 10.1109/CVPR.2016.91

Redmon, J., & Farhadi, A. (2017). YOLO9000: Better, Faster, Stronger. *Proceedings - 30th IEEE Conference on Computer Vision and Pattern Recognition.* DOI: 10.1109/CVPR.2017.690

Redmon, J., & Farhadi, A. (2018, April 8). *YOLOV3: an incremental improvement.* arXiv.org. https://arxiv.org/abs/1804.02767v1

Roboflow: Computer vision tools for developers and enterprises. (n.d.). https://roboflow.com/

Rout, N. K., Dutta, G., Sinha, V., Dey, A., Mukherjee, S., & Gupta, G. (2023, November 10). *Improved pothole detection using YOLOV7 and ESRGAN.* arXiv.org. https://arxiv.org/abs/2401.08588v1

Scm, D. S. S. (2015). *A comparative study on web scraping.* http://ir.kdu.ac.lk/handle/345/1051

She, X., Hong-Wei, Z., Wang, Z., & Yan, J. (2021). Feasibility study of asphalt pavement pothole properties measurement using 3D line laser technology. *International Journal of Transportation Science and Technology*, 10(1), 83–92. DOI: 10.1016/j.ijtst.2020.07.004

Skalski, P. (n.d.). *GitHub - SkalskiP/make-sense: Free to use online tool for labelling photos.*https://makesense.ai. GitHub. https://github.com/SkalskiP/make-sense/

Suong, L. K., & Jangwoo, K. (2018). Detection of potholes using a deep convolutional neural network. *Journal of Universal Computer Science*, 24(9), 1244–1257.

Wang, C. Y., Bochkovskiy, A., & Liao, H. Y. M. (2023). YOLOv7: Trainable bag-of-freebies sets new state-of-the-art for real-time object detectors. In Proceedings of the IEEE/CVF conference on computer vision and pattern recognition (pp. 7464-7475).

Wang, H., Lu, B., Li, J., Liu, T., Xing, Y., Lv, C., Cao, D., Li, J., Zhang, J., & Hashemi, E. (2022). Risk assessment and mitigation in local path planning for autonomous vehicles with LSTM based predictive model. *IEEE Transactions on Automation Science and Engineering*, 19(4), 2738–2749. DOI: 10.1109/TASE.2021.3075773

Wang, H. W., Chen, C., Cheng, D. Y., Lin, C., & Lo, C. (2015). A Real-Time pothole detection approach for intelligent transportation system. *Mathematical Problems in Engineering*, 2015, 1–7. DOI: 10.1155/2015/813507

Woo, S., Youtie, J., Ott, I., & Scheu, F. (2021). Understanding the long-term emergence of autonomous vehicles technologies. *Technological Forecasting & Social Change. Technological Forecasting and Social Change*, 170, 120852. DOI: 10.1016/j.techfore.2021.120852

Ye, W., Jiang, W., Tong, Z., Yuan, D., & Xiao, J. (2021). Convolutional neural network for pothole detection in asphalt pavement. *Road Materials and Pavement Design*, 22(1), 42–58. DOI: 10.1080/14680629.2019.1615533

Chapter 11
Unraveling Data Complexity in the Metaverse for Anomaly Detection With Python on NYC Taxi

Özen Özer
Kirklareli University, Turkey

Nadir Subasi
Kirklareli University, Turkey

ABSTRACT

In the dynamic environment of the Metaverse, where virtual interactions and transactions thrive, detecting data anomalies becomes imperative for maintaining integrity and security. This paper explores the application of Pythonbased anomaly detection models, including Inter Quartile Range (IQR), Median Absolute Deviation (MAD), and Local Outlier Factor (LOF), in identifying anomalies within NYC Taxi data. Through comprehensive analysis and experimentation, we investigate the effectiveness and comparative performance of these models in detecting outliers amidst the complex and diverse data landscape of the Metaverse. In the NYC Taxi Data, which contains 10320 data, it was analyzed with the mentioned algorithms and 2 anomalies (0.019%) were found with IQR. In the same data set, 1 anomaly (0.009%) was found with MAD model and 1032 anomalies (10%) were found with LOF

DOI: 10.4018/979-8-3373-1399-3.ch011

1. INTRODUCTION

The Metaverse is an evolving digital realm where virtual interactions and transactions are becoming increasingly prevalent. As individuals continue to immerse themselves in this dynamic environment, the need for maintaining data integrity and security has become paramount. One of the key challenges in this rapidly changing landscape is the detection of data anomalies, which can have significant implications for the reliability and trustworthiness of virtual interactions.

Anomalies in data can arise due to various factors, such as errors, fraud, or unusual patterns that deviate from the norm. Detecting and addressing these anomalies in a timely manner is crucial for ensuring the quality and accuracy of data within the Metaverse. In this paper, we focus on the application of Python-based anomaly detection models to identify anomalies within NYC Taxi data, a dataset that captures a wide range of interactions and transactions in the virtual realm.

Python has emerged as a powerful tool for data analysis and machine learning, offering a wide range of libraries and frameworks that facilitate the development of sophisticated anomaly detection algorithms. In this study, we explore the use of three popular anomaly detection models - Inter Quartile Range (IQR), Median Absolute Deviation (MAD), and Local Outlier Factor (LOF) - to detect anomalies within NYC Taxi data. By leveraging these models, we aim to showcase the effectiveness of Python-based anomaly detection techniques in uncovering anomalous patterns within complex datasets.

Through this research, we seek to contribute to the growing body of knowledge on anomaly detection in the Metaverse, highlighting the importance of data integrity and security in virtual environments. By gaining insights into the detection of anomalies within NYC Taxi data, we aim to provide valuable guidance for developers, researchers, and stakeholders working within the virtual realm.

2. RELATED WORK

In 2019, the book “Multivariate Data Analysis” by Hair Jr. et al., published by Cengage Learning, became widely used in academic settings to teach multivariate data analysis. It covers various multivariate techniques and their applications, providing a comprehensive introduction to the subject. The book's strength lies in its practical approach, making complex statistical methods accessible to students and practitioners.

Another popular textbook in the field of multivariate statistics is “Using Multivariate Statistics” by Tabachnick and Fidell. This book offers comprehensive coverage of multivariate techniques, including their theoretical underpinnings and practical applications. It serves as an essential resource for students and researchers who need to understand and apply multivariate statistical methods in their work.

The book "Methods of Multivariate Analysis" by Rencher and Christensen offers a rigorous treatment of multivariate statistical methods. It covers a wide range of topics, including the multivariate normal distribution, principal component analysis, factor analysis, and discriminant analysis. This book is particularly valued for its detailed explanations and mathematical rigor.

Field et al. focus on statistical analysis using the R programming language in their book "Discovering Statistics Using R". It includes discussions on both univariate and multivariate analysis, highlighting their importance in data analysis. This book is a valuable resource for those who prefer learning statistics through practical examples and coding in R.

The book "An Introduction to Applied Multivariate Analysis with R" by Everitt and Hothorn provides an introduction to applied multivariate analysis using the R programming language. It covers topics such as multivariate data visualization, clustering, and classification. This book is designed for readers who wish to apply multivariate techniques in their research using R.

John Tukey's classic book "Exploratory Data Analysis" introduces the concept of the interquartile range (IQR) and its application in exploratory data analysis. Tukey's work has been foundational in the field, emphasizing the importance of understanding data distributions and identifying outliers.

The paper by Leys et al. discusses the advantages of using the Median Absolute Deviation (MAD) over the standard deviation for identifying outliers. This paper provides a compelling argument for the robustness of MAD in detecting outliers in various datasets.

The seminal paper by Breunig et al. introduces the Local Outlier Factor (LOF) algorithm for identifying local outliers in high-dimensional datasets. This algorithm has become a key tool in data mining and machine learning for outlier detection.

Hawkins' book "Identification of Outliers" provides a comprehensive overview of outlier detection methods, including techniques based on the interquartile range and other robust statistics. This book is a valuable reference for researchers and practitioners dealing with outlier detection in their data.

The paper by Rousseeuw and Croux discusses alternative measures of dispersion to the median absolute deviation and their applications in outlier detection. This paper highlights the importance of robust statistical measures in improving the accuracy of outlier detection.

The book "Outlier Analysis" by Aggarwal et al. provides a comprehensive overview of outlier analysis techniques, including methods based on statistical measures like IQR and MAD, as well as more advanced approaches like LOF. This book is an essential resource for those studying data mining and anomaly detection.

The paper by Hyndman and Fan discusses the computation of sample quantiles, including quartiles and the interquartile range, in statistical software packages. This paper is important for understanding how different software implementations can affect the results of statistical analyses.

Additionally, several other notable works contribute to the field:

- "Applied Multivariate Statistical Analysis" by Johnson and Wichern offers a broad introduction to multivariate statistical methods with practical applications.
- "Multivariate Statistical Methods: A Primer" by Manly and Navarro Alberto provides an accessible introduction to multivariate techniques for beginners.
- "Modern Multivariate Statistical Techniques" by Izenman covers a wide range of multivariate methods, from classical to modern approaches.
- "Multivariate Analysis for the Behavioral Sciences" by Kim and Mueller focuses on applications of multivariate techniques in social and behavioral sciences.
- "Analyzing Multivariate Data" by Lattin, Carroll, and Green presents multivariate methods with a strong emphasis on practical data analysis.

3. METHODOLOGY

3.1 Univariate and Multivariate Analysis

Univariate and multivariate analysis are two fundamental approaches used in statistics and data analysis, each serving distinct purposes but sharing some commonalities. We can delve into their definitions and explore their similarities and differences as follows:

Univariate Analysis

Definition: Univariate analysis focuses on analyzing a single variable at a time. It explores the distribution and characteristics of that variable without considering the relationships with other variables.

Purpose of the it may be explained as "Descriptive which describes the distribution, central tendency, and dispersion of a single variable" or "Inferential that helps make inferences and test hypotheses about the population parameters based on sample statistics."

Examples are given as follows basically:

- *Descriptive Statistics*: Mean, median, mode, standard deviation, variance, etc.
- *Inferential Statistics:* One-sample t-tests, Chi-square tests, etc.

Similarity with Multivariate Analysis:

- Both univariate and multivariate analyses are statistical methods used to understand data.
- They both employ statistical techniques to summarize and analyze datasets.

Differences:

- Univariate analysis deals with one variable, while multivariate analysis deals with multiple variables simultaneously.
- Univariate analysis is simpler since it focuses on a single variable, while multivariate analysis can be more complex due to the consideration of interactions and correlations among multiple variables.
- Univariate analysis provides insights into the distribution of individual variables, while multivariate analysis explores relationships among variables and their collective impact on outcomes.
- While both types of analysis use statistical techniques, the methods employed in multivariate analysis, such as regression analysis, factor analysis, and cluster analysis, are often more advanced and sophisticated compared to those used in univariate analysis.

Multivariate Analysis.

Definition: Multivariate analysis involves the simultaneous analysis of multiple variables to understand the relationships and dependencies among them. It aims to uncover patterns, associations, and structures within complex datasets.

Purpose of the it may be explained as "Identify Patterns that helps identify patterns and relationships among multiple variables" or "Predictive Modeling which enables the development of predictive models by considering the joint effects of multiple variables" with "Dimensionality Reduction Techniques like factor analysis and principal component analysis (PCA) help reduce the dimensionality of data while retaining important information."

The followings can be given as examples too:

- Regression Analysis: Linear regression, logistic regression.
- Factor Analysis: Exploratory factor analysis (EFA), confirmatory factor analysis (CFA).
- Cluster Analysis: K-means clustering, hierarchical clustering.
- Discriminant Analysis: Linear discriminant analysis (LDA), quadratic discriminant analysis (QDA).

Similarity with Univariate Analysis:

- Both univariate and multivariate analyses aim to extract meaningful insights from data.
- They both utilize statistical techniques and methodologies for analysis.

Differences:

- Multivariate analysis deals with multiple variables simultaneously, making it more complex than univariate analysis.
- While univariate analysis focuses on the characteristics of individual variables, multivariate
- Multivariate analysis requires interpreting complex relationships among variables, whereas univariate analysis provides simpler interpretations focused on individual variables.

In summary, univariate analysis is concerned with the characteristics of single variables, while multivariate analysis explores the relationships among multiple variables. Both approaches play crucial roles in data analysis, offering insights into different aspects of datasets and facilitating informed decision-making.

3.2 The History of the Inter Quartile Range (IQR)

Median Absolute Deviation (MAD), and Local Outlier Factor (LOF)

. The history of statistical measures such as the Interquartile Range (IQR), Median Absolute Deviation (MAD), and Local Outlier Factor (LOF) in identifying anomalies can be traced back to the evolution of statistical methods for data analysis and anomaly detection.

Inter Quartile Range (IQR)

The concept of quartiles and the interquartile range has its roots in descriptive statistics and the graphical representation of data. The idea of dividing a dataset into quartiles to understand its spread and central tendency dates back to the early development of statistical methods.

IQR is often used to detect outliers in a dataset by considering the spread of the middle 50% of the data. Anomalies are identified as observations that fall below Q1 - 1.5 * IQR or above Q3 + 1.5 * IQR, where Q1 and Q3 represent the first and third quartiles, respectively.

Median Absolute Deviation (MAD)

. The MAD is a robust measure of variability that captures the dispersion of data points around the median. It has roots in the development of robust statistics, which aim to provide reliable estimates even in the presence of outliers.

MAD is used as a measure of variability that is less influenced by extreme values compared to the standard deviation. Anomalies can be identified using MAD by considering observations with values beyond a certain threshold, typically a multiple of MAD away from the median.

Local Outlier Factor (LOF)

LOF is a relatively recent development in the field of anomaly detection, specifically designed to identify outliers in high-dimensional datasets. It emerged from the need to address the limitations of traditional methods in detecting anomalies in complex data structures, such as those encountered in data mining and machine learning applications.

LOF assesses the local density deviation of a data point with respect to its neighbors, identifying outliers as data points with significantly lower local density compared to their neighbors. It is particularly useful in detecting anomalies in datasets where traditional methods based on distance metrics may fail to capture the underlying data structure.

3.3 Evolution in Identifying Anomalies

The development of these measures reflects a progression in statistical methods for identifying anomalies, from traditional descriptive statistics to robust measures and, more recently, specialized techniques tailored for complex data structures.

As datasets become larger and more complex, there is a continued emphasis on developing advanced anomaly detection methods that can effectively handle diverse data types and structures.

4. ANALYSIS OF NEW YORK CITY TAXI DATA

Taxis in New York come in two varieties, yellow and green. Traditional yellow taxis can pick up passengers anywhere in the five boroughs, although street pickups tend to be largely concentrated in lower Manhattan and at airports.

Green taxis began to enter service with the launch of the Boro Taxi programme in August 2013. In an effort to increase access to street transport in all boroughs, green taxis have restrictions on picking up on Manhattan streets below West 110th Street and East 96th. The New York City Taxi and Limousine Commission provides public web access to records of trips made by the city's yellow taxis since 2009 [NYC].

Green taxi records have been available since the programme's inception in 2013. Each journey record is identified by cut-off and medallion numbers and contains information on pick-up and drop-off times, pick-up and drop-off longitude and latitude, number of passengers, measured time and distance, and fare data.

One of our objectives is the detection of abnormal traffic events, so we decided to focus yellow taxi data to find outliers.

In its original form, the 2013 yellow taxi data contains records of about 170 million journeys. As it is quite noisy and contains a large number of entries with missing, incomplete or erroneous information, our first step was to clean the data during pre-processing.

We also remove journeys with inconsistent or impossible values, such as journeys with zero duration or journeys whose measured distance is less than the straight distance between the listed pick-up and drop-off locations. In this section, we detail the application of outlier analysis approach to analyze taxi traffic in New York City.

5. EXPERIMENTS

To examine the performance of outlier detection methods, we tested the Interquartile Range (IQR), Median Absolute Deviation (MAD), and Local Outlier Factor (LOF) on both manually generated datasets and a real-world traffic dataset called NYC-taxi. The NYC-taxi dataset comprises comprehensive records of taxi trips in New York City, providing a rich source of data for testing various outlier detection techniques.

We began by generating outliers and outlier scores manually using an algorithm designed to introduce anomalies into the dataset systematically. This step allowed us to create a controlled environment in which the performance of different outlier detection methods could be objectively assessed. In Table 1, we list the manually generated outliers and their corresponding outlier scores. These scores were derived from a range of variables, each containing a variable number of outliers, to ensure a thorough evaluation of each method's effectiveness.

For the IQR method, we calculated the interquartile range for each variable and identified data points that fell outside 1.5 times the IQR above the third quartile or below the first quartile as outliers. This traditional approach is widely used due to its simplicity and effectiveness in identifying univariate outliers.

The MAD method involved calculating the median absolute deviation for each variable. Data points that deviated more than a certain threshold from the median were flagged as outliers. This method is particularly robust to the presence of extreme values, making it a reliable choice for datasets with heavy-tailed distributions.

The LOF algorithm was employed to detect local outliers in the dataset. LOF measures the local density deviation of a given data point with respect to its neighbors, identifying points that have significantly lower density compared to their surroundings as outliers. This method is especially useful for detecting outliers in high-dimensional datasets.

Our evaluation of these methods on the NYC-taxi dataset revealed insights into their respective strengths and weaknesses. The real-world traffic data provided a complex, dynamic environment that challenged each method's ability to accurately identify outliers.

In addition to our own experiments, several key works in the field have informed and guided our approach to outlier detection. ***For example:***

- ✓ "Identification of Outliers" by Hawkins offers a foundational understanding of various outlier detection techniques, providing detailed explanations and examples.
- ✓ The paper "Robust Statistics" by Huber and Ronchetti discusses the robustness of different statistical methods, including MAD, highlighting their importance in outlier detection.
- ✓ "The Elements of Statistical Learning" by Hastie, Tibshirani, and Friedman includes a comprehensive section on anomaly detection, covering algorithms like LOF and their applications in machine learning.
- ✓ "Outlier Analysis" by Aggarwal presents a broad overview of modern outlier detection methods, integrating both statistical approaches and machine learning techniques.
- ✓ "An Introduction to Statistical Learning" by James, Witten, Hastie, and Tibshirani provides practical insights into implementing outlier detection methods using R, with examples and case studies relevant to real-world data.

To further enhance our understanding, we reviewed additional literature focusing on the practical application and theoretical underpinnings of these methods:

"Applied Multivariate Statistical Analysis" by Johnson and Wichern offers an in-depth look at multivariate techniques, including outlier detection in high-dimensional spaces.

"Multivariate Statistical Methods: A Primer" by Manly and Navarro Alberto provides an accessible introduction to these techniques, making it a useful resource for beginners.
"Modern Multivariate Statistical Techniques" by Izenman covers a wide range of methods, from classical approaches to cutting-edge techniques in the field.
"Multivariate Analysis for the Behavioral Sciences" by Kim and Mueller focuses on applications in social and behavioral sciences, emphasizing practical data analysis.
"Analyzing Multivariate Data" by Lattin, Carroll, and Green emphasizes practical implementation and interpretation of multivariate methods, including outlier detection.

These references collectively enhance our understanding and application of outlier detection methods, providing a robust foundation for examining the performance of IQR, MAD, and LOF on diverse datasets like NYC-taxi. Our study contributes to the broader field by offering practical insights and empirical evidence on the effectiveness of these methods in identifying outliers in real-world scenarios.

To examine the performance of Outlier methods, we tested IQR, MAD, LOF with both manually generated by an algorithm for real-world traffic data sets called NYCtaxi. We first test the manually generated outliers and outliers scores listed in Table 1, which have a variable number of outliers and outlier scores.

Table 1. Number of outliers and scores

Method	Outliers Scores (Mean)	Number of Outliers	Outlier Rate(%)
IQR	-272.425 – 30373.175	2	0.019%
MAD	0.9066	1	0.009%
LOF	1.0240	1032	10%

5.1 Comparison of IQR, MAD and LOF

IQR, MAD and LOF methods are compared. These three methods are statistical methods used to detect and analyze outliers in a dataset. The advantages, disadvantages and usage areas of each method will be analyzed.

IQR refers to the difference between quartiles in the data set. Calculate the difference between the third quarter (Q3) and the first quarter (Q1). The IQR method considers the center and amplitude of the distribution in the data set. Therefore, it can be used to detect values far from the mean in the data set. Result of IQR outlier detector for NYCtaxi dataset is shown in Figure 1.

Figure 1. IQR outlier results

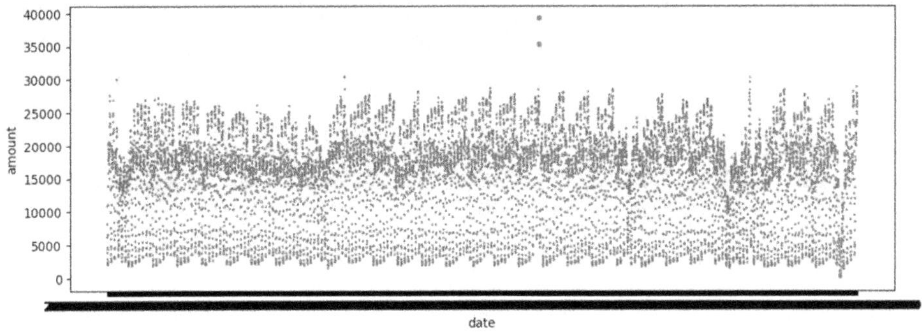

MAD refers to the median of the absolute distances of data points from the median value. The MAD method can better reflect the variability and dispersion of the dataset and identify outliers more accurately.

Figure 2. MAD outlier results

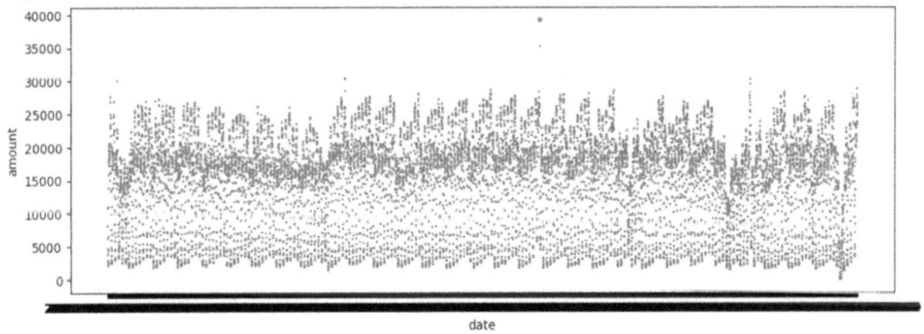

LOF determines the local density of each data point by comparing it with the density of its neighbours. The LOF method is used to measure the typical value of each data point in the dataset and is effective in detecting hidden structures in the dataset.

Figure 3. LOF outlier results

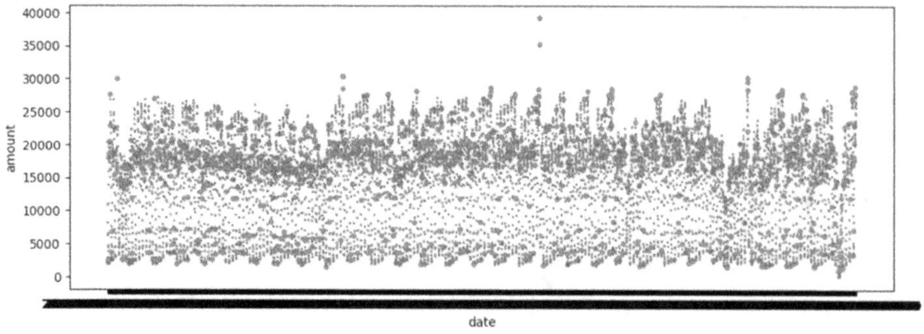

Comparing the IQR, MAD and LOF methods for outlier detection is an important step to identify the advantages and disadvantages of each method.

The IQR method has the advantage of detecting outliers using the interquartile range of the data distribution. In this way, it can uncover statistically significant outliers independent of the overall distribution of the data. On the other hand, the MAD method is used to identify outliers using a measure of the spread of the median and has the advantage that it is generally not sensitive to outliers affecting the median. MAD can produce more reliable results as it favours the median to avoid overshadowing the central tendency in the dataset. The LOF method determines the degree of outlier of each data point by calculating the local outlier factor. Thus, it can detect outliers by evaluating the intensity differences between data points.

Comparing these methods provides information about their ability to deal with different data structures and outliers, providing an important understanding of which method may be more effective in which situations.

In outlier detection, choosing the right method depending on the characteristics of the data set is a critical step to minimize undesirable results and obtain reliable results. Therefore, the advantages and disadvantages of different methods such as IQR, MAD and LOF should be considered, the appropriate one should be selected according to the purpose of the analysis and data characteristics. For the data set used, IQR seems to be a more successful outlier detection method compared to other methods.

6. CONCLUSION

In this study, the IQR, MAD and LOF methods for outlier detection are compared. Experimental observations show that it does not matter which statistics are used to threshold outlier scores; the performance of the methods can be improved, i.e. by first excluding scores above the original threshold and then recalculating the statistics, the actual elimination of outliers can be done using the revised threshold. According to the experimental tests shown in this work, all the techniques tested (IQR, MAD, LOF) were developed in the Python environment.

The performance of each method was compared on the same data set. r. This comparison is important to better understand the advantages and disadvantages of each method and to determine when it is preferable. It is also evaluated in terms of its use in instantaneous and outlier detection.

DECLARATION OF COMPETING INTEREST

The authors declare that they have NO known competing financial interests or personal relationships that could have appeared to influence the work reported in this paper.

ACKNOWLEDGEMENT

We thank referees for assistance with their suggestions and comments that greatly improved the manuscript.

REFERENCES

Aggarwal, C. C., & Yu, P. S. (Eds.). (2018). *Outlier Analysis* (2nd ed.). Springer.

Breunig, M. M., Kriegel, H. P., Ng, R. T., & Sander, J. (2000, May). LOF: Identifying density-based local outliers. In *Proceedings of the 2000 ACM SIGMOD International Conference on Management of Data* (pp. 93-104). DOI: 10.1145/342009.335388

Everitt, B. S., & Hothorn, T. (2011). *An Introduction to Applied Multivariate Analysis with R*. Springer Science & Business Media. DOI: 10.1007/978-1-4419-9650-3

Field, A., Miles, J., & Field, Z. (2012). *Discovering Statistics Using R*. SAGE Publications Ltd.

Hair, J. F.Jr, Black, W. C., Babin, B. J., & Anderson, R. E. (2019). *Multivariate Data Analysis* (8th ed.). Cengage Learning.

Hawkins, D. M. (1980). *Identification of Outliers*. Chapman and Hall. DOI: 10.1007/978-94-015-3994-4

Hyndman, R. J., & Fan, Y. (1996). Sample quantiles in statistical packages. *The American Statistician*, 50(4), 361–365. DOI: 10.1080/00031305.1996.10473566

Leys, C., Ley, C., Klein, O., Bernard, P., & Licata, L. (2013). Detecting outliers: Do not use standard deviation around the mean, use absolute deviation around the median. *Journal of Experimental Social Psychology*, 49(4), 764–766. DOI: 10.1016/j.jesp.2013.03.013

NYC. The Official Website of the City of New York, www.nyc.gov/html/tlc/html/about/triv_record_data.shtml. (up to date).

Rencher, A. C., & Christensen, W. F. (2012). *Methods of Multivariate Analysis* (3rd ed.). Wiley. DOI: 10.1002/9781118391686

Rousseeuw, P. J., & Croux, C. (1993). Alternatives to the median absolute deviation. *Journal of the American Statistical Association*, 88(424), 1273–1283. DOI: 10.1080/01621459.1993.10476408

Tabachnick, B. G., & Fidell, L. S. (2019). *Using Multivariate Statistics* (7th ed.). Pearson.

Tukey, J. W. (1977). *Exploratory Data Analysis*. Addison-Wesley.

Chapter 12

The Synergistic Power of AIoT in Enhancing EFL Student CCT Skills in Higher Education

Muthmainnah Muthmainnah
https://orcid.org/0000-0003-3170-2374
Universitas Al Asyariah Mandar, Indonesia

Misdi Misdi
https://orcid.org/0000-0002-3543-0823
Universitas Swadaya Gunung Jati, Indonesia

Muliati Muliati
https://orcid.org/0009-0007-5859-5480
Universitas Bosowa, Makassar, Indonesia

Ahmad Al Yakin
https://orcid.org/0000-0003-3170-2374
Universitas Al Asyariah Mandar, Indonesia

Dalwinder Kaur
Manipal GlobalNXT University, Kuala Lumpur, Malaysia

Eka Apriani
Institut Agama Islam Negeri, Curup, Indonesia

V. Vasantha Kumar
Sourashtra College Autonomous, Madurai, India

ABSTRACT

The study monitored undergraduate students in English language education courses as they utilized artificial intelligence-enhanced learning resources. We created this course by integrating GenAI and IoT from various applications like Google (IoT), YouTube, and AI-ChatGPT into the learning plan, with a focus on modernity and interaction through AIoT. We aim to enhance students' communication and critical

DOI: 10.4018/979-8-3373-1399-3.ch012

thinking skills by incorporating this technology into a dynamic learning environment that fosters 21st-century learning skills. The integration of AI-ChatGPT, YouTube, Mind mapping and Google (IoT) into the English education curriculum enhances students' critical thinking and communication skills, making learning more engaging and enjoyable. Insights like these highlights how the Internet of Things (IoT) can transform classroom methods and improve student achievement in the modern era.

INTRODUCTION

Generative Artificial intelligence (GenAI) is revolutionizing many parts of human existence in this age of heightened societal digitization. Artificial intelligence is reshaping numerous industries and its ability to automate processes, handle massive volumes of data, and offer predictive insights (Ullah, et al. 2024; Durach and Gutierrez 2024), traditional AI definitions emphasized the capability of AI to mimic human cognitive functions, including reasoning, decision-making, and goal setting. However, the integration of AIoT has enhanced these capabilities, resulting in smart applications that serve as excellent instructional tools.

AIoT technology, which integrates AI with the Internet of Things, can bolster many different approaches to education. This is especially true of project-based and problem-oriented learning models. According to Park and Kwon (2024), students can improve their problem-solving and collaborative skills, gain a deeper understanding of technology, and learn to adapt to AI systems efficiently with the help of these models. This integration is crucial for today's schools because it creates a dynamic and engaging classroom where students can learn and grow. Using digital apps and tools to encourage creative thinking, the “maker movement” exemplifies the “learning by doing” concept. Traditional education, on the other hand, tends to centre on theoretically grounded, hands-on courses. The focus of maker education is on getting students to work together on real-world projects, with an emphasis on social engagement. When studying English, this approach is especially useful since it encourages students to think creatively and realistically by giving them opportunities to convey their thoughts (Nguyen et al., 2024; Rajamohan et al. 2024).

According to Fernández et al., (2024) and Onesi et al. (2024), the Internet of Things (IoT) marks a worldwide movement in education that seeks to foster students' creative abilities to better equip them for the demands of the future. To improve national competitiveness, mold students' information literacy, and foster their creative and critical thinking abilities, educational reforms now incorporate courses on AI, IoT, and AIoT. The line with the technical developments of Industry 5.0, the focus on computational thinking highlights the significance of these abilities in educational and occupational settings.

University course preparation has a long history of prioritizing professors' interests and pedagogical goals over students' needs and current workforce developments. A misalignment between academic understanding and industry expectations is forcing graduates entering the job to undergo retraining. To close this achievement gap, we must assess instructors' knowledge and practices to ensure that students receive high-quality instruction in cutting-edge fields such as AIoT. Because of the rapid development of new technologies, the traditional classroom is no longer the primary site of education. Education 5.0's focus is "maker learning," which encourages students to be actively involved in their own education and to work independently, as pointed out by An Darwish, et al. (2024). This method considers the changing needs of the industry, streamlines course delivery, and adapts to the changing learning environment of the digital age.

We need to adapt to a changing educational landscape, fulfill industry expectations, and improve course delivery efficiency is driving the shift towards Education 5.0. By integrating sensing technologies across disciplines and focusing on problem-solving, English classes that use the AIoT teaching model cultivate an atmosphere that encourages students to think critically and communicate effectively. Despite the intricacy of AIoT applications, this technology has the potential to greatly inspire learning and deliver valuable educational experiences through the exploration of ICT, social media, and AI.

This study examines how the Internet of Things (IoT) and artificial intelligence (AI-ChatGPT) affect the growth of CT and communication abilities. We present an AIoT pedagogical approach that integrates AI-ChatGPT with internet of things (IoT) social media platforms such as Google and YouTube, and we test how well it improves students' performance in English classes and their CCT (communication and critical thinking) skills. The study's overarching goals are to improve educational research, bring attention to the value of strong communication and critical thinking abilities, lay the groundwork for comprehending technological ideas, and ultimately facilitate more effective learning through the integration of AIoT into English classes. In this English lesson, students will learn how to plan and execute communication projects, communicate effectively, and reflect on their own ideas and actions.

The aim of this study is to investigate the effects of incorporating CCT pedagogical strategies into English language classes on students' learning intentions and CCT competences, as well as to gradually introduce the essential AIoT pedagogical models. Schools, universities, and other appropriate educational environments can apply the AIoT teaching model for English courses due to its comprehensive approach to modern education.

LITERATURE REVIEW

In today's lightning-fast digital age, the convergence of artificial intelligence (AI) and the Internet of Things (IoT) gave rise to a revolutionary paradigm known as AIoT. This confluence is transforming many aspects of our lives, including schooling (Wahshat, et al. 2024). The term "AIoT" describes a mutually beneficial partnership between artificial intelligence (AI) and internet of things (IoT) devices, which enhances the performance of both systems. Learning with AIoT entails using AI-powered apps and devices to gather, process, and analyze data from objects and sensors connected to the Internet of Things (IoT). This environment allows for interactive and individualized learning experiences that go beyond the typical classroom setting (Mamatnabiyev et al. 2024).

1. Enhance CCT Skills

For second language acquisition, long periods of immersion in the target language are necessary. Researchers have shown that chatbots powered by AI can be a great tool for language learners. They can help students learn new words and improve their conversational skills every day (Duong and Suppasetseree, 2024). Additionally, they have the potential to stimulate students' interest and encourage their growth (Jeon., 2024). Teachers can also help students recognize and understand linguistic features in language learning by emphasizing relevant information and responding to student questions (Cooray et al. 2024). Because these chatbots are accessible at all times, students can enhance their language abilities regardless of where they are (Ebadi and Amini et al., 2024). This type of real-time feedback can help students improve their English learning results by adjusting their communication output (Fathi, et al. 2024). Chatbots powered by AI can personalize the learning experience for undergraduates, tailor the delivery of course material to each student's needs, and provide individualized lesson plans if students encounter challenges (Muthmainnah et al. 2024). Using this chatbot in the classroom may create an engaging and natural learning environment (Zhang et al., 2024). It has the ability to offer expressions, queries, and vocabulary that human language communication partners could lack (Hand and Lee 2024). Chatbots powered by artificial intelligence can greatly improve English language acquisition by administering formative evaluations and offering immediate feedback. Research by Koç, (2024) and and Alshewiter et al. (2024) suggests that these technical developments are critical for effective language learning. Multiple studies have demonstrated the effectiveness of chatbots in language learning. Mohamed (2024) and Alrajhi (2024) found that these tools were effective in retaining students' attention, which in turn led to higher levels of learning and motivation. It is well known that chatbots can establish a welcoming classroom climate where students feel safe enough to speak up during class discussions. They inspire pupils to speak up more and have faith in their language skills, according to studies conducted by Alabbas and Alomar (2024).

According to Koç, (2024) research chatbots also help students and teachers connect, which leads to improved communication and more assistance throughout the learning process. In a nutshell, chatbots driven by artificial intelligence have been a boon to language learning, providing advantages including formative assessment, quick feedback, higher engagement, better learning outcomes, and more interaction between students and teachers. Improvement of Analytical Capacity Critical thinking involves relying on evidence to guide one's actions. With the advent of AIoT, educational institutions can collect vast quantities of data from a variety of sources. With practice, students can learn to make sense of this data, spot trends, and extract useful information. Students develop their ability to think critically through participating in data-driven activities, which challenge them to ask questions, analyze data, and draw valid investigation by Muthmainnah et al. (2022). Because AIoT teaches students about systems and devices that are interdependent, it is possible to use IoT to address issues. Students can effectively address real-world problems, come up with creative solutions, and understand the interplay between ChatGPT and the Internet of Things (IoT) Khlaisang and Teo, (2024). This practical experience inspires them to critically analyze problems and devise innovative solutions.

The AIoT landscape raises questions of privacy, security, and biases in user data, requiring ethical concerns as a form of critical thinking. According to Borkovska et al. (2024); Essel et al. (2024) students who take part in these talks are also motivated to maximize their technological literacy and think critically about the societal ramifications of AIoT. By breaking down these complicated situations, students improve their analytical and problem-solving skills, as well as their capacity to think creatively and independently. The goal of optimizing adaptive learning is to enhance critical thinking abilities within the Hyplex learning environment. In order to maximize learning in intelligent settings, AIoT platforms can adjust learning routes according to student involvement and performance. This approach customizes the material's complexity to match each student's current academic level, enabling them to hone their problem-solving and critical-thinking skills as they progress through the curriculum (Tang, 2024). One area where technology has proven to be revolutionary is in education, where AIoT has sparked a revolution. According to Kostikova, et. al (2024), AIoT helps students succeed in today's highly connected society by improving their communication and critical thinking abilities. If teachers and students embrace this paradigm shift, the next generation will be confident and competent in all communication forms, including technology, to solve future problems.

2. The Effects of AIoT on Language Acquisition

As pointed out by Yuan (2024), the advent of AI has caused widespread modernization and disruption across many industries, including education. Of all the revolutionary AI tools, the conversational AI interface known as OpenAI Chat Generative Pre-Trained Transformer (ChatGPT) is unparalleled. Data scientists can greatly benefit from this cutting-edge tool, which uses natural language processing to comprehend and adequately answer human inquiries (Monika, and Suganthan 2024). ChatGPT's strength is in the high-quality responses it can produce for a variety of inquiries; this is all down to its deep learning algorithm base. According to the model description by Bekou et al. (2024), this approach uses a combination of machine learning and natural language processing algorithms to generate text that is both grammatically correct and relevant to the context. Notably, ChatGPT is now a cultural phenomenon, not just a technological one. According to Steiss et al. (2024), this shows that AI is having an increasing effect on how humans communicate and engage with machines. Consequently, ChatGPT has the potential to revolutionize and enhance numerous facets of our educational system, among others.

It is widely believed among technological professionals that natural events serve as inspiration for technological advancement, according to many technological professionals. Rafique, et al. (2024) argues that while developing AI system technologies, humans offer the finest example to follow. From this perspective, we must emulate language, as it represents one of the most sophisticated tools humans have ever devised. Kondurkar et al. (2024) posits that ChatGPT, a generative AI system, can mimic language as a soft technology, employing machine and deep learning techniques to acquire a new language. ChatGPT and related generative AI systems primarily learn to produce and interpret human language to comprehend the complex patterns and structures of human-like language (Tuan et al., 2024). ChatGPT, as a generative AI technology, is a groundbreaking development in natural language processing. Both the answers and the processing of natural language input are areas where this revolutionary system truly shines in its replication of human language. However, keep in mind that ChatGPT is still only a machine and lacks human intelligence and context awareness (Clark, 2024). Despite this, it remains a strong tool for NLP. In contrast to people, ChatGPT doesn't have any preexisting information or comprehension of its surroundings. The way it learns from its training data, however, is the only basis for its linguistic interpretation (Haleem et al., 2023). On the other hand, this study will use ChatGPT—which is built for two-way user interaction—to teach English on smartphones in the classroom (Jefferson, 2019; Norouzi et al., 2020).

Acquiring a second language has long been considered an essential competency for working across cultural boundaries and competing on a global scale. A new paradigm in language instruction is emerging as a result of the merging of AIoT and AI-powered language models such as ChatGPT, made possible by the exponential growth of computing power Rezaei and Ansary (2024). This potent mix is revolutionizing language learning and providing new opportunities to learners from all over the world. It is well known that AIoT-personalized learning paths add a human element to language acquisition. Internet of Things (IoT) devices with AI capabilities can analyze students' progress, preferences, and habits to personalize their learning path. This customization of information and tasks can accommodate students' skill levels, learning styles, and pace. As a result, the student maintains interest and drive, both of which are critical to language learning. ChatGPT is able to mimic realistic and engaging discussions because it uses a sophisticated AI language model. Students frequently use AI to help them engage in discourse with ChatGPT motivate to practice speaking, listening, and understanding (Muthmainnah, 2024) study. This method of instruction encourages students to practice speaking in a controlled environment, which helps them gain self-assurance and refine their language abilities. AIoT devices such as smart speakers or language learning apps can integrate with ChatGPT to offer interactive language practice opportunities.

Internet of Things (IoT) devices that can recognize and translate spoken languages in real-time provide a more engaging linguistic environment (Li, et al. 2024). Students can use these gadgets to communicate in a foreign language in their daily lives and have programmed smart home equipment to speak the target language, enabling learners to consistently engage in the native language. Similar Dhivvya and Karnati, (2024) argue to how people pick up their native tongue, this kind of immersion helps students pick up new languages quickly. Timely feedback and corrections are typically necessary for language learners to enhance their skills while learning in a classroom setting. Because it is an AI language model, Chatbot can provide you with immediate comments on your grammar mistakes, and pronunciation, sentence structure, and vocabulary (Shin et al.) Jinming and Daniel, (2024). The student is able to quickly recognize and fix their mistakes, which reinforces the correct use of language. ChatGPT capabilities, when integrated into language learning systems enabled by the Internet of Things, can enhance the efficiency and effectiveness of learning loops.

AIoT tools enable language learning to transcend physical location and time constraints. Through their Internet of Things (IoT) devices, students can access language learning materials at any time. According to Boonchom et al. (2024) learning a new language becomes more accessible to more people because of this adaptability, which takes into account different learning styles and timetables. In addition, students can make the most of their study time by practicing their language abilities while engaging in common activities such as traveling, playing sports, or cooking. Lehman, (2024) stated by understanding other cultures intrinsically relates to learning a new language, which is why AIoT therapies enhance cultural awareness and context. By supplying cultural insights, colloquial idioms, and real-life events, AIoT solutions can put language instruction into perspective. The Internet of Things (IoT) can teach students to appreciate linguistic diversity. According to Kim and Su (2024), this will aid students in politely and successfully interacting with native speakers. Learners are better prepared for meaningful cross-cultural encounters when they use a comprehensive approach to language learning.

According to Al Yakin (2024), AIoT can help remove obstacles to language learning in a variety of settings. The ability of Internet of Things (IoT) devices to translate text or speech into another language, including English, allows for effortless communication between people who speak various languages. The effects on international cooperation and connections are substantial and far-reaching. An innovative breakthrough with far-reaching consequences is the incorporation of AIoT into language instruction. Learners can engage in dynamic discussions, immerse themselves in authentic language use, practice to the max, and accurately imitate native speakers—all while embarking on a personalized language learning path. Language learning becomes more accessible, efficient, and culturally enriching by utilizing the capabilities of AI-powered IoT devices and language model in a new era of global communication and comprehension.

METHODOLOGY

This study set out to assess how ChatGPT worked in enhancing students' critical thinking and communication abilities (CCT) in English as a Foreign Language (EFL) courses. We randomly selected 46 fourth-semester students from a population of 350 students from Indonesian universities, who participated in the level 4 English program, for this study. Questionnaires and in-person interviews were the tools of choice in this mixed-methods study. Following eight weeks of English lessons using AIoT, we asked students to complete a questionnaire measuring their communication and critical thinking abilities. The questionnaire consisted of ten questions about communication and ten questions about critical thinking. We use SPSS version 26 to test the data and produce numbers, percentages, and averages using descriptive statistics.

We sent a survey to students in the fourth-level English Sunset program who took eight weeks of courses in public health, agribusiness, and AIoT (ChatGPT, Google Sites, and YouTube) to better understand the structure of the university's EFL classrooms. Based on the features listed, it appears that not all EFL classrooms are implementing AIoT. Next, we'll go over the eight weeks of instruction that cover the topic of artificial intelligence (AIoT) and its potential applications in English as a Foreign Language (EFL) classrooms. This study lays the groundwork for future research on the role of AIoT in improving critical thinking and communication abilities by utilizing 21st century learning strategies in English as a Foreign Language (EFL) courses at the university level.

Next step, we conducted in-depth interviews with students to gain a better understanding of the use of AIoT in the classroom. The study interviewed six participants on the topic of using AIoT-ChatGPT to enhance critical thinking and communication abilities, selecting them based on their availability and positive survey responses. The interviewer, an AIoT user, has been using chatbots to improve English conversation skills for at least two months. We gave participants a five-point scale to rate their communication proficiency, chatbot comfort, AIoT-assisted critical thinking, and motivation to learn with AIoT. Lower scores indicated less perceived AIoT impact, while higher scores indicated significant impact and more substantial. We do this to ensure that the results are credible. According to Flick (2009), the interviewees satisfied the study's inclusion criteria because their average questionnaire score was 4 and 3 Individuals Involved

Participants

The study population consisted of 350 undergraduate students. Of 46 EFL students in total filled out this survey from agriculture department. Following their introduction to AIoT, participants in this study voluntarily contributed their time. The vast majority of EFL students have never utilized AIoT in their coursework. All participants in the English Sunset level 3 program had an intermediate level of English proficiency; the study included students from Al Asyariah Mandar University in West Sulawesi, Indonesia. We used a random selection approach to select the research sample, ensuring an equal chance of inclusion for each group of potential participants.

Figure 1. AIoT tools for CCT students practices

The Procedure of Collecting Data

An organized method that provides useful insights into numerous aspects of research is involved in analyzing data received from surveys employing a Likert scale. In this study, researchers in the field of English as a foreign language use the Likert scale to assess undergraduate students' attitudes towards AIoT interventions. The researcher used a preset 5-point scale (strongly agree 5, agree 4, neutral 3, disagree 2, strongly disagree 1) to gauge the participants' levels of agreement or disagreement with a sequence of statements in the questionnaires.

Table 1. Questionnaire scores categorized according to students' replies.

Score		**Classification**
5	**85-100**	**Very good**
4	75-84	Good
3	65-74	Fairly
2	45-64	Low
1	0-44	Very low

Data Analysis Procedure

Using a quantitative systematic approach, this study analysed data using Likert scales. After data collection is complete, the next step is to analyse and interpret the findings using SPSS, or the Statistical Package for the Social Sciences. The first stage involves assigning a numerical number to each answer choice—often ranging from 1 for strongly disagreeing to 5 for strongly agreeing—and entering the obtained data into the software. The researcher then determined the standard deviation, median, and mean for each Likert scale item. These data, which summarize the participants' responses, reveal key patterns and the distribution of their answers. We administered questionnaires to collect data for the Likert scale data analysis, and then processed and analysed the data using software like SPSS. Through this systematic methodology, researchers can gain useful insights into participants' attitudes and views to better understand the AIoT phenomenon under study.

Findings

The results showed that 46 participants were eligible for inclusion in the analysis; this was due to their eagerness to share their thoughts on how AIoT will affect students' ability to communicate and think critically in English classes, as well as their general interest in the language. In addition, the assignment required analysing the survey responses provided by each student. The researcher might use the following table, which summarizes the questionnaire evaluation results, to make sense of the students' scores.

Table 2. Descriptive statistics of students CCT skills

Statements	N	Minimum	Maximum	Mean	Std. Deviation
Communication Skills					
1. I highly recommend Google Sites, YouTube, Mind mapping and ChatGPT for English language study	46	4.00	5.00	4.6087	.49344
2. I believe that my English-speaking skills have improved after utilizing AIoT.	46	4.00	5.00	4.6087	.49344
3. ChatGPT, AI Cicibot, and friends are great ways for me to hone my English communication abilities.	46	4.00	5.00	4.6087	.49344
4. To hone my English communication skills, I use ChatGPT, my AI friends Cici, and my human friends. Look at some videos on YouTube	46	4.00	5.00	4.2391	.43127
5. The Internet of Things (IoT) enhances all of modes of English communication.	46	4.00	5.00	4.6739	.47396
6. . In my opinion, using AIoT for English language learning improves communication skills.	46	4.00	5.00	4.6522	.48154

continued on following page

Table 2. Continued

Statements	N	Minimum	Maximum	Mean	Std. Deviation
7. When studying English, I frequently make use of AIoT devices or apps.	46	4.00	5.00	4.8261	.38322
8. I confidence to speak in English now	46	3.00	5.00	4.3043	.59140
9. The use of AIoT enhances the desire to study English and the self-assurance to speak.	46	2.00	5.00	3.7826	1.07317
10. In my opinion, AIoT has been a great help to me as I've worked to enhance my English communication abilities.	46	3.00	5.00	4.0652	.49000
Critical Thinking Skills.					
11. Using AI-driven language learning tools sparked my interest in studying English.	46	4.00	5.00	4.4130	.49782
12. The benefit of learning English using AIoT technology is that it improves my ability to analyze and comprehend lengthy English texts.	46	3.00	5.00	3.8043	.80608
13 I am more invested in my English language studies when I use AI chatbots and IoT solutions.	46	3.00	5.00	4.3261	.70093
14. Using AI-powered language apps Internet of Things (IoT) chatbots force me to use critical thinking skills whenever I resolve issues using language	46	3.00	5.00	4.1739	.85126
15. The AI-assisted language learning system adapts to my style and speed.	46	3.00	5.00	4.1739	.85126
16. AIoT promotes group projects and conversations as a means of English language acquisition, which in turn encourages critical thinking	46	3.00	5.00	4.1739	.85126
17. The AI-driven language tool provides me with targeted English activities that test my cognitive abilities	46	3.00	5.00	4.0000	.55777
18. AIoT technology positively affects the capacity to relate concepts when learning English	46	3.00	5.00	4.1522	.66558
19. I was able to overcome the challenges of critically evaluating language learning research and come to my own conclusions	46	4.00	5.00	4.4565	.50361
20. AIoT tools can help to improve critical thinking skills and information source analysis is possible with the use of AIoT tools.	46	4.00	5.00	4.4565	.50361
Valid N (listwise)	46				

Table 2 shown the study reveals that students highly appreciate the role of Artificial Intelligence (AIoT) technology in enhancing their English language studies. Students endorsed Google Sites, YouTube, and ChatGPT for English language study, with a mean score of 4.6087. They also believed that AIoT tools improved their English-speaking skills. The use of ChatGPT, AI-Cicibot, and peer interactions for honing communication abilities also had a high mean score of 4.6087. When combined with traditional methods, the mean score was slightly lower at 4.2391. Overall, students felt that AIoT significantly aided their English communication abilities, with a mean score of 4.0652.

Critical thinking skills were also positively impacted by AIoT tools. Students showed interest in studying English, improved comprehension and analysis of lengthy texts, increased engagement with AI chatbots, and improved critical thinking in problem-solving. AIoT also promoted group projects and discussions as learning methods, targeted cognitive activities, and the ability to relate concepts in learning English. Overall, AIoT tools were seen to improve critical thinking skills and the ability to analyze information sources.

Table 3. Descriptive statistics of the students mean score

	N	Minimum	Maximum	Mean	Std. Deviation
CCTSkills	46	76.00	96.00	86.5000	5.98424
Valid N (listwise)	46				

Based on the data in Table 3 described the mean scores of 46 students' Communication and Critical Thinking (CCT) skills, revealing a high level of perceived pro ficiency. The average mean score was 86.5000, with a standard deviation of 5.98424. The results suggest that students generally perceive their CCT skills positively, with mean scores indicating high confidence and ability. The AIoT-enhanced learning environment effectively fosters essential communication and critical thinking skills. The data of the students mean score also presenting in the graphic in figure 1 below.

Figure 2. The histogram and Q-QPlot of students CCT skills

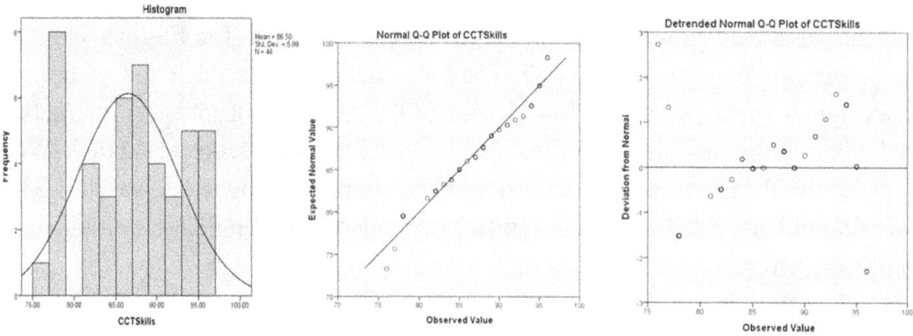

The histogram and Q-Q plot of students' Communication and Critical Thinking (CCT) skills scores in Figure 2 provide a detailed analysis of their distribution. The histogram shows a symmetric distribution, with a peak at 86.50, suggesting a high mean score. The Q-Q plot, on the other hand, shows a normal distribution with few extreme deviations. The combined analysis indicates that the AIoT-enhanced learning environment effectively supports the development of students' communication and critical thinking skills, resulting in a high and relatively consistent level of proficiency across the sampled students. The normal distribution of scores further suggests that educational interventions using AIoT are equitably beneficial, with most students experiencing similar levels of improvement.

Table 4. Reliability statistics

Cronbach's Alpha	N of Items
.815	20

Table 4 displays a 20-item scale that evaluates students' CCT skills and provides reliability statistics. There is strong item-to-item correlation and excellent internal consistency (Cronbach's alpha =.815). An acceptable Cronbach's alpha value is over 0.7, and a high score is above 0.8. The 20-item scale offers a strong evaluation of internal consistency since it is based on a broad set of items.

The high reliability score (.815) of the 20-item scale, which assesses students' CCT skills, is crucial for ensuring the accuracy of the study's conclusions and its application in future studies or educational assessments. This reliable finding indicates that AIoT tools positively impact students' communication and critical thinking skills. In addition to serving as a consistent tool for researchers and educators, the reliability of the scale makes it suitable for use in future studies or educational programs that aim to test CCT skills.

Table 5. ANOVA

	Sum of Squares	df	Mean Square	F	Sig.
Between Groups	215.451	10	21.545	4.735	.000
Within Groups	159.267	35	4.550		
Total	374.717	45			

Table 5 displays the analysis of variance (ANOVA) results comparing the CCT skills of various student groups. The ANOVA findings indicate a significant variation in CCT skills scores between groups, with the sum of squares between groups (215.451) significantly larger than the sum of squares within groups (159.267). Given the total number of observations and the number of groups, the degrees of freedom values are suitable. The fact that the mean square between groups is 21.545 and the mean square within groups is 4.550 shows that there is a significant difference in variability between the two sets of data. It is quite improbable that the observed differences in the groups' mean CCT skills scores happened by coincidence, given the statistical significance of the F-value (4.735) and the p-value (.000).

The findings of the analysis of variance show that the various groups of pupils tested had significantly different CCT abilities. The significant group differences warrant additional investigation into their origins because of the high F-value and extremely low p-value. The study's implications include the need for more research into the factors that differentiate the groups, the identification of more effective educational interventions or teaching methods for CCT skill development, and the use of these findings to create targeted programs that improve CCT skills for students from all backgrounds, ultimately leading to better educational outcomes.

Table 6. Correlations communication and critical thinking

		Communication	Critical Thinking
Communication	Pearson Correlation	1	.419**
	Sig. (2-tailed)		.004
	N	46	46
Critical Thinking	Pearson Correlation	.419**	1
	Sig. (2-tailed)	.004	
	N	46	46

**. Correlation is significant at the 0.01 level (2-tailed).

Table 6 demonstrates the correlation between the students' critical thinking and communication abilities. The correlation coefficients indicate a positive association between improving one's critical thinking abilities and improving one's communication abilities. At the 0.01 level, there is a statistically significant association, as indicated by the correlation's significance value of.004. Results show a moderately good correlation between communication and critical thinking abilities; the study included 46 participants. Enhancing students' communication skills also appears to enhance their critical thinking skills. A p-value of.004 indicates that the observed link is statistically significant and not the result of chance alone.

Because of the positive relationship, it stands to reason that courses that teach students to communicate effectively may also help them become better critical thinkers, and the reverse is also true. Teachers can help students' cognitive development in all areas. It is possible that teaching students to think critically and communicate well would be more effective if done in a more integrated fashion. To further understand this connection, other studies should investigate its mechanisms, such as which forms of communication exercises have the greatest impact on improving critical thinking. The correlation study highlights the interrelated nature of students' communication and critical thinking skills, demonstrating a substantial positive association between the two.

Additionally, semi-structured interviews will strengthen the data obtained about the influence of AIoT on developing communication and critical thinking skills for learning English in the context of the English Sunset program in a hybrid classroom environment. The following is a summary of the interview findings:

> Q1: How do you think students' communication abilities, particularly their English language acquisition, will change because of AIoT integration into language learning?

Extract 1: December 19th, 2023.
"I believe that using AIoT in language study could completely transform the way I improve my English-speaking abilities." With the help of AIoT-enabled devices and applications, I can create my own unique learning routes that will put my

speaking and listening skills to the test in real-world scenarios while providing me with immersive language experiences and real-time feedback".

Q2: "How are specific AIoT tools or applications affecting your English communication skills?" and the question asks you to provide examples. The second excerpt is from December 19th, 2023.

"There is no doubt. A language learning software that utilizes AIoT exemplifies ChatGPT's shortcomings in evaluating and correcting pronunciation. One great thing about this software, though, is that it can quickly identify where my pronunciation is lacking by listening to their speech and comparing it to native patterns".

Q 3: Is it based on technology or more conventional approaches to language study?

This is the third extract from the document dated December 19th, 2023..
"Hmm... AIoT can offer useful feedback and practice, but there are subtleties in human communication like body language and cultural background that machines can't capture."

Q4 is: What is your take on the rapidly developing area of artificial intelligence-assisted language learning?

Extract 4. December 19th, 2023.
" AIoT tools can create compelling language learning experiences, in my opinion. Stay updated on the latest technological advancements and their potential applications in the classroom. While AIoT can be a useful tool for students, they also need to make an effort to practice having natural conversations in order to hone their communication abilities. I believe that parents may help their children develop a well-rounded linguistic skill set by exposing them to both human interaction and activities that utilize artificial intelligence and the internet of things"

Q 5: How does the use of AIoT in a language classroom influence your capacity for critical analysis?

The extracts 5 and 2 are from December 19th, 2023.
"In my opinion, the integration of AIoT into language learning brings interactive and adaptive elements that stimulate students' critical thinking," the author writes. "We are motivated to analyze and make decisions about what and how

they learn because we believe that intelligent language learning apps may personalize content based on individual development and preferences."

Q6: What kind of artificial intelligence (AI) applications exist that help students develop their critical thinking skills while studying English?

Exctract 6. December 19th, 2023.
"There is no doubt. A language learning platform that analyzes students' reading habits, vocabulary retention, and comprehension levels using AIoT is one example. With this information, we can better tailor reading recommendations to each student, inspire them to critically examine a variety of books, and help them synthesize knowledge for improved analytical and critical reading abilities".

Q 7: The application of technology in this manner intrigues me. Could you talk about the effects of AIoT on group critical thinking skills in English language learners, as well as how collaborative learning fits into this?

Extract 3: December 19th, 2023.
"There is no doubt. By allowing students to practice language and communication in real-time, regardless of their geographical location, AIoT can promote collaborative learning.

In my opinion, students can practice critical thinking skills through debate, argument analysis, and the development of coherent responses to other viewpoints in virtual language exchange programs or group conversations facilitated by artificial intelligence".

Q 8: We heard that virtual language exchange is a powerful tool. Equally important, are there any possible problems or downsides to integrating AIoT in this setting?

Exctract 1 and Extract 5: December 19th, 2023.
"Obviously, relying too much on AI-powered platforms may compromise students' ability to think critically on their own. Students may not have the opportunity to learn from a variety of sources or practice independent evaluation of course content if we only encourage them to use personalized content recommendations. I believe it's crucial to strike a balance between assisting AI and encouraging independent exploration".

Q9. We really appreciate your elaboration on that equilibrium. Just one last question before we wrap up: how can schools and teachers make the most of AIoT to help English language learners hone their critical thinking skills?

Extract 1-6 December 19th, 2023.
"Our recommendation is to use AIoT as an adjunct to more conventional forms of education rather than a replacement for them. Encourage your students to think critically about the recommendations that emerge".

DISCUSSION

The fact that AIoT significantly affects students' capacity for critical thinking and effective communication is one of the most groundbreaking new aspects of this research. This article delves into the ways in which the Internet of Things (IoT) is revolutionizing education by enhancing students' capacity for critical thinking and effective communication. AI-assisted communication capabilities create a personalized learning environment. With the assistance of AIoT, it is possible to design a classroom specifically for each student. AIoT can cater to students' learning preferences and communication methods. Because of its flexibility, educational information can be more interesting, helping pupils improve their communication skills. According to this study, AIoT allows students at the undergraduate level to improve their critical thinking abilities and communicate with human-machine simulations. In this study, students get the chance to interact in real-world settings through AIoT-based simulation. In this study, students learn English as a second language using immersive experiences that include conversations with language bots driven by artificial intelligence or interactions with Internet of Things devices, use mid mapping, Cici bot and YouTube to improve their communication skill and critical thinking skill (CCT) in Table 2.

AIoT allows students to collaborate with peers worldwide, enabling cross-border collaboration. In ChatGPT classrooms powered by the Internet of Things (IoT), students from all over the world may work together to solve issues, share ideas, and improve their communication skills in a multicultural setting. Professors and institutions in this study have not adequately supported undergraduates to participate in international collaborations aimed at enhancing their English language skills, thereby hindering the development of their communication abilities through these interactions in Figure 1. The core of effective science education lies in this AIoT research, which engages students in learning through experiments, activities, and projects based on their own questions. This strategy enhances students' ability to apply their knowledge to real-world problems. In class, students use ChatGPT for experimental projects under the premise that it is a reference tool and an assistant robot that facilitates responsiveness by giving efficient feedback. Table 1 clearly shows that critical thinking abilities are required for the anticipated participation and feedback.

The key to successful education is helping students make connections between what they learn in science class and their real lives. As a result, undergraduates will find the content more engaging and relevant. Teachers can see a considerable improvement in their students' English communication skills when they use ChatGPT to create dialogues and summaries of course material, which they then use to practice real-world conversation with their classmates.

Adapting lessons to the specific requirements of each student is a hallmark of high-quality language education. Teachers employ a range of strategies to cater instruction to the unique requirements of their students, including delivering content in many forms and administering multiple-choice exams through ChatGPT. Both ChatGPT and the development of the Internet of Things are integral parts of a set of educational activities. Incorporating Google, YouTube, Cicibot and ChatGPT into the classroom helps with instruction. Improving one's capacity for both oral and written expression and analysis is the goal of this set of AIoT exercises based on interview data. The hallmark of an effective English class in this study similar with Aeni et al. (2024) findings when the students working together to solve issues and share ideas. Because of this, undergraduate group projects are a lot more successful in fostering students' analytical, collaborative, and interpersonal skills. They collaborate by sharing resources, implementing AIoT in the classroom, and receiving and offering feedback on ideas.

According to this study, students may benefit from science classes that make good use of technology resources to strengthen their analytical, logical, and communicative abilities, as well as their grasp of more advanced scientific topics. Various online resources and apps, including simulations of the AIoT and virtual labs, can be considered "digital tools" in this sense. Furthermore, successful science teachers utilize a battery of assessments to gauge their pupils' development and guide their instruction. This broad category includes many kinds of evaluations. Effective scientific knowledge dissemination in the classroom requires teachers to stay abreast of and make use of the most recent research, best practices in writing, EFL competencies, metacognitive skills and technological developments in their professions also highlighted by the scholars Yuen and Schlote, (2024); Wan and Moorhouse, (2024); Lee and Lim, (2024); Apriani et al, (2024); Al Yakin et al. (2024) findings.

We infer from AIoT's comments and our research on ESL students' communication and critical thinking abilities that it would be helpful for these students. While AIoT can be beneficial, instructors should supervise its use in class before students use it independently, ensuring they consult with their professors to ensure optimal utilization. Maximizing the benefits of AIoT requires adapting lessons to individual students' English skill levels. Chatting with AIoT might not be the best idea for a beginner English learner, for instance. Alternatively, we can provide students with fundamental reading, writing, and vocabulary exercises using Google, AI bot and ChatGPT.

Students have different learning styles; some remember more from reading, while others do better with auditory or physical learning. AIoT can accommodate these differences. In order for AIoT to be effective in the classroom, its implementation must take into account individual students' learning styles and requirements. Students can use ChatGPT to find relevant resources based on what interests them. For example, students who are struggling with American or British culture or language can use ChatGPT, Google Scholar, or YouTube to locate videos, articles, and podcasts about AIoT.

CONCLUSION

The findings demonstrate a positive relationship between AIoT usage, communication skills, and critical thinking abilities in English as a Foreign Language (EFL) students. Increased engagement with AIoT tools enhances students' understanding and proficiency in English. The gradual implementation of AIoT approach across various educational levels has a significant impact on learning activities, particularly in the context of EFL. The study highlights the multifaceted benefits of integrating AIoT technology into English courses, such as vocabulary expansion, listening skills, collaboration and teamwork, discussion and critical idea sharing, and visualization of activities. The AIoT program also offers the capability to automatically grade videotaped project-based spoken presentations uploaded to YouTube, introducing AIoT literacy to educators and tailoring learning instructions to meet undergraduate students' specific needs. The implications of this study extend beyond immediate educational benefits, calling for broader involvement of educators, practitioners, educational authorities, and decision-makers in exploring the potential benefits and challenges of AIoT technology in higher education. By emphasizes the transformative potential of AIoT in enhancing educational outcomes, preparing students for the demands of the modern, technologically driven world.

This study provides valuable insights into the effectiveness of artificial intelligence (AIoT) in education, particularly in English as a Foreign Language (EFL). It demonstrates that AIoT enhances students' critical thinking, communication, listening, collaboration, and vocabulary skills. The research also introduces novel evaluation approaches, such as automated grading for speaking presentations in filmed projects also presents a framework for AIoT integration, providing a roadmap for successful technology integration. It also promotes AIoT literacy among educators and students, emphasizing its importance in navigating new technology in educational settings.

For future research we recommended to examine AIoT tools' performance in diverse educational contexts and compare conventional classroom instruction with AIoT-based ones. We should conduct comparative studies to gain a comprehensive understanding of the advantages and limitations of AIoT solutions.We need more research to comprehend how AIoT tools affect student engagement and motivation. We should investigate teacher training and support for successful AIoT tool implementation. We need to address ethical considerations such as data privacy, algorithmic bias, and the digital divide. Finally, we should design AIoT tools with students in mind, enabling teachers and programmers to better meet their needs. In conclusion, this study underscores the potential of AIoT in improving educational outcomes and offers numerous avenues for future research.

REFERENCES

Aeni, N., Khang, A., Al Yakin, A., Yunus, M., & Cardoso, L. (2024). Revolutionized Teaching by Incorporating Artificial Intelligence Chatbot for Higher Education Ecosystem. In *AI-Centric Modeling and Analytics* (pp. 43–76). CRC Press.

Al Yakin, A. M., Obaid, A. J., Abdul, L., Warsah, I., Muthmainnah, M., & Elngar, A. A. (2024). Harnessing the Power of ChatGPT to Explore Student Metacognitive Skills in Learning Sociology Education. In *Advanced Applications of Generative AI and Natural Language Processing Models* (pp. 405–423). IGI Global.

Alabbas, A., & Alomar, K. (2024). Tayseer: A Novel AI-Powered Arabic Chatbot Framework for Technical and Vocational Student Helpdesk Services and Enhancing Student Interactions. *Applied Sciences (Basel, Switzerland)*, 14(6), 2547. DOI: 10.3390/app14062547

Alrajhi, A. S. (2024). Artificial intelligence pedagogical chatbots as L2 conversational agents. *Cogent Education*, 11(1), 2327789. DOI: 10.1080/2331186X.2024.2327789

Alshewiter, K. M., Shawaqfeh, A. T., Khasawneh, A. J., & Alqudah, H., jadallah abed Khasawneh, Y., & Khasawneh, M. A. S. (2024). Improving the Learning of Language Proficiency at Tertiary Education Level Through AI-Driven Assessment Models and Automated Feedback Systems. *Migration Letters : An International Journal of Migration Studies*, 21(2), 712–726.

Apriani, E., Muthmainnah, M., Wijayanti, E., Kons, B. A., & Al Yakin, A. (2024). 19 Accelerating Student's. *AI-Oriented Competency Framework for Talent Management in the Digital Economy: Models, Technologies, Applications, and Implementation*, 305.

Bekou, A., Mhamed, M. B., & Assissou, K. (2024). Exploring opportunities and challenges of using ChatGPT in English langua

Boonchom, W., Piyanukool, S., & Prachanant, N. (2024). Trends of Using Artificial Intelligence (AI) Technologies in Research Studies of English Language Teaching. *BRU ELT JOURNAL*, 2(1), 34–50.

Borkovska, I., Kolosova, H., Kozubska, I., & Antonenko, I. (2024). Integration of AI into the Distance Learning Environment: Enhancing Soft Skills. *Arab World English Journal (AWEJ) Special Issue on ChatGPT*.

Clark, D. (2024). *Artificial Intelligence for Learning: Using AI and Generative AI to Support Learner Development*. Kogan Page Publishers.

Cooray, S., Hettiarachchi, C., Nanayakkara, V., Matthies, D., Samaradivakara, Y., & Nanayakkara, S. (2024, April). Kavy: Fostering Language Speaking Skills and Self-Confidence Through Conversational AI. In *Proceedings of the Augmented Humans International Conference 2024* (pp. 226-236). DOI: 10.1145/3652920.3652944

Darwish, O., Al-shorman, A., AlSobeh, A., & Tashtoush, Y. (2024, April). A Survey Analysis of Internet of Things (IoT) Education Across the Top 25 Universities in the United States. In *International Conference on Advanced Information Networking and Applications* (pp. 497-510). Cham: Springer Nature Switzerland. DOI: 10.1007/978-3-031-57931-8_47

Dhivvya, J. P., & Karnati, S. B. (2024, March). BuddyBot: AI Powered Chatbot for Enhancing English Language Learning. In *2024 IEEE International Conference on Interdisciplinary Approaches in Technology and Management for Social Innovation (IATMSI)* (Vol. 2, pp. 1-6). IEEE.

Duong, T., & Suppasetseree, S. (2024). The Effects of an Artificial Intelligence Voice Chatbot on Improving Vietnamese Undergraduate Students' English Speaking Skills. *International Journal of Learning. Teaching and Educational Research*, 23(3), 293–321.

Durach, C. F., & Gutierrez, L. (2024). "Hello, this is your AI co-pilot"–operational implications of artificial intelligence chatbots. *International Journal of Physical Distribution & Logistics Management*, 54(3), 229–246. DOI: 10.1108/IJPDLM-01-2024-0031

Ebadi, S., & Amini, A. (2024). Examining the roles of social presence and human-likeness on Iranian EFL learners' motivation using artificial intelligence technology: A case of CSIEC chatbot. *Interactive Learning Environments*, 32(2), 655–673. DOI: 10.1080/10494820.2022.2096638

Essel, H. B., Vlachopoulos, D., Essuman, A. B., & Amankwa, J. O. (2024). ChatGPT effects on cognitive skills of undergraduate students: Receiving instant responses from AI-based conversational large language models (LLMs). *Computers and Education: Artificial Intelligence*, 6, 100198. DOI: 10.1016/j.caeai.2023.100198

Fathi, J., Rahimi, M., & Derakhshan, A. (2024). Improving EFL learners' speaking skills and willingness to communicate via artificial intelligence-mediated interactions. *System*, 121, 103254. DOI: 10.1016/j.system.2024.103254

Fernández-Batanero, J. M., Montenegro-Rueda, M., Fernández-Cerero, J., & López Meneses, E. (2024). Adoption of the Internet of Things in higher education: Opportunities and challenges. *Interactive Technology and Smart Education*, 21(2), 292–307. DOI: 10.1108/ITSE-01-2023-0025

Han, J., & Lee, D. (2024). Research on the development of principles for designing elementary English-speaking lessons using artificial intelligence chatbots. *Humanities & Social Sciences Communications*, 11(1), 1–15. DOI: 10.1057/s41599-024-02646-w

Jeon, J. (2024). Exploring AI chatbot affordances in the EFL classroom: Young learners' experiences and perspectives. *Computer Assisted Language Learning*, 37(1-2), 1–26. DOI: 10.1080/09588221.2021.2021241

Jinming, D. U., & Daniel, B. K. (2024). A Systematic Review of AI-Powered Chatbots in EFL Speaking Practice: Transforming Language Education. *Computers and Education: Artificial Intelligence*, 100230.

Khlaisang, J., & Teo, T. (2024). An innovation-based virtual flipped learning system in a ubiquitous learning environment the 21st century skills of higher education learners. *Journal of Educational Technology & Society*, 27(1), 100–116.

Kim, A., & Su, Y. (2024). How implementing an AI chatbot impacts Korean as a foreign language learners' willingness to communicate in Korean. *System*, 122, 103256. DOI: 10.1016/j.system.2024.103256

Koç, F. Ş. (2024). The development of listening and speaking skills in EFL via an artificially intelligent chatbot application: A quasi-experimental design study.

Koç, F. Ş. (2024). The development of listening and speaking skills in EFL via an artificially intelligent chatbot application: A quasi-experimental design study.

Kondurkar, I., Raj, A., & Lakshmi, D. (2024). Modern Applications With a Focus on Training ChatGPT and GPT Models: Exploring Generative AI and NLP. In *Advanced Applications of Generative AI and Natural Language Processing Models* (pp. 186–227). IGI Global.

Kostikova, I., Holubnycha, L., Besarab, T., Moshynska, O., Moroz, T., & Shamaieva, I. (2024). ChatGPT for Professional English Course Development. *International Journal of Interactive Mobile Technologies*, 18(2).

Lee, K. A., & Lim, S. B. (2023). Designing a leveled conversational teachable agent for English language learners. *Applied Sciences (Basel, Switzerland)*, 13(11), 6541. DOI: 10.3390/app13116541

Lehman, W. (2024, May). Chat (bot) ting Up a Storm: How AI Is Changing the Game in Foreign Language Teaching. In *World Languages and Cultures in the Public Sphere: Selected Proceedings of the 25th Southeast Conference on Languages, Literatures, and Film* (p. 165). BrownWalker Press.

Li, B., Bonk, C. J., Wang, C., & Kou, X. (2024). Reconceptualizing self-directed learning in the era of generative AI: An exploratory analysis of language learning. *IEEE Transactions on Learning Technologies*, 17, 1489–1503. DOI: 10.1109/TLT.2024.3386098

Mamatnabiyev, Z., Chronis, C., Varlamis, I., & Zhaparov, M. (2024). Development of an Educational Robot for Exploring the Internet of Things. *International Journal of Advanced Computer Science and Applications*, 15(4). Advance online publication. DOI: 10.14569/IJACSA.2024.01504120

Mohamed, A. M. (2024). Exploring the potential of an AI-based Chatbot (ChatGPT) in enhancing English as a Foreign Language (EFL) teaching: Perceptions of EFL Faculty Members. *Education and Information Technologies*, 29(3), 3195–3217. DOI: 10.1007/s10639-023-11917-z

Monika, M., & Suganthan, C. (2024). A Study on Analyzing the Role of Chatgpt in English Acquisition among ESL Learners during English Language Classroom. *language, 8*(2).

Muthmainnah, M., Apriani, E., Seraj, P. M. I., Obaid, A. J., & Al Yakin, A. M. (2024). Nudging Motivation to Learn English Through a ChatGPT Smartphone-Based Hybrid Model. In *Advanced Applications of Generative AI and Natural Language Processing Models* (pp. 26–42). IGI Global.

Muthmainnah, P. M. I. S., & Oteir, I. "Playing with AI to Investigate Human-Computer Interaction Technology and Improve Critical Thinking Skills to Catch Up with the 21st Century Era", Education Research International, vol. 2022, Article ID 6468995, 17 pages, 2022. DOI: 10.1155/2022/6468995

Nguyen, A., Kremantzis, M., Essien, A., Petrounias, I., & Hosseini, S. (2024). Enhancing Student Engagement Through Artificial Intelligence (AI): Understanding the Basics, Opportunities, and Challenges. *Journal of University Teaching & Learning Practice*, 21(06). Advance online publication. DOI: 10.53761/caraaq92

Onesi-Ozigagun, O., Ololade, Y. J., Eyo-Udo, N. L., & Ogundipe, D. O. (2024). Revolutionizing education through AI: A comprehensive review of enhancing learning experiences. *International Journal of Applied Research in Social Sciences*, 6(4), 589–607. DOI: 10.51594/ijarss.v6i4.1011

Park, W., & Kwon, H. (2024). Implementing artificial intelligence education for middle school technology education in Republic of Korea. *International Journal of Technology and Design Education*, 34(1), 109–135. DOI: 10.1007/s10798-023-09812-2 PMID: 36844448

Rafique, H., Nazeer, I., & Rehman, J. (2024). The Impact of ChatGPT on Language Evolution: A Linguistic Analysis. *Journal of Education and Social Studies*, 5(1), 56–68. DOI: 10.52223/jess.2024.5106

Rajamohan, K., Rangasamy, S., Manoj, C. N., Mary, L., Sabu, R., & Jose, S. S. (2024). Internet of Things (IoT) as a Game Changer to the Education Sector. In *Advanced IoT Technologies and Applications in the Industry 4.0 Digital Economy* (pp. 251–274). CRC Press. DOI: 10.1201/9781003434269-14

Rezaei, S., & Ansary, A. (Eds.). (2024). *Artificial Intelligence of Things (AIoT) for Productivity and Organizational Transition.* IGI Global. DOI: 10.4018/979-8-3693-0993-3

Shin, D., Lee, J. H., & Noh, W. I. (2024). Realizing Corrective Feedback in Task-Based Chatbots Engineered for Second Language Learning. *RELC Journal*, •••, 00336882231221902. DOI: 10.1177/00336882231221902

Steiss, J., Tate, T., Graham, S., Cruz, J., Hebert, M., Wang, J., & Olson, C. B. (2024). Comparing the quality of human and ChatGPT feedback of students' writing. *Learning and Instruction, 91*, 101894. ge teaching (ELT) in Morocco. *Focus on ELT Journal*, 6(1), 87–106.

Tang, F. (2024). Understanding the role of digital immersive technology in educating the students of english language: Does it promote critical thinking and self-directed learning for achieving sustainability in education with the help of teamwork? *BMC Psychology*, 12(1), 144. DOI: 10.1186/s40359-024-01636-6 PMID: 38481260

Tuan, N. T., Moore, P., Thanh, D. H. V., & Pham, H. V. (2024). A Generative Artificial Intelligence Using Multilingual Large Language Models for ChatGPT Applications. *Applied Sciences (Basel, Switzerland)*, 14(7), 3036. DOI: 10.3390/app14073036

Ullah, I., Khan, I. U., Ouaissa, M., Ouaissa, M., & El Hajjami, S. (Eds.). (2024). *Future Communication Systems Using Artificial Intelligence, Internet of Things and Data Science*. CRC Press. DOI: 10.1201/9781032648309

Wahshat, H., Khalifeh, A., Taha, A., Wahsheh, F., Amayreh, K., & Matalka, M. (2024). Individual, technological, organizational, and environmental factors impact of the internet of things on e-learning adoption in higher education institutions in Jordan. *International Journal of Data and Network Science*, 8(3), 1451–1462. DOI: 10.5267/j.ijdns.2024.3.020

Wan, Y., & Moorhouse, B. L. (2024). Using Call Annie as a Generative Artificial Intelligence Speaking Partner for Language Learners. *RELC Journal*, •••, 00336882231224813. DOI: 10.1177/00336882231224813

Yuan, L. (2024). Where does AI-driven Education, in the Chinese Context and Beyond, go next? *International Journal of Artificial Intelligence in Education*, 34(1), 31–41. DOI: 10.1007/s40593-023-00341-6

Yuen, C. L., & Schlote, N. (2024). Learner Experiences of Mobile Apps and Artificial Intelligence to Support Additional Language Learning in Education. *Journal of Educational Technology Systems*, 52(4), 00472395241238693. DOI: 10.1177/00472395241238693

Zhang, C., Meng, Y., & Ma, X. (2024). Artificial intelligence in EFL speaking: Impact on enjoyment, anxiety, and willingness to communicate. *System*, 121, 103259. DOI: 10.1016/j.system.2024.103259

Chapter 13

Transforming Civic Education With ChatGPT to Boost Student Interaction and Social learning in the GenAI Era

Ahmad Al Yakin
https://orcid.org/0000-0003-3170-2374
Universitas Al Asyariah Mandar Sulawesi Barat, Indonesia

A. Ramli Rasjid Rasyid
Universitas Negeri Makassar, Indonesia

Ali Said Al-Matari
https://orcid.org/0000-0002-8124-4730
A'Sharqiyah University, Oman

Muthmainnah Muthmainnah Muthmainnah
https://orcid.org/0000-0003-3170-2374
Universitas Al Asyariah Mandar, Indonesia

Ahmed J Obaid
https://orcid.org/0000-0003-0376-5546
University of Kufa, Iraq

Luis M. Cardoso
https://orcid.org/0000-0003-3748-5593
Polytechnic Institute of Portalegre, Portugal & Centre for Comparative Studies, University of Lisbon, Portugal

Ahmad A. Elngar
https://orcid.org/0000-0001-6124-7152
Faculty of Computers and Artificial Intelligence, Beni-Suef University, Egypt

ABSTRACT

To increase social interaction and encourage independent learning in Pancasila and citizenship education classes, this project investigates the use of ChatGPT as a new teaching tool. The main goal is to test how well ChatGPT encourages student

DOI: 10.4018/979-8-3373-1399-3.ch013

participation and group work in different classroom environments. By offering empirical evidence of the efficacy of ChatGPT as a teaching tool, this research adds to the growing body of literature on AI in education. We have highlighted the importance of AI in facilitating social interactions and self-directed learning, which is essential for the growth of analytical and communicative abilities. Educators and lawmakers can use this research's practical insights on how to incorporate AI tools into school courses to improve student engagement and learning outcomes. To better equip their students to face the problems of the digital age, instructors should participate in ongoing professional development opportunities to learn how to use AI tools in the classroom.

INTRODUCTION

Sociology is very necessary in studying the current 5.0 society, including education. Additionally, sociology, as a subfield of social science, highlights many aspects of human civilization. It explores topics including social stratification, interaction, conflict, social norms, and deviance (Khalil, 2024). According to sociologists, "social interaction" is defined as a two-way exchange between people, groups, and groups themselves. Because social interaction is essential for the survival of humans and communities, it is an intrinsic and inevitable aspect of human beings. Humans tend to institutionalize and internalize behaviours that they engage in repeatedly to satisfy their needs. Therefore, communication and reciprocal interaction are very important for the establishment of bonds between people, communities, and other social entities (Snell et al., 2023).

In today's digital era, more and more people interact with each other's smartphones or devices. Although face-to-face communication is still important, technological advances have made it easier to communicate through other channels, such as messaging and the telephone (Oba and Berger, 2024). These changes demonstrate the power of human relationships to adapt to new situations. Associative contacts are more positive, and dissociative encounters are more negative. These are the two main types of social interaction. These interactions form social institutions. Humans' incessant pursuit of various demands in a manner that conforms to accepted norms gives rise to social structure and various institutions have emerged to meet different human needs Vom et al. (2024) Sociology plays a crucial role in the study of society, and sociology, as a subfield of the social and societal sciences, highlights many aspects of human civilization, including civilization in the digital era. According to Moretti Traver, et al. (2024), they studied topics such as social change, stratification, deviance, social norms, and interaction.

According to Gray and Dunn (2024), “social interaction” is defined as two-way exchanges between people, groups, and groups themselves. Since social interaction is essential to the survival of both people and communities, this is an inevitable and intrinsic aspect of being human. Humans tend to institutionalize and internalize behaviours that they engage in repeatedly to meet their requirements. Therefore, connecting with others is essential for building relationships in society. Social interaction and engagement are on the rise in today's digital age. Although face-to-face communication is still vital, technology has made it easier to communicate via other channels, like messaging and the phone (Shannon et al., 2023). This change demonstrates the power of human relationships to adapt to new situations. Associative contacts are more positive, and dissociative encounters are more negative. These are the two primary types of social interactions. These interactions shape social institutions. Humans' incessant pursuit of a variety of demands in a way that is consistent with accepted norms gives rise to social structures. A wide range of human needs have given rise to a wide range of institutions (Carlyle, 2024). Social engagement in the classroom is also crucial and classroom discussions foster growth in students' social and personal lives, in addition to their academic knowledge. It equips students with academic knowledge and important interpersonal skills, preparing them for a well-rounded and successful future. Less social engagement in the classroom can lead to a number of problems, such as less feedback and less student motivation, Okada et al. (2024); Agrahari, (2024).

Class group discussions are a great way to get students pumped up for class. Hughes (2024) study found that active discussion with instructors and peers may boost student engagement. When students are actively immersed in their learning, they may be more inclined to speak out, ask questions, and go more deeply into class topics. Without chances to make friends, students may lose interest in school and lose their will to learn. Furthermore, classroom social ties function as a feedback system. Students gain a lot when they get immediate feedback on their work from instructors and peers. This feedback loop is vital for their growth and comprehension. Less in-person interaction between students, instructors, and peers may reduce the amount of time students spend receiving comments on their work. Given the documented low level of social interaction among pre-service teachers in tertiary institutions' new technology courses in education, specifically in the Pancasila and Citizenship Education study programs, conducting this research is crucial. This could lead to students not actively sharing ideas and disrupting discussion activities both in and out of the classroom. For example, in pre-service teacher training, students might think about the social aspect of education by discussing what they've learned, expanding on one another's views, and asking questions all through class.

Lack of collaboration and interaction in the classroom can negatively impact students' learning and overall well-being. Only formal education can hone crucial social skills such as effective communication, teamwork, and empathy. The future of every student's career and personal success depends on their growth in these areas. Students might not have many chances to practice these abilities unless they talk to each other often; this study suggests using ChatGPT to encourage more social interactions while learning. To address this problem, teachers need to think about the strategy to incorporate social connections into the classroom if they want to solve this issue by facilitating chances for peer cooperation, online forums, class debates, and group activities. Teachers can aid their students' motivation, feedback, discussion participation, and skill development by promoting social contact. Therefore, this project seeks to fill the gaps by utilizing ChatGPT as a media and technology tool for university-level Pancasila and citizenship education classes, with the goal of increasing students' social interactions in class.

LITERATURE REVIEW

Achieving learning goals relies heavily on teachers and students interacting in the classroom. The art of effective and natural communication between educators and their students is a powerful force in education (Leng, 2024). Students have an easier time engaging with and understanding course material when instructors are excellent communicators. On the other side, learning can become complicated and difficult for pupils to embrace if teachers are unable to engage successfully (Marginson, 2024). Consequently, it is crucial for educators to have strong communication and engagement skills in order to excel in the classroom. Moreover, educational institutions frequently exhibit a diverse array of perspectives, Xu et all (2024). This includes disparities in religion, racial/ethnic background, socioeconomic status, and other areas. When students come from a variety of backgrounds and experiences, it encourages them to think critically, helps them support one another, and teaches them to value one another's viewpoints. In an inclusive classroom where diversity enhances the educational experience, we expect students to develop skills in critical thinking, effective collaboration, and respect for the diverse world around them. Dempere, et al. (2023) argue ChatGPT increased digital engagement, human-machine conversations, educational influence, recognition of social and cultural diversity, and ethical discourse about the changing nature of human relationships in the AI era are just a few ways in which ChatGPT is changing social interactions.

Taking part in social activities, whether they take place in person or online, is beneficial to one's mental health. This alleviates feelings of isolation, despair, and worry. Social contact is critical for an individual to learn and grow. Because of this, people can work together, share information, and gain new insights, all of which improve critical thinking. According to Rajaei (2024) and Alanezi (2024), by building professional relationships through social media is crucial in today's digital age for maintaining and expanding one's social circle. This facilitates opportunities, employment, and professional connections. By bringing people from all over the world together, digital platforms foster an environment that is more accepting of many cultures. Creativity and invention flourish in online communities where users are more likely to work together. Innovation in both solutions and goods can result from the free flow of ideas and criticism. In addition, social interaction is crucial to the success of online communities, which serve to unite people who share common interests, aspirations, or ambitions. In online education, social interactions are crucial because they promote student engagement, discussion, and peer learning (Tao, 2024; Muthmainnah et al. 2024). Digital interactions are valuable despite their differences from face-to-face encounters. The digital era has created a world that seamlessly blends physical and digital interactions. To live a full life, it is vital to balance these connections. People still require face-to-face communication, even in this technological age. Even though technology opens new ways to communicate, it shouldn't take the place of people's innate urge to interact with one another.

Social Interaction Model in the Classroom

Teachers, classmates, and the classroom setting all have a role in students' personal and academic growth, according to social interaction theory Saxer et al. (2024). Social constructivism, a learning theory that emphasizes the importance of social interaction in knowledge creation, closely aligns with social interaction theory. Hill et al. reported this in 2009. Students gain a comprehension of ideas and practice problem-solving abilities via collaborative learning, according to this notion. Zahra et al (2024) cites Lev Vygotsky's work on ZPD as a foundational concept in the field of social interaction theory. Here, we discuss the distinction between students' abilities under guidance and those under their own initiative. To help their pupils close this achievement gap, teachers play a crucial role in improving their students' learning. Social interaction theory relies heavily on classroom interactions like student-teacher and peer-to-peer dialogue.

Using techniques such as group work, peer instruction, and collaborative learning in lesson plans is a great way to tap into students' natural curiosity and enthusiasm for learning. Student collaboration allows for the free flow of ideas, the testing of hypotheses, and the development of body knowledge, Vandermaas (2024). Effective teacher-student relationships are just as important as interactions with peers when it comes to promoting learning. To aid their students in navigating the learning process, teachers offer direction, criticism, and encouragement. A classroom may be a safe and supportive place when teachers and students have good relationships. That is why talking to one another is fundamental to interacting with others. Effective verbal and nonverbal communication are crucial for idea transmission, concept clarification, and misunderstanding resolution. As a student's learns a language through interaction with others, there is a strong correlation between social engagement and language development (Zola and Zola, 2024).

Classroom social interactions foster collaborative problem-solving, which in turn improves students' critical thinking and problem-solving abilities. Students can draw from a variety of experiences and viewpoints when they collaborate on challenging assignments Al Yakin et al. (2024). Teachers give students comments to help them improve their metacognitive skills, and students often get feedback from their classmates when they work in groups, (Tzimas and Demetriadis, 2024). According to Al Sasi et al (2024) based on social interaction theory, the effective classroom practices include student-teacher collaboration, peer-to-peer communication, and positive relationships between instructors and their students. Asad and Malik, (2024) stated by promotes participation and knowledge building through interaction, acknowledging that learning is a social process affected by cultural and contextual influences. Because of its potential to boost student achievement, social engagement in the classroom is crucial, students can hone their communication and interpersonal abilities through classroom social interaction exercises. Students will gain a more thorough grasp of the subject matter through active participation in class discussions, group projects, and peer instruction (Lin and Wang, 2024; Saleem et al. 2024).

Collaborating with coworkers to discuss and understand newly created content and ideas is an example of a social interaction activity. Dispute resolution, cooperation, and teamwork are just a few of the important social skills that students can acquire in the classroom (Dragnić- et al., 2024). These skills are critical to academic and professional success. When given the chance to actively connect with one another while learning, students are more likely to form meaningful relationships, (Ouyang, et al. 2023). As a result, students are more likely to make lasting connections, which in turn can lead to a more welcoming and accepting classroom environment. Students may gain a sense of belonging through these types of connections. (Fasting, and Breilid, 2024; Chen et al. (2024) found that practice sessions with instructors and peers enhance stud Fasting, R. B., & Breilid, ents' verbal and nonverbal communication abilities. Learning to communicate effectively will benefit students in many areas of life beyond the classroom.

Cerasa, et al. (2024) and Flores et al., (2024) claim that the emergence of social relationships improves the mental health of students. Because of this, feelings of isolation, worry, and loneliness have less of a negative effect on school performance. Being around positive individuals has a multiplicative influence on one's mental health and academic performance; engaging in social interaction is one way to counteract this. As demonstrated by Tuan et al. (2024), the development of language (communication) and social skills also relies on interactions with teachers and classmates. Through practice and interaction, students can develop their social and language abilities, resulting in better language communication. Students' interactions with one another in class are valuable for more than just piqued interest in studying. According to Chang et al. (2024), it promotes students' intellectual, social, and emotional growth, which in turn helps them flourish both in and out of the classroom. The combination of technology and interpersonal communication in ChatGPT-mediated social engagement is intriguing. There has been a lot of buzz lately about OpenAI's ChatGPT, a cutting-edge language model that might have widespread applications in fields like education and society (Farazouli (2024), ChatGPT serves as a text-based discussion companion for its users. This opens the door to any kind of conversation, from deep dives into a topic to casual small talk.

As a kind of social interaction, colleagues exchange fresh ideas and content in order to better grasp it. Dispute resolution, cooperation, and teamwork are just a few of the important social skills that students can acquire in the classroom (Guerrero et al., 2024). These skills are critical to academic and professional success. When given the chance to actively connect with one another while learning, students are more likely to form meaningful relationships. Kansok et al. (2024) and Mattson (2024) claim students are more likely to make lasting connections, which in turn can lead to a more welcoming and accepting classroom environment. These types of connections can help students gain a sense of purpose and belonging (Dulfer, et al. 2024). According to studies by Dost (2024) students can develop a sense of purpose and belonging through these types of connections. Othman, et al. (2024) research reveals that practice sessions with instructors and peers enhance students' verbal and nonverbal communication abilities. Learning to communicate effectively will benefit students in many areas of life beyond the classroom.

Utilizing ChatGPT as a Platform for Social Interaction

Senter (2024) claim that the emergence of social relationships improves the mental health of students. Because of this, feelings of isolation, worry, and loneliness have less of a negative effect on school performance. Engaging in social interaction is one activity that can mitigate the negative impacts of being with positive people on mental health and academic achievement. Lover and Webber (2024) state that the development of social and linguistic competence depends on interactions with both instructors and classmates. Through practice and interaction, students can develop their social and language abilities, resulting in better language communication. Students' interactions with one another in class are valuable for more than just piqued interest in studying. According to Zammit (2024),social interaction its promotes students' intellectual, social, and emotional growth, which in turn helps them flourish both in and out of the classroom.

The combination of technology and interpersonal communication in ChatGPT-mediated social engagement is intriguing. There has been a lot of buzz lately about OpenAI's ChatGPT, a cutting-edge language model that might have widespread applications in fields like education and society. According to Chan and Colloton, (2024). ChatGPT serves as a text-based discussion companion for its users. This opens the door to any kind of conversation, from deep dives into a topic to casual small talk. At the same time, ChatGPT can serve as a learning medium and technology, according to Yu (2023) and Liu and Ma, (2024), making it easier for students to grasp complex ideas and utilize them as learning materials in educational settings. Answering questions, elaborating on subjects, summarizing, and offering learning aids can improve the educational experience. Furthermore, by providing explanations and examples, ChatGPT contributes to the creation of new knowledge. Educators and researchers can have meaningful discussions with ChatGPT about a wide range of subjects.

Artificial intelligence's impact and moral implications are examined within the framework of HRI exercises. ChatGPT's societal and educational use may result in both positive and negative outcomes. There are also concerns about prejudice, privacy, and the limitations of AI, even while it provides access to knowledge and support (Elbanna and Armstrong, 2024). As technology improves, ChatGPT will be able to mimic human emotions in social interactions, further blending the gap between humans and AI. ChatGPT has quickly become a popular platform for people to connect, learn, and share what they know. There are continuous debates over the social, emotional, and ethical consequences of human-AI interactions, and its usefulness and adaptability are growing. This research subsequently highlights the importance of ChatGPT in bridging knowledge gaps that develop in Pancasila and citizenship education classes.

Many benefits accrue from incorporating ChatGPT into classrooms, but two of the most important are the enhancement of students' social relationships and the improvement of their mental health. Academic performance suffers when students experience isolation, anxiety, and loneliness; nevertheless, the social bonds developed via these exchanges can greatly alleviate these symptoms (Kolade et al., 2024). Positive social connections, made easier by ChatGPT, can boost mental health and academic performance in a multiplicative way.

Interactions with instructors and classmates are crucial for the growth of social and linguistic competence, as pointed out by Tai (2024). With ChatGPT, students can work on their social and linguistic skills, which in turn improves their communication abilities. According to Kelly (2024), this kind of engagement is vital for students' emotional, social, and intellectual development in addition to their academic success. Without this kind of development, students can't succeed in school or in life.

ChatGPT is an interesting combination of technology and human communication; it's a conversational companion that can handle anything from serious philosophical debates to lighthearted small talk (Jo and Park, 2024). This feature enables ChatGPT to facilitate meaningful conversations and group projects in various ways. Fahad et al. (2024) assert that these encounters play a crucial role in cultivating important social skills like collaboration, conflict resolution, and academic and professional success.

Park and Ahn, (2024) research suggests that students who engage in class discussions and foster relationships with their classmates via online platforms like ChatGPT are more likely to achieve academic and social success. These relationships further enhance students' well-being and academic performance by fostering a sense of purpose and belonging (Ng, et al. 2024). ChatGPT-facilitated practice sessions can enhance students' verbal and nonverbal communication skills, which are advantageous in many areas of life (Liu et al., 2024).

Nevertheless, there are several restrictions associated with ChatGPT. Even though AI has the potential to be a game-changer in the classroom, many are still worried about privacy invasion, bias, and the technology's ethical ramifications. Concerns about AI's ability to propagate biases or improperly handle sensitive data are substantial. The intricacies of human emotions and social situations are difficult for AI to grasp and respond to, which limits its ability to completely mimic human interactions. The distance between human and AI interactions may get smaller as ChatGPT's capacity to imitate human emotions and behaviors improves with the advancement of technology. However, this raises ethical questions about the validity of such exchanges and their impact on interpersonal relationships. The ongoing discussions surrounding these topics highlight the importance of thoroughly examining and regulating the social, emotional, and ethical implications of interactions between humans and AI.

METHODOLOGY

This quantitative study conducted at teacher professional education at Universitas Al Asyariah Mandar. During their first semester of student teaching, aspiring educators have their social interaction practices observed and evaluated. Furthermore, support staff receive training to assist instructors in efficiently implementing the curriculum in their classrooms. It is worth mentioning that the preservice teachers program follows the guidelines set out by Eshelman and Hogue (2023) and places an emphasis on working together and getting their hands dirty. We expect teachers to actively participate in curriculum implementation by providing the necessary materials and resources. Achieving this principle relies heavily on classroom interaction's social input.

Research Participants

Participants in this research were pre-service teachers from Al Asyariah Mandar University who were the subjects of this research, totaling 20 pre-service teachers for the 2023 academic year. A mixed methods strategy that integrates quantitative and qualitative methods is part of the study plan. Statistical analysis was used to find pre-service teachers' responses in the quantitative section regarding the impact of ChatGPT in social interaction and social learning, which requires structured data collection using questionnaires and classroom observations. Qualitative data collected from the observation of pre-service teachers' social interactions in the classroom and documented them using field notes, recordings of classroom conversations, and audio and video transcripts.

The use of quantitative and qualitative data allows for a thorough examination of prospective teachers' social and reflective practices in educational group discussions and idea-sharing activities. Prospective educators who take Pancasila and citizenship education classes with new technology will find this learning approach very useful. The systematic approach of using ChatGPT as a teaching tool highlights the importance of thorough data collection and analysis in this study. The main aim of this research is to improve teaching methods by explaining the intricacies of social interactions and reflective activities of pre-service teachers by maximizing the benefits of ChatGPT.

Research Setting

This research uses ChatGPT in the classroom and finds that it can produce useful educational materials such as reading, problem-solving exercises, summaries, and sharing ideas or discussions. ChatGPT is a useful tool for encouraging communication between students in class and maximizing their interactions together during the learning process. The organized content provided by ChatGPT helps simplify difficult subjects and fosters a more engaging and interactive environment for learning.

Prospective teachers also use ChatGPT as a metasearch engine. To streamline the research process, we use ChatGPT to uncover credible sources, references, and detailed explanations about various issues. When put into practice in the classroom by post-service teachers, it is known that ChatGPT can start a conversation about anything, get students to think critically and debate controversial issues, and teach them how to discover, evaluate, and synthesize knowledge from AI-generated sources.

Pre-service teachers criticized ChatGPT's information, which also supported their understanding of the subject matter.

During the collaborative project stage, we discovered how ChatGPT simplified group conversations among pre-service teachers. They actively discuss, connect, and reflect on their knowledge. When these activities are planned, lecturers play a big part in keeping an eye on how students use ChatGPT to learn. They make sure that the content created by AI is correct and useful, and they teach future teachers how to think critically and analyze things. For example, they teach future teachers how to judge the reliability of ChatGPT output. Then pre-service teachers can ensure the correctness of the information from ChatGPT, they discuss it in class and also use additional credible resources such as Google Scholar and Google Play Books. ChatGPT designs this technique to complement its content by ensuring its accuracy and validity. Integrating these resources enables pre-service teachers to provide a more comprehensive education while increasing information reliability.

The curriculum in this study incorporates ChatGPT to enhance social interaction and social learning. Lecturers use ChatGPT as a teaching resource to monitor and ensure that pre-service teachers develop critical thinking, research skills, and dynamic and engaging class discussions. Instructors supervise their students to ensure they learn to critically assess AI-generated content while keeping their knowledge accurate and intact. This research highlights the importance of integrating technological resources with more conventional resources to build effective and engaging learning environments.

Figure 1. Introducing ChatGPT in the class

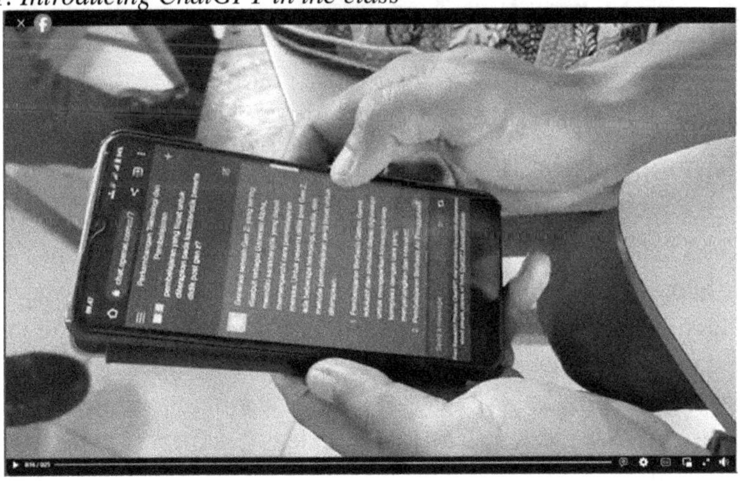

Figure 2. Social Interaction in the classroom

Figure 3. Discussion and demonstration the materials

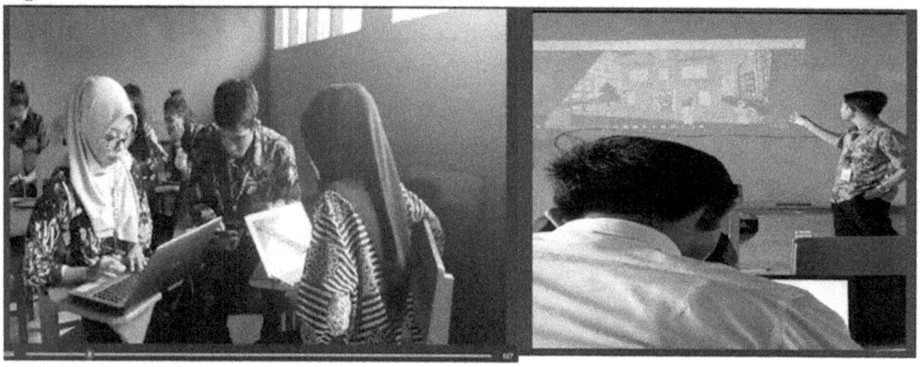

Procedures for Collecting Data

Surveys and direct observations of ChatGPT teaching were the primary tools for gathering information in this study. This research's goal can be better understood by using quantitative and qualitative data, which is what this approach aims to provide. One of the most important tools for collecting quantitative data is a questionnaire. Following Suárez et al. (2024); Aybar et al. (2024) instructions, this survey used a Likert scale for data collection. The Likert scale is a numerical system that allows measuring attitudes, opinions, or perceptions in an organized manner. The scale typically ranges from 1 (strongly disagree) to 5 (strongly agree). This strategy allows researchers to collect measurable, uniform data that can be easily analyzed statistically. Participants were asked to complete a survey measuring their experience with ChatGPT and its impact on their learning.

In addition to the quantitative data obtained from surveys, qualitative data was also taken from direct reports. This method is used to observe social interaction activities that occur during ChatGPT learning. Discussions, exchange of ideas, group concept formation, reflection processes, and other markers of social interaction are recorded and closely monitored. This qualitative method enriches the research by bringing out the intangible and nuanced aspects of human relationships. To give more meaning to the quantitative data collected from surveys, researchers make sure to record and document individual behavior and group dynamics.

This research used a mixed-methods approach, combining formal quantitative data collection with in-depth participant observation. A Likert scale questionnaire was used for the first one. To fully understand the impact of ChatGPT on the learning process, a comprehensive analysis was conducted using numerical data and detailed observational insights. This dual approach allows for a thorough examination.

Analyzing Data

A thorough analysis was carried out on the data collected from both surveys and observations. The statistical package for the social sciences, SPSS, was used to perform the quantitative analysis. Numerical data obtained from the survey were processed and analyzed using SPSS, a comprehensive program for handling and interpreting statistical data. The data is analyzed using various statistical methods to find trends, patterns, and correlations. Theme analysis was carried out on qualitative data obtained from observations. Part of this procedure entails looking for similarities and trends in students' interactions with each other as they learn. By bringing together statistical validation and a deep contextual understanding of workplace social dynamics, this analytical tool builds a powerful framework for understanding the complex impact of using ChatGPT as a teaching tool.

Findings

At this stage, the researcher scrutinizes the data obtained from the research instrument, specifically from questionnaires and observations. This questionnaire contains important indicators regarding pre-service teachers' opinions about ChatGPT-based learning media, which they can use to enhance their learning experience for social interactions.

Table 1. Descriptive statistics of the preservice teachers survey

Statements	N	Minimum	Maximum	Mean	Std. Deviation
1. The lecturer introduces ChatGPT for leaning activities	20	4.00	5.00	4.7000	.47016
2. I already know how to use ChatGPT for leaning	20	2.00	5.00	4.2000	.95145
3. I believe ChatGPT can make students and teachers more social in the classroom.	20	3.00	5.00	4.0000	.72548
4. I find that ChatGPT is a great tool for classroom instruction and academic pursuits.	20	3.00	5.00	4.3500	.74516
5. I think student-teacher interactions With ChatGPT easier	20	3.00	5.00	4.4500	.68633
6. Class discussions and idea sharing can be facilitated by ChatGPT.	20	4.00	5.00	4.8500	.36635
7. The ideas generated by ChatGPT are intriguing.	20	4.00	5.00	4.3000	.47016
8. I use ChatGPT to provide each student specific feedback.	20	3.00	5.00	4.1500	.81273
9. ChatGPT is a great tool for improving your social interaction and learning abilities.	20	3.00	5.00	3.9500	.51042
10. As a learning resource, ChatGPT can be seamlessly integrated into group projects or collaborative learning.	20	3.00	5.00	4.2500	.55012
11. ChatGPT is a great tool for enhancing student engagement in class discussions.	20	3.00	5.00	3.8500	.48936
12. Using ChatGPT, students can be encouraged to think critically while they study.	20	3.00	5.00	4.2500	.63867
13. I have no problem communicating using ChatGPT.	20	3.00	5.00	4.3000	.65695
14. ChatGPT has the potential to foster or enhance social relationships within the classroom.	20	4.00	5.00	4.4000	.50262
15. ChatGPT in class is great since it encourages more participation through conversation activities, idea sharing, and learning resource utilization	20	3.00	5.00	3.9500	.51042
16. I think ChatGPT helps students have better social relationships while they learn.	20	3.00	5.00	4.3500	.58714
17. ChatGPT increase my learning skills and easy for me to understand.	20	4.00	5.00	4.4000	.50262
18. Using ChatGPT makes classroom engagement much more efficient.	20	4.00	5.00	4.4000	.50262
19. ChatGPT is designed to increase students interest in learning civic education and pancasila	20	3.00	5.00	3.9500	.51042

continued on following page

Table 1. Continued

Statements	N	Minimum	Maximum	Mean	Std. Deviation
20. Using ChatGPT can make learning activities in the classroom more fun, enjoyable and relax.	20	3.00	5.00	4.3500	.58714
Valid N (listwise)	20				

Table 1 shows the pre-service teachers' survey revealed that they generally perceive ChatGPT positively, particularly in facilitating discussions, enhancing social interactions, and improving classroom engagement. The survey was conducted among 20 pre-service teachers and the responses were measured on a Likert scale. The results showed a high mean of 4.70, indicating that lecturers introduced ChatGPT for learning activities. The respondents had varying levels of familiarity with ChatGPT, with a positive view of its ability to enhance social interactions. The survey also showed strong agreement on its effectiveness for classroom instruction, ease of interaction, facilitation of discussions, intriguing ideas, providing feedback, improving interaction and learning, integration into group projects, enhancing engagement, encouraging critical thinking, ease of communication, fostering social relationships, encouraging participation, improving social relationships, increasing learning skills, efficiency in engagement, interest in civic education, and enjoyable learning. The high means and relatively low standard deviations suggest a consensus among respondents on the benefits of using ChatGPT in educational settings. Overall, the data indicates that pre-service teachers generally perceive ChatGPT positively, particularly in facilitating discussions, enhancing social interactions, and improving classroom engagement.

Table 2. Mean score of the social interaction survey

	N	Minimum	Maximum	Mean	Std. Deviation
ChatGPT and social interaction	20	80.00	94.00	85.4000	4.25997
Valid N (listwise)	20				

Twenty pre-service teachers under observation (N) formed the data set on which the descriptive statistics were provided. The value of the Table 2 variable could range from eighty to ninety-four. The standard deviation of this variable is 4.25997 and the mean score is 85.40.

Figure 4. The histogram and P-P plot of respondent's survey

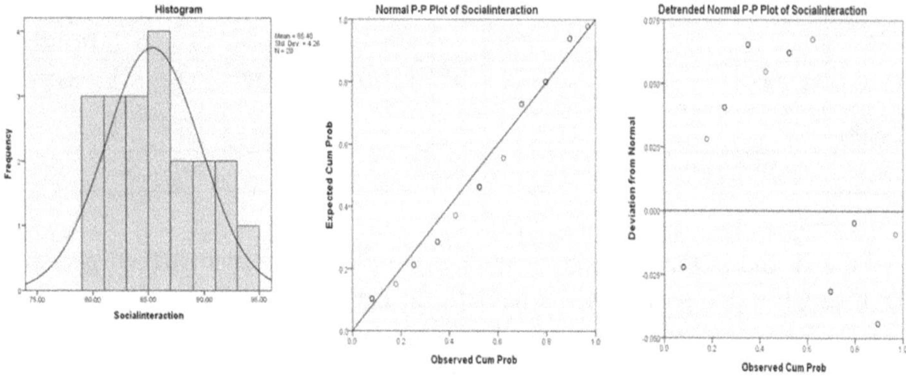

Figure 1 displays the distribution of preservice teacher responses from the online questionnaire results, with scores ranging from 75 to 95. The evidence supports the idea that students strongly believe that ChatGPT enhances their classroom learning experiences across a variety of domains. Students believe that ChatGPT will enhance their social interactions, encourage them to think critically, lead them in fruitful debates, spark their imaginations, and provide them with many possibilities for information exchange with their classmates. This strong support for ChatGPT highlights its effectiveness as a useful tool for creating interactive and collaborative classrooms. Figure 1 shows that students have positive attitudes towards ChatGPT, as well as the ability of ChatGPT to enhance students' educational experiences by encouraging them to actively participate, share what they know, and engage their cognitive abilities.

DISCUSSION

According to the findings, pre-service teachers generally have a positive impression of ChatGPT and its ability to enhance classroom social interaction as students learn. They also have a decent grasp of how to include AI into their lessons and are highly motivated to make this vision a reality. The outcomes are consistent with expectations as teachers have been guided through professional development. Figure 2 shows that, in theory, instructors are more likely to embrace and, more importantly, use ChatGPT in the classroom when it supports social interactions during classroom learning. People who took the survey use ChatGPT to supplement their education.

The AI-ChatGPT technology was deemed a valuable tool for the learning process and its ease of application by pre-service instructors. Figure 3 shows that group members were more motivated to talk to each other and engage in group activities when they had easy access to and meaningful replies to learning subjects. They concluded that organizations and technicians had shown enough support for using ChatGPT in the classroom. Their understanding of AI-TK, AI-TPK, and AI-TPACK, the acronyms for "technology knowledge," "technology pedagogical knowledge," and "technology pedagogical content knowledge," was limited to that. A lot of people think that they will keep learning ChatGPT and utilize it for other things related to education, such making quizzes, resumes, and questions. Teachers in Pancasila and citizenship education programs continue to make better use of the ChatGPT app when teaching students about new technologies, according to the findings of this study.

Based on the results, the social interactions that instructors have using ChatGPT are most affected by their performance goals. According to table 1, questions 10 and 12, high-performing educators firmly believe that incorporating ChatGPT will enhance their teaching abilities and provide them with easier access to relevant information. According to the survey takers, ChatGPT can help them hone their abilities. engage in critical thinking while evaluating data. It is evident that to promote the use of ChatGPT as a medium and technology for learning at the university level, professors must assist teachers in comprehending its efficacy in the classroom. Whenever and wherever they need to practice genuine conversation, pre-service teachers can use the ChatGPT app similar with Strzelecki, (2024), Elbanna and Armstrong (2024); Acosta, et al. (2024) investigation. In addition, this has the potential to remove the primary barriers to communication at Al Asyariah Mandar University, which are the limitations on the exchange of ideas and information. As more and more educators see the value of technology in the classroom, we can only hope that their social interactions on ChatGPT will grow. We also hope that this growth will ripple out into their daily lives, impacting their families, communities, and the school where they work.

In Figure 3, we can observe that pre-service teachers and citizenship education educators are likely to be interested in using ChatGPT due to its convenience, the fact that the information it provides can spark discussions, and the fact that it helps keep scientific ethics in check by providing access to credible sources through Google Scholar. This pattern is consistent with earlier studies that investigated the effects of the ChatGPT intervention on raising discourse (Stahl and Eke 2024). The results demonstrate that ChatGPT can enhance classroom discussions and interactions. Past studies have shown that educators are worried that students may become overly dependent on digital resources due to their heavy use of ChatGPT (Wang, 2024);

Sun, et al. (2024). This could threaten the survival of more traditional methods of instruction and make teachers work harder and less efficiently. The study's authors concluded that instructors' social interactions appeared lively due to the accessible nature of ChatGPT technology, which served as a useful tool for instruction.

Similarly, the lecturer's chosen teaching model determines the accuracy of AI predictions about teachers' social interactions. Teachers are more inclined to use ChatGPT if they have faith in their abilities to integrate the tool with their subject matter expertise and pedagogical practices. Yu, (2024); Al Shloul, et al. (2024) all found that using ChatGPT technology in the classroom increased students' enthusiasm to study. Following the implementation of ChatGPT, it is evident that they frequently utilize it to generate ideas for team discussions (Figure 2). That people are self-assured and knowledgeable about how ChatGPT might enhance education and social interaction in general is evident from the results.

Although the majority of pre-service teachers in this study had never heard of ChatGPT before the pilot, they were receptive to the idea and eager to find methods to integrate course material with relevant technological abilities into their lesson plans. Consequently, educators should review and update course materials to provide pre-service teachers with supplementary learning resources that can be accessed through ChatGPT. This will enhance learning efficiency, empower students to actively participate in feedback and social interactions, and ultimately alleviate their anxiety associated with new information. Beyond that, the instructor still has the pre-service instructors make connections between the Google Scholar information they consulted.

The findings of this study highlight the growing responsibility of pre-service teachers to take an active role in their own education, going beyond the traditional function of acquiring knowledge through creating learning materials based on ChatGPT. Instead of spending time searching for relevant resources on Google Scholar and ChatGPT, pre-service teachers can save time during learning activities by collaborating with peers to generate new ideas, such as making connections between different pieces of material. This, in turn, will improve students' engagement, performance, and quality of learning in comparison to a purely theoretical approach. Through the utilization of ChatGPT as a multimodal educational experiment, this research equips future educators with the technological know-how to experiment with AI-integrated learning.

The results of this study show that ChatGPT is beneficial for education and has a good effect on future educators. ChatGPT can enhance social connections in the classroom and is generally well-received by pre-service teachers. They are really committed to using ChatGPT and other forms of artificial intelligence in the classroom, and they have a solid grasp on how to do it. The usage of ChatGPT by educators has been directed by professional development opportunities, which have helped to align outcomes with expectations. We find that future educators see the AI-ChatGPT technology as a tool that will enhance their education. They are more motivated to participate in discussions and exchanges with their peers since ChatGPT is easy to use and can give substantial answers to a variety of issues (Figure 2). There is enough support for ChatGPT integration in the classroom, according to educational institutions and engineers. As evidence of their dedication to learning, pre-service teachers have researched AI-related elements and used ChatGPT for a variety of instructional applications. How instructors engage with ChatGPT is heavily influenced by their expectations. Incorporating ChatGPT is seen by high-performing educators to enhance their lessons and provide students with more resources. Educators view ChatGPT as a tool that can help them develop their abilities, foster collaboration, and promote critical thinking.

In both formal and informal settings, ChatGPT can facilitate more communication amongst future educators. It is believed that ChatGPT will promote discussion and uphold ethical standards in education because to its user-friendliness and wealth of information. Although some worry about students being too dependent on technology in the classroom, ChatGPT can improve classroom discussions and interactions. Teachers are more inclined to use ChatGPT in the classroom if they feel comfortable incorporating it into their pedagogical paradigm. After its launch, ChatGPT remains in use for the same primary function: to facilitate idea generation. Despite their lack of familiarity with ChatGPT, pre-service teachers were eager to incorporate it into their lessons. Teachers are encouraged to update course materials to include more learning resources that can be accessed through ChatGPT, which will help students learn more efficiently and feel less anxious when confronted with new material.

The Limitation of this Study

Due to its exclusive emphasis on using the ChatGPT deep learning course to survey social interactions in the classroom, this survey research did not examine other general courses. Several contextual variables, such as geography, socioeconomic level, school size, and academic year, were considered when making conclusions regarding these findings because the research participants were prospective teachers at Universitas Al Asyriah Mandar. This research methodology and set of beliefs are specific to this class of survey with a limited number of respondents. Due to the short duration of ChatGPT-based learning (only four weeks and one hundred minutes), there may be some temporary obstacles in drawing conclusions from the research. Our findings suggest that professional teacher education programs can incorporate ChatGPT courses, with a combination of classroom theory and practical experience being the most effective way to teach these concepts. While online learning platforms make it easy to teach courses on new technologies in education, not all pre-series instructors will be comfortable using them. As part of the ChatGPT classroom experiment, the researchers demonstrated the operation of ChatGPT to prospective educators and then submitted it to them to run under the guidance of their mentor teachers. In this research, the author considers the influence of the time he spends teaching using ChatGPT as a technology on his students' social interactions. The findings of this study—both instructive and useful—can help future educators think critically about the ideas they will teach and give group presentations on the topics they have covered. In this activity, teacher candidates follow their lecturers as they demonstrate how to hold meetings under close supervision. After that, pre-service teachers work in groups to complete the assignment. ChatGPT is a great tool for pre-service teachers to review material before, during, and after class. It also increases social contacts, which is beneficial for their theoretical and practical development as educators.

CONCLUSION

In conclusion, there is an urgent need to resolve the constraints and ethical concerns surrounding ChatGPT, even though it presents encouraging benefits for improving educational environments' social interaction and learning. Future research should continue to explore these relationships to ensure the integration of AI technologies like ChatGPT in a way that maximizes their benefits while minimizing hazards. This is of utmost importance in areas like citizenship education and Pancasila, where encouraging critical thinking and teamwork is fundamental. By bolstering social relationships, improving learning experiences, and providing support to instructors and pre-service teachers in various educational settings, this research makes a constructive contribution to education. Using ChatGPT as a teaching tool and source of information, this article aims to analyze the anticipated social interactions connected to learning activities. Discussion activities, idea sharing, concept construction, knowledge connection, and proper material reflection are all greatly aided by ChatGPT, according to a review of research on its potential and benefits in the context of student learning. Also included in this article is an analysis of all the ChatGPT tools that are out there and can be accessed through mobile devices and computers. Group projects are great for practicing both human-to-human and human-to-machine social interactions, which will help you feel more confident when presenting your ideas and thoughts on the subject. When everyone's roles and duties are clear, everyone feels more empowered.

Furthermore, the results imply that being a part of a group is linked to having more social interactions. Using ChatGPT technology, which combines generalization with learning motivation to encourage pre-service teachers to communicate with one another while they learn, is likely to enhance their social interaction thinking. This paper's pedagogical goal is to provide pre-service teachers with the tools they need to design and implement classroom learning activities that promote social interaction and the development of the skills. Additionally, we suggest that teaching methods based on the findings of this research might be developed to promote robust development of each social skill in ChatGPT-based courses.

REFERENCES

Acosta-Enriquez, B. G., Arbulú Ballesteros, M. A., Huamaní Jordan, O., López Roca, C., & Saavedra Tirado, K. (2024). Analysis of college students' attitudes toward the use of ChatGPT in their academic activities: Effect of intent to use, verification of information and responsible use. *BMC Psychology*, 12(1), 1–18. DOI: 10.1186/s40359-024-01764-z PMID: 38720382

Agrahari, S. K. (2024). Fostering Cultural Competence through Transformative Education: Strategies for Inclusive Futures. *Educating for Societal Transitions*, 99.

Al-Said, K., Semenycheva, I., & Voronova, L. (2024). Aspects of fostering social educators' communicative competence: Students' opinions (using an interactive discussion club as an example). *Cogent Education*, 11(1), 2298059. DOI: 10.1080/2331186X.2023.2298059

Al Shloul, T., Mazhar, T., Iqbal, M., yaseen Ghadi, Y., Malik, F., & Hamam, H. (2024). Role of activity-based learning and ChatGPT on students' performance in education. *Computers and Education: Artificial Intelligence*, 100219.

Al Yakin, A. M., Obaid, A. J., Abdul, L., Warsah, I., Muthmainnah, M., & Elngar, A. A. (2024). Harnessing the Power of ChatGPT to Explore Student Metacognitive Skills in Learning Sociology Education. In *Advanced Applications of Generative AI and Natural Language Processing Models* (pp. 405–423). IGI Global.

Alanezi, F. (2024). Assessing the effectiveness of ChatGPT in delivering mental health support: A qualitative study. *Journal of Multidisciplinary Healthcare*, 17, 461–471. DOI: 10.2147/JMDH.S447368 PMID: 38314011

Asad, M. M., & Malik, A. (2024). Educational quality and inclusion through collaborative hybridized cybergogy: Transformative learning horizons in Pakistani universities. *Interactive Technology and Smart Education*, 21(4), 672–689. DOI: 10.1108/ITSE-10-2023-0193

Carlyle, D. (2024). *Researching Child-Dog Relationships and Narratives in the Classroom: Rhythms of Posthuman Childhoods*. Taylor & Francis.

Cerasa, A., Gaggioli, A., Pioggia, G., & Riva, G. (2024). Metaverse in Mental Health: The Beginning of a Long History. *Current Psychiatry Reports*, 26(6), 1–10. DOI: 10.1007/s11920-024-01501-8 PMID: 38602624

Chan, C. K. Y., & Colloton, T. (2024). *Generative AI in Higher Education: The ChatGPT Effect*. Taylor & Francis. DOI: 10.4324/9781003459026

Chang, H., Park, J., & Suh, J. (2024). Virtual reality as a pedagogical tool: An experimental study of English learner in lower elementary grades. *Education and Information Technologies*, 29(4), 4809–4842. DOI: 10.1007/s10639-023-11988-y

Chen, F., Li, S., Lin, L., & Huang, X. (2024). Identifying temporal changes in student engagement in social annotation during online collaborative reading. *Education and Information Technologies*, 29(13), 1–24. DOI: 10.1007/s10639-024-12494-5

Dempere, J., Modugu, K., Hesham, A., & Ramasamy, L. K. (2023, September). The impact of ChatGPT on higher education. [). Frontiers Media SA.]. *Frontiers in Education*, 8, 1206936. DOI: 10.3389/feduc.2023.1206936

Dost, G. (2024). Students' perspectives on the 'STEM belonging' concept at A-level, undergraduate, and postgraduate levels: An examination of gender and ethnicity in student descriptions. *International Journal of STEM Education*, 11(1), 1–33. DOI: 10.1186/s40594-024-00472-9

Dragnić-Cindrić, D., Lobczowski, N. G., Greene, J. A., & Murphy, P. K. (2024). Exploring the Teacher's Role in Discourse and Social Regulation of Learning: Insights from Collaborative Sessions in High-School Physics Classrooms. *Cognition and Instruction*, 42(1), 92–123. DOI: 10.1080/07370008.2023.2266847

Dulfer, N., Gowing, A., & Mitchell, J. (2024). Building belonging in online classrooms: Relationships at the core. *Teaching in Higher Education*, •••, 1–17. DOI: 10.1080/13562517.2024.2349993

Elbanna, S., & Armstrong, L. (2024). Exploring the integration of ChatGPT in education: Adapting for the future. *Management & Sustainability: An Arab Review*, 3(1), 16–29. DOI: 10.1108/MSAR-03-2023-0016

Elbanna, S., & Armstrong, L. (2024). Exploring the integration of ChatGPT in education: Adapting for the future. *Management & Sustainability: An Arab Review*, 3(1), 16–29. DOI: 10.1108/MSAR-03-2023-0016

Fahad, S. A., Salloum, S. A., & Shaalan, K. (2024). The Role of ChatGpt in Knowledge Sharing and Collaboration Within Digital Workplaces: A Systematic Review. *Artificial Intelligence in Education: The Power and Dangers of ChatGPT in the Classroom*, 259-282.

Farazouli, A., Cerratto-Pargman, T., Bolander-Laksov, K., & McGrath, C. (2024). Hello GPT! Goodbye home examination? An exploratory study of AI chatbots impact on university teachers' assessment practices. *Assessment & Evaluation in Higher Education*, 49(3), 363–375. DOI: 10.1080/02602938.2023.2241676

Fasting, R. B., & Breilid, N. (2024). Cross-professional collaboration to improve inclusive education. *Scandinavian Journal of Educational Research*, 68(4), 661–676. DOI: 10.1080/00313831.2023.2175248

Flores-Piñero, M. D. C., Valdivia-Moral, P., Ramos-Mondejar, L., & González-Hernández, J. (2024). Motivational Climate, Physical Self-Concept, and Social Relationships in Adolescents in Physical Education Classes: A Systematic Review. *Education Sciences*, 14(2), 199. DOI: 10.3390/educsci14020199

Gray, L. E., & Dunn, S. D. (Eds.). (2024). *Incorporating the Human Element in Online Teaching and Learning*. IGI Global. DOI: 10.4018/979-8-3693-4131-5

Guerrero Elecalde, R., Contreras García, J., Bonilla Martos, A. L., & Serrano Arnáez, B. (2024). Digital and Social-Civic Skills in Future Primary Education Teachers: A Study from the Didactics of Social Sciences for the Improvement of Teacher Training in Competences. *Education Sciences*, 14(2), 211. DOI: 10.3390/educsci14020211

Hughes, S. (2024). Teaching assistants as facilitators of learning and engagement: Promoting social justice for students with special educational needs and disabilities (SEND). In *Social Justice in Practice in Education* (pp. 156-174). Routledge.

Jo, H., & Park, D. H. (2024). Effects of ChatGPT's AI capabilities and human-like traits on spreading information in work environments. *Scientific Reports*, 14(1), 7806. DOI: 10.1038/s41598-024-57977-0 PMID: 38565880

Kansok-Dusche, J., Wettstein, A., Castellanos, M., Schwab, C., Subramaniam, A., Wachs, S., & Bilz, L. (2024). Associations between social competence, perceived parents' prosocial educational goals and adolescents' hate speech perpetration in school. *Victims & Offenders*, 19(3), 419–446. DOI: 10.1080/15564886.2023.2189191

Kelly, M. (2024). Re-imagining pedagogical approaches to support student engagement in secondary schools. In *Re-imagining Teaching Improvement: From Early Childhood to University* (pp. 47–70). Springer Nature Singapore.

Khalil, S. (2024). Understanding Society: An Introduction to Social Science. *Pakistan Journal of Social Science Review*, 2(01), 57–68.

Kolade, O., Owoseni, A., & Egbetokun, A. (2024). Is AI changing learning and assessment as we know it? Evidence from a ChatGPT experiment and a conceptual framework. *Heliyon*, 10(4), e25953. DOI: 10.1016/j.heliyon.2024.e25953 PMID: 38379960

Leng, L. (2024). Challenge, integration, and change: ChatGPT and future anatomical education. *Medical Education Online*, 29(1), 2304973. DOI: 10.1080/10872981.2024.2304973 PMID: 38217884

Lin, J., & Wang, Y. (2024). Unpacking the mediating role of classroom interaction between student satisfaction and perceived online learning among Chinese EFL tertiary learners in the new normal of post-COVID-19. *Acta Psychologica*, 245, 104233. DOI: 10.1016/j.actpsy.2024.104233 PMID: 38522350

Liu, G., & Ma, C. (2024). Measuring EFL learners' use of ChatGPT in informal digital learning of English based on the technology acceptance model. *Innovation in Language Learning and Teaching*, 18(2), 125–138. DOI: 10.1080/17501229.2023.2240316

Liu, Z., Zhu, Z., Zhu, L., Jiang, E., Hu, X., Peppler, K. A., & Ramani, K. (2024, May). ClassMeta: Designing Interactive Virtual Classmate to Promote VR Classroom Participation. In *Proceedings of the CHI Conference on Human Factors in Computing Systems* (pp. 1-17). DOI: 10.1145/3613904.3642947

Lovell, J., & Webber, M. (2024). Adaptation of Connecting People to address loneliness and social isolation in university students: A feasibility study. *Journal of Higher Education Policy and Management*, 46(1), 32–47. DOI: 10.1080/1360080X.2023.2242079

Mansouri, M., & Taylor, H. (2024). Does Cultural Robotics Need Culture? Conceptual Fragmentation and the Problems of Merging Culture with Robot Design. *International Journal of Social Robotics*, 16(2), 385–401. DOI: 10.1007/s12369-023-01085-y

Marginson, S. (2024). Student self-formation: An emerging paradigm in higher education. *Studies in Higher Education*, 49(4), 748–762. DOI: 10.1080/03075079.2023.2252826

Mattson, K. (2024). *Digital citizenship in action: empowering students to engage in online communities*. International Society for Technology in Education.

Muthmainnah, M., Darmawati, B., Khang, A., Al Yakin, A., Obaid, A. J., & Elngar, A. A. Shaping Artificial Intelligence-Perceived Hybrid Learning Environment at University Toward the Global Talent Development Strategy. In *AI-Oriented Competency Framework for Talent Management in the Digital Economy* (pp. 219–243). CRC Press. DOI: 10.1201/9781003440901-14

Ng, D. T. K., Tan, C. W., & Leung, J. K. L. (2024). Empowering student self-regulated learning and science education through ChatGPT: A pioneering pilot study. *British Journal of Educational Technology*, 55(4), 1328–1353. DOI: 10.1111/bjet.13454

Oba, D., & Berger, J. (2024). How communication mediums shape the message. *Journal of Consumer Psychology*, 34(3), 406–424. DOI: 10.1002/jcpy.1372

Okada, A., Panselinas, G., Bizoi, M., Malagrida, R., & Torres, P. L. (2024). Fostering Transversal Skills through Open Schooling with the CARE-KNOW-DO Framework for Sustainable Education. *Sustainability (Basel)*, 16(7), 2794. DOI: 10.3390/su16072794

Othman, R. M., Abushunar, M. S., & Al-Shboul, S. S. (2024). Investigating the Effectiveness of Synchronous E-Learning Tools on Developing EFL Learners' Language Communication Skills Among the Hashemite University Students. *Theory and Practice in Language Studies*, 14(2), 597–605. DOI: 10.17507/tpls.1402.33

Ouyang, F., Zhang, L., Wu, M., & Jiao, P. (2024). Empowering collaborative knowledge construction through the implementation of a collaborative argument map tool. *The Internet and Higher Education*, 62, 100946. DOI: 10.1016/j.iheduc.2024.100946

Park, H., & Ahn, D. (2024, May). The Promise and Peril of ChatGPT in Higher Education: Opportunities, Challenges, and Design Implications. In *Proceedings of the CHI Conference on Human Factors in Computing Systems* (pp. 1-21). DOI: 10.1145/3613904.3642785

Rajaei, A. (2024). Teaching in the Age of AI/ChatGPT in Mental-Health-Related Fields. *The Family Journal (Alexandria, Va.)*, 32(1), 6–10. DOI: 10.1177/10664807231209721

Saleem, F., Chikhaoui, E., & Malik, M. I. (2024, February). Technostress in students and quality of online learning: Role of instructor and university support. []. Frontiers Media SA.]. *Frontiers in Education*, 9, 1309642. DOI: 10.3389/feduc.2024.1309642

Saxer, K., Schnell, J., Mori, J., & Hascher, T. (2024). The role of teacher–student relationships and student–student relationships for secondary school students' well-being in Switzerland. *International Journal of Educational Research Open*, 6, 100318. DOI: 10.1016/j.ijedro.2023.100318

Scuotto et al. (2024) stated despite the prevalence of digital communication, humans still have an innate desire for face-to-face interaction, making social interaction the most important item in the digital age. This helps people feel more connected, empathic, and at home in their community.

Scuotto, V., Manlio, D. G., Malhotra, A., & Pereira, V. (2024). The evolution of human-machine interaction, from Taylorism to human-centered approach. *Journal of Management History*, 30(2), 141–146. DOI: 10.1108/JMH-04-2024-305

Senter, M. S. (2024). The impact of social relationships on college student learning during the pandemic: Implications for sociologists. *Teaching Sociology*, 52(1), 39–54. DOI: 10.1177/0092055X231178505

Shannon, L. (2023). *Interconnected Realities: How the Metaverse Will Transform Our Relationship with Technology Forever*. John Wiley & Sons.

Snell, S. A., Swart, J., Morris, S., & Boon, C. (2023). The HR ecosystem: Emerging trends and a future research agenda. *Human Resource Management*, 62(1), 5–14. DOI: 10.1002/hrm.22158

Stahl, B. C., & Eke, D. (2024). The ethics of ChatGPT–Exploring the ethical issues of an emerging technology. *International Journal of Information Management*, 74, 102700. DOI: 10.1016/j.ijinfomgt.2023.102700

Strzelecki, A. (2024). Students' acceptance of ChatGPT in higher education: An extended unified theory of acceptance and use of technology. *Innovative Higher Education*, 49(2), 223–245. DOI: 10.1007/s10755-023-09686-1

Suárez-García, A., Álvarez-Hernández, M., Arce, E., & Ribas, J. R. (2024). Exploring the Efficacy of Binary Surveys versus Likert Scales in Assessing Student Perspectives Using Bayesian Analysis. *Applied Sciences (Basel, Switzerland)*, 14(10), 4189. DOI: 10.3390/app14104189

Sun, D., Boudouaia, A., Zhu, C., & Li, Y. (2024). Would ChatGPT-facilitated programming mode impact college students' programming behaviors, performances, and perceptions? An empirical study. *International Journal of Educational Technology in Higher Education*, 21(1), 14. DOI: 10.1186/s41239-024-00446-5

Tai, T. Y. (2024). Comparing the effects of intelligent personal assistant-human and human-human interactions on EFL learners' willingness to communicate beyond the classroom. *Computers & Education*, 210, 104965. DOI: 10.1016/j.compedu.2023.104965

Tao, A. (2024). The Dynamics of Community Engagement in Distance Education: A Sociological Analysis Based on Online Learning Platforms. *International Journal of Interactive Mobile Technologies*, 18(7), 4–18. DOI: 10.3991/ijim.v18i07.48599

Traver, A. E., Shin, H., & Bakalarz-Duverger, M. (2024). Introduction: Teaching and Learning a Humanistic Sociology. *Teaching Sociology*, 52(2), 107–114. DOI: 10.1177/0092055X241233882

Tuan, P. T., Phuong, N. T. H., & Nam, N. T. (2024). Engaging Students in The Learning Process with Game-Based Learning. *International Journal of English Language Studies*, 6(1), 54–60. DOI: 10.32996/ijels.2024.6.1.5

Tzimas, D. E., & Demetriadis, S. N. (2024). Impact of Learning Analytics Guidance on Student Self-Regulated Learning Skills, Performance, and Satisfaction: A Mixed Methods Study. *Education Sciences*, 14(1), 92. DOI: 10.3390/educsci14010092

Vandermaas-Peeler, M., Byrd, S. K., Reiness, C. G., Frederick, K. A., Holmes, B. E., Bates, S. C., & Wooley, C. A. (2024). Transformation through a Disciplinary Lens. In *Transforming Academic Culture and Curriculum* (pp. 72–93). Routledge. DOI: 10.4324/9781003448327-5

vom Lehn, D., Gibson, W., & Ruiz-Junco, N. (2024). *People, technology, and social organization*. Routledge.

Wang, K. D., Burkholder, E., Wieman, C., Salehi, S., & Haber, N. (2024, January). Examining the potential and pitfalls of chatgpt in science and engineering problem-solving. [). Frontiers Media SA.]. *Frontiers in Education*, 8, 1330486. DOI: 10.3389/feduc.2023.1330486

Xu, T., Gao, Q., Ge, X., & Lu, J. (2024). The relationship between social media and professional learning from the perspective of pre-service teachers: A survey. *Education and Information Technologies*, 29(2), 2067–2092. DOI: 10.1007/s10639-023-11861-y PMID: 37361740

Yu, H. (2024). The application and challenges of ChatGPT in educational transformation: New demands for teachers' roles. *Heliyon*, 10(2), e24289. DOI: 10.1016/j.heliyon.2024.e24289 PMID: 38298626
Yu, H. (2024). The application and challenges of ChatGPT in educational transformation: New demands for teachers' roles. *Heliyon*, 10(2), e24289. DOI: 10.1016/j.heliyon.2024.e24289 PMID: 38298626
Zahra, A., Waheed, Z., Fatima, T., & Khong, K. W. (2024). Leveraging Technology for Environmental Awareness: Insights from Experimental Research with Middle School Students in Malaysia. *RMLE Online: Research in Middle Level Education*, 47(4), 1–16. DOI: 10.1080/19404476.2024.2322046

Zammit, A. (2024). Exploring the Influence of Social Emotional Development on Academic Achievement in Vocational Education: Insights from Student Mentor Interventions. *MCAST Journal of Applied Research & Practice*, 8(1), 80–104. DOI: 10.5604/01.3001.0054.5096

Zola, J., & Zola, J. (2024). *Teaching as if Students Matter: A Guide to Creating Classrooms Based on Relationships and Engaged Learning*. State University of New York Press.

Chapter 14
Unveiling Academic Success:
Harnessing Graph Machine Learning for Student Performance Prediction

Polisetty Sri Hari Sai Saran
Department of Data Science and Business Systems, Faculty of Engineering and Technology, College of Engineering and Technology, SRM Institute of Science and Technology, Kattankulathur, India

R. Krishna Kumari
https://orcid.org/0000-0002-1802-628X
Department of Mathematics, Faculty of Engineering and Technology, College of Engineering and Technology, SRM Institute of Science and Technology, Kattankulathur, India

ABSTRACT

In this chapter, we delve into the utilization of graph machine learning techniques to forecast student academic performance. By harnessing graph-based representations of educational data, our study endeavors to unearth underlying patterns and connections that impact student success. Through a fusion of feature engineering, graph analytics, and predictive modeling, we aim to investigate the efficacy of graph-based methodologies in improving the precision and interpretability of student performance prediction systems. This paper investigates the effectiveness of logistic regression, K-nearest neighbors (KNN), and a custom Graph Neural Network (GNN) model for predicting student performance in exams. Our analysis reveals that the custom GNN model outperforms both logistic regression and KNN, achieving higher accuracy and efficiency in student performance prediction. The custom GNN model leverages

DOI: 10.4018/979-8-3373-1399-3.ch014

the graph-based representation of educational data, which enhances its ability to capture complex relationships and dependencies among students.

1. INTRODUCTION

Predicting student academic performance is a crucial task in educational institutions worldwide, as it allows educators to identify struggling students early and provide them with the necessary support to succeed (Balcioglu et al.,2023). However, traditional predictive models based on tabular data often fall short in capturing the intricate interplay of factors that influence student success (Subramanian et al.,2023). These models typically rely on individual attributes such as demographics, prior grades, and standardized test scores, overlooking the rich relational context inherent in educational data (Brooks et al.,2023).

In recent years, the advent of graph-based machine learning techniques has opened up new avenues for analyzing complex relational data (Dervenis et al.,2022). By representing educational data as a graph, wherein nodes correspond to entities such as students, courses, and assignments, and edges encode relationships such as enrollment, collaboration, and academic performance, we can capture the multidimensional interactions and dependencies that shape student outcomes (Gajwani et al.,2021, Kaur et al.,2023). This graph- based representation offers a holistic view of the educational ecosystem, enabling us to uncover hidden patterns and insights that traditional models may overlook.

One of the key challenges in student performance prediction lies in the early identification of at-risk students who may benefit from targeted interventions. Early intervention strategies have been shown to significantly improve student retention and success rates (Ojajuni et al.,2021, Soyoye et al.,2023). However, conventional models often lack the granularity and interpretability necessary to pinpoint students who are most in need of support. By leveraging graph-based representations, which inherently encode structural information and relational context, we can enhance the accuracy and interpretability of predictive models, thereby facilitating timely interventions and personalized support strategies (Patil et al.,2023, Sekeroglu et al.,2019).

In this research article, we aim to explore the potential of graph machine learning techniques for predicting student academic performance. We will begin by discussing the limitations of traditional predictive models and the motivations for adopting a graph-based approach. We will then outline our objectives, which include:

(i) Constructing a comprehensive graph representation of educational data, incorporating various entities and relationships relevant to student performance prediction.

(ii) Investigating feature engineering techniques tailored to graph-based representations, aiming to extract informative features that capture both individual characteristics and relational dynamics.
(iii) Employing advanced graph analytics and predictive modeling techniques to uncover hidden patterns and relationships that influence student outcomes.
(iv) Evaluating the performance of graph-based predictive models in comparison to traditional approaches, with a focus on accuracy, interpretability, and scalability.
(v) Discussing the implications of our findings for educational practice and the potential for enhancing early intervention and support strategies.

Through this research, we seek to advance our understanding of how graph-based machine learning can be leveraged to address the challenges of predicting student academic performance. By harnessing the power of relational data and advanced analytics, we aim to contribute to the development of more effective and interpretable predictive models, ultimately improving student outcomes and promoting success in educational settings.

Student performance prediction has been a subject of extensive research in the field of education, driven by the imperative to improve learning outcomes and support student success (Benkhalfallah et al.,2023, Das et al.,2019). A plethora of studies have investigated the multifaceted factors that contribute to academic achievement, ranging from socio-economic background and demographics to individual learning behaviors and engagement. Understanding these factors is crucial for developing accurate predictive models that can identify at-risk students and guide intervention efforts. Several seminal studies have highlighted the significance of socio-economic status (SES) as a determinant of student success. Research has consistently shown that students from disadvantaged backgrounds face greater academic challenges due to limited access to resources, lack of parental support, and exposure to adverse environmental factors (Soyoye et al.,2023). Moreover, disparities in access to quality education exacerbate existing inequalities, perpetuating cycles of underachievement and disadvantage. Therefore, incorporating SES-related variables into predictive models is essential for addressing equity issues and ensuring inclusive educational practices.

In addition to socio-economic factors, prior academic performance emerges as a robust predictor of future success (Brooks et al.,2023, Hennebelle et al.,2024). Numerous studies have demonstrated the predictive power of historical grades, standardized test scores, and GPA in forecasting student outcomes. However, it is essential to recognize that academic performance is not solely determined by past achievements but also influenced by various contextual factors, including the learning environment, teacher quality, and peer interactions. Thus, predictive models must account for both individual characteristics and relational dynamics to accurately assess student trajectories (Kaur et al.,2022). Further- more, recent advancements in machine learning have spurred interest in graph-based techniques for representing and analyzing educational data (Kumar et al.,2023). Graphs offer a natural framework for capturing the complex relational structures inherent in educational environments, wherein nodes represent entities such as students, courses, and concepts, and edges denote relationships such as enrollment, collaboration, and prerequisite dependencies. By modeling educational data as a graph, researchers can leverage powerful graph-based algorithms to uncover hidden patterns and insights that traditional methods may overlook (Radhya et al.,2022).

Graph convolutional networks (GCNs) and graph attention networks (GATs) have emerged as prominent approaches for predictive modeling tasks on graph-structured data. GCNs extend the concept of convolutional neural networks to irregular graph domains, enabling the propagation of information across nodes while preserving local neighborhood structure (Sahlaoui et al.,2021). GATs, on the other hand, leverage attention mechanisms to selectively aggregate information from neighboring nodes, allowing for adaptive feature weighting and enhanced expressiveness. These graph-based methods have demonstrated promising results in various applications, including node classification, link prediction, and recommendation systems (Sharma et al.,2023). Moreover, researchers have explored diverse graph representations of educational data, including student social net- works, course prerequisite graphs, and knowledge graphs. Social network analysis techniques have been employed to analyze patterns of collaboration and information flow among students, highlighting the role of peer interactions in shaping learning outcomes (Zeineddine et al.,2021). Course prerequisite graphs provide insights into the hierarchical structure of academic curricula, facilitating course recommendation and curriculum planning. Knowledge graphs, constructed from semantic relationships between concepts and learning objectives, support personalized learning pathways and adaptive assessment strategies (Subramanian et al.,2023).

In the realm of student performance prediction and educational data analytics, the k-nearest neighbors (KNN) algorithm stands as a fundamental yet effective method for classification and regression tasks. The essence of KNN lies in its simplicity and intuitive approach to prediction based on similarity metrics. KNN operates under the principle that similar instances in feature space tend to exhibit similar outcomes. Graph Neural Networks (GNNs) are a class of neural networks specifically designed to work with graph-structured data. Unlike traditional neural networks that operate on fixed-size, grid-like structures (e.g., images), GNNs can effectively process data represented as graphs, where nodes represent entities and edges represent relationships between these entities. This makes GNNs particularly powerful for applications involving complex relational data. In the context of student performance prediction, this means that students with similar characteristics, academic histories, or learning behaviors are likely to achieve comparable results. In our study, we employed a custom Graph Neural Network (GNN) model to predict student performance and compared its effectiveness against traditional models such as logistic regression and K-nearest neighbors (KNN). The custom GNN model leverages graph-based representations of educational data to capture complex relationships and dependencies among students. This approach allows the GNN to provide more accurate and interpretable predictions compared to logistic regression and KNN. The custom GNN model exhibited superior accuracy, robustness, and efficiency in student performance prediction. By incorporating local context awareness and graph structures, the custom GNN demonstrated improved performance, particularly in handling dynamic and evolving datasets. This capability to model intricate relational dynamics provided significant advantages over traditional models that rely on individual attributes and overlook the rich relational context inherent in educational data.

2. DATASET DESCRIPTION

The dataset used for this study is designed to predict whether a student will pass or fail an exam based on two main features: the number of study hours for the upcoming exam and the student's score in the previous exam. Additionally, the dataset includes the target variable, where a value of 1 represents a pass and 0 represents a fail in the current exam.

2.1 Features, Target Variable and Dataset Size

(i) **Study Hours (Numeric)**: This feature represents the number of hours a student spent studying for the upcoming exam. It serves as an indicator of the student's level of preparation and dedication towards academic success.

(ii) **Previous Exam Score (Numeric)**: This feature indicates the student's score in the previous exam. It provides insight into the student's academic performance and proficiency in the subject matter.

Target Variable: The target variable in the dataset is binary and represents whether the student passed or failed the current exam. A value of 1 denotes a pass, indicating that the student met the required criteria for success in the exam. Conversely, a value of 0 indicates a fail, signifying that the student did not achieve the necessary level of proficiency in the exam.

Dataset Size: The dataset consists of data for 500 students, ensuring a diverse range of study patterns and previous exam performances. This ample sample size allows for robust analysis and modeling, capturing variations in study habits, academic backgrounds, and exam outcomes among students.

Table 1. First few rows of the dataset

	Study Hours	Previous Exam Score	Pass/Fail
0	4.370861	81.889703	0
1	9.556429	72.165782	1
2	7.587945	58.571657	0
3	6.387926	88.827701	1
4	2.404168	81.083870	0
..	...	...	...
495	4.180170	45.494924	0
496	6.252905	95.038815	1
497	1.699612	48.209118	0
498	9.769553	97.014241	1
499	9.875897	66.760346	1

[500 rows x 3 columns]

By leveraging this dataset, we aim to explore the relationship between study hours, previous exam scores, and the likelihood of exam success. Through predictive modeling techniques, we seek to develop a model that can accurately classify students into pass or fail categories based on their study behavior and academic history. This analysis has the potential to provide valu- able insights for educators and policymakers to support student success and enhance academic outcomes.

3. PREDICTING STUDENT PERFORMANCE IN EXAMS USING LOGISTIC REGRESSION

Logistic regression is a popular statistical method used for predicting binary outcomes, making it suitable for predicting student performance in exams, where the outcome is typically pass or fail. The logistic regression model estimates the probability that a given student will pass the exam based on one or more predictor variables, such as study hours, previous exam scores, demographic factors, etc.

3.1 Data Preparation

Before building the logistic regression model, the dataset needs to be prepared. This involves:

- **Data Cleaning**: Handling missing values, outliers, and inconsistencies in the dataset.
- **Feature Selection**: Identifying relevant predictor variables that are likely to influence stu- dent performance.
- **Data Splitting**: Splitting the dataset into training and testing sets to evaluate the model's performance.

3.2 Model Building

Once the dataset is prepared, the logistic regression model can be built using the training data. The logistic regression model estimates the probability p of a student passing the exam as a function of the predictor variables. Mathematically, the logistic regression model can be represented as:

$p = 1 + e^{-z}$,

Where $z = \beta_0 + \beta_1 x_1 + \beta_2 x_2 + \ldots + \beta_n x_n$ is the linear combination of predictor variables and their

respective coefficients $\beta_0, \beta_1, \ldots, \beta_n$.

In this equation:

- p represents the probability of a student passing the exam.
- e is the base of the natural logarithm.
- z is the logic function, which is a linear combination of predictor variables.
- $x_1, x_2, \ldots, x_n$ represent the predictor variables.
- $\beta_0, \beta_1, \ldots, \beta_n$ are the coefficients of the predictor variables.

The logistic regression model computes the log-odds of the probability of the event of interest (in this case, passing the exam) and then applies the sigmoid function to map the log-odds to a probability value between 0 and 1.

3.3 Model Evaluation

After training the logistic regression model, it needs to be evaluated using the testing data to assess its performance and generalization ability. Common evaluation metrics for logistic regression models include:

- **Accuracy**: The percentage of correctly classified instances.
- **Precision**: The proportion of true positive predictions among all positive predictions.
- **Recall**: The proportion of true positive predictions among all actual positive instances.
- **F1 Score**: The harmonic mean of precision and recall, providing a balance between the two metrics.
- **ROC Curve and AUC**: Receiver Operating Characteristic (ROC) curve and Area under the Curve (AUC) measure the model's ability to discriminate between positive and negative classes across different thresholds.

Algorithm 1 Logistic Regression Algorithm for Predicting Student Performance

Input: Training features X_{train}, training labels y_{train}, test features X_{test}
Output: Predicted labels for test set y_{pred}
Initialization: Initialize logistic regression classifier
Feature Scaling: Scale features using StandardScaler
Training: Fit logistic regression model to training data
Prediction: Predict labels for test set using trained model
function LOGISTICREGRESSION(X_{train}, y_{train}, X_{test})
Feature Scaling:
Initialize StandardScaler *sc*
$X_{train} \leftarrow sc.fit\_transform(X_{train})$
$X_{test} \leftarrow sc.transform(X_{test})$
Training:
Initialize logistic regression classifier *classifier* with random state 42
$classifier.fit(X_{train}, y_{train})$
Prediction:
$y_{pred} \leftarrow classifier.predict(X_{test})$
returny_{pred}
end function=0

Figure 1. Students' performance prediction using logistic regression

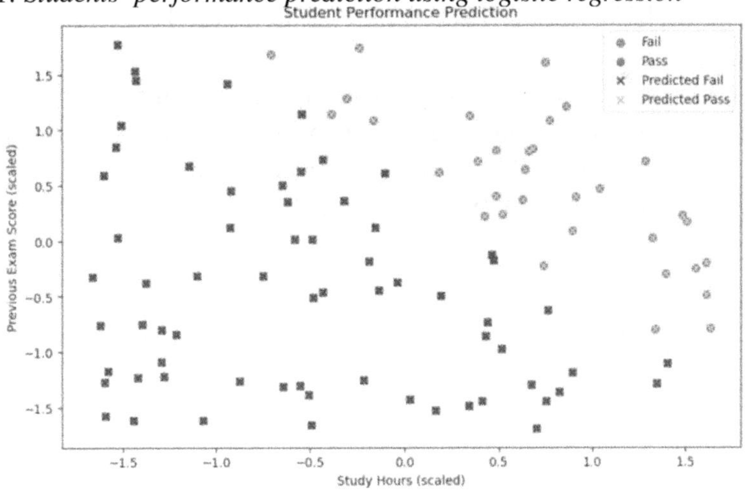

3.4 Analysis of Logistic Regression Model Output

Logistic regression is a powerful tool for predicting student performance in exams based on relevant predictor variables. By estimating the probability of passing the exam, educators and stakeholders can identify at-risk students and implement targeted intervention strategies to improve academic outcomes. The logistic regression model achieved an accuracy of 0.86 on the testing set, indicating its effectiveness in predicting student performance. The decision boundary of the logistic regression model was visualized to provide a graphical representation of the classification process. The decision boundary separates the feature space into regions corresponding to pass and fail outcomes.

The results demonstrate the potential of logistic regression as a tool for predicting student performance in exams. The model's accuracy suggests that study hours and previous exam scores are significant predictors of exam outcomes. The decision boundary visualization provides insights into the regions of feature space where students are more likely to pass or fail the exam. Thus, the logistic regression model shows promise in accurately predicting student performance in exams based on study hours and previous exam scores. Further research could explore additional features and more advanced machine learning algorithms to improve prediction accuracy.

4. K-NEAREST NEIGHBORS (KNN) ALGORITHM FOR STUDENT PERFORMANCE PREDICTION

The K-nearest neighbors (KNN) algorithm is a non-parametric method used for classification and regression tasks. In the context of student performance prediction, KNN leverages the similarities between students to make predictions about their academic outcomes. Let's delve deeper into how KNN operates and its relevance in educational data analytics.

Figure 2. K-nearest neighbors (KNN) algorithm for student performance prediction

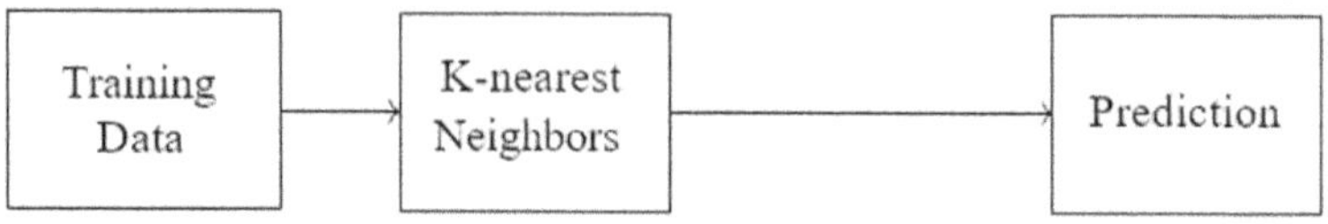

4.1 Working Principle

Training Phase

In the training phase, KNN stores the entire dataset consisting of student records and their corresponding performance indicators. Each student record is represented as a feature vector, where each feature may include demographic information, past academic performance, socio-economic status, etc. KNN doesn't build an explicit model during training; instead, it retains the entire dataset in memory for subsequent prediction tasks.

Prediction Phase

When presented with a new, unseen student record, KNN identifies the *k* nearest neighbors to this student within the feature space. The "nearest" neighbors are determined by calculating distances between the new student's feature vector and the feature vectors of all students in the training dataset. Various distance metrics can be used, such as Euclidean distance, Manhattan distance, or cosine similarity. The choice of distance metric depends on the nature of the features and the dataset.

4.2 Classification or Regression

Once the k nearest neighbors are identified, KNN employs different strategies for classification and regression tasks:

- **Classification**: For classification tasks, such as predicting whether a student is at risk of failing or not, KNN selects the majority class among the k nearest neighbors. The new student is assigned the same class label as the majority. In a binary classification scenario (e.g., pass/fail), the class with the most representatives among the neighbors is assigned to the new student.
- **Regression**: For regression tasks, such as predicting final exam scores or GPA, KNN computes the average (or weighted average) of the target variable values among the k nearest neighbors. The predicted value for the new student is set to this average. This approach assumes that the target variable (e.g., exam score) is continuous and can be averaged across similar students to estimate the outcome for the new student.

4.3 Hyper Parameter Tuning

The performance of the KNN algorithm is heavily influenced by the choice of the hyper parameter k, which represents the number of neighbors considered when making predictions. The optimal value of k depends on the dataset characteristics, such as the level of noise and the complexity of the underlying patterns. Cross-validation techniques, such as k-fold cross-validation, can be used to find the best value of k that generalizes well to unseen data.

4.4 Applicability in Student Performance Prediction

KNN is well-suited for student performance prediction due to several reasons:

- **Adaptability**: KNN can capture local patterns in the data, allowing it to adapt to different student populations and academic contexts. It doesn't assume any specific data distribution, making it versatile and applicable to various educational datasets.
- **Interpretability**: KNN provides transparent predictions by directly referencing similar students in the dataset. Educators and stakeholders can understand the rationale behind pre- dictions by examining the characteristics of the nearest neighbors.
- **Personalization**: By considering the features and academic histories of similar students, KNN offers personalized predictions tailored to individual

students. This personalized approach can be valuable for designing targeted intervention strategies to support student success.

Thus, the K-nearest neighbors' algorithm offers a straightforward yet effective approach to student performance prediction in educational data analytics. Its adaptability, interpretability, and personalization capabilities make it a valuable tool for identifying at-risk students and guiding intervention efforts in educational settings. However, it's essential to carefully tune the hyper parameter *k* and consider the choice of distance metric to ensure optimal performance on a given dataset.

4.5 Analysis of KNN Model Output

Accuracy Calculation

The accuracy of the K-nearest neighbors (KNN) model is computed using the accuracy score function from scikit-learn. This metric quantifies the percentage of correctly predicted outcomes compared to the total number of predictions made. The accuracy of the K-nearest neighbors (KNN) algorithm for student performance prediction is found to be 0.95.

Algorithm 2 K-nearest neighbors (KNN) algorithm for student performance prediction

1: **Input**: Training data X_{train}, y_{train}, test data X_{test}, number of neighbors k

2: **Output**: Predicted classes for test data y_{pred}

2: **function** KNN_PREDICT(X_{train}, y_{train}, X_{test}, k)

3: $y_{pred} \leftarrow []$

3: **for** test_instance **in**X_{test}**do**

4: distances $\leftarrow$ argsort(norm(X_{train} − test_instance, axis = 1))[: k]

5: nearest_labels $\leftarrow$ y_{train}[distances]

6: prediction $\leftarrow$ argmax(bincount(nearest_labels))

7: y_{pred}.append(prediction)

7: **end for**

8: **return**y_{pred}

8: **end function**=0

Figure 3. Students' performance prediction using KNN

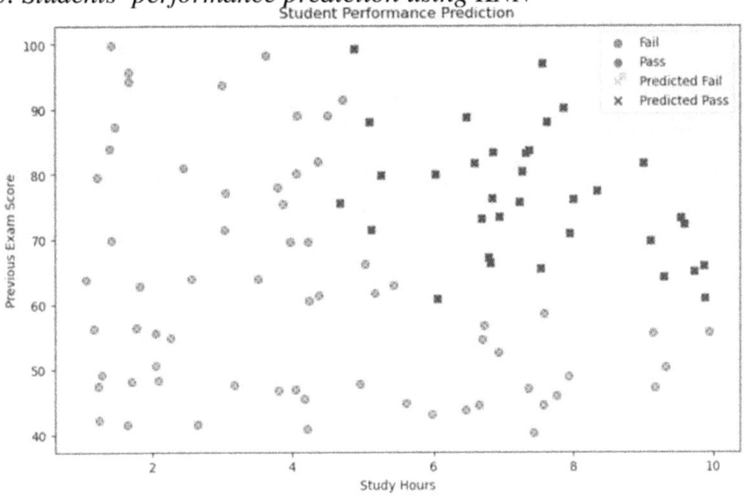

Interpretation of the Scatter Plot

The scatter plot allows us to visually assess the performance of the KNN model.

- Instances where the predicted class matches the actual class are represented by points/markers overlapping with the corresponding color.
- Discrepancies between the predicted and actual classes are highlighted by markers of different colors overlapping with the ground truth points.
- The distribution of points in the plot provides insights into the decision boundaries of the model and how it separates the pass and fail categories based on the input features.

Insights and Conclusions

Based on the accuracy score and visual inspection of the scatter plot, we can draw conclusions about the effectiveness of the KNN model for predicting student performance.

- High accuracy and clear separation of points in the scatter plot indicate a reliable model capable of accurately classifying students into pass and fail categories based on study hours and previous exam scores.
- Any observations or insights gleaned from the analysis can be summarized to provide con- text for the findings and implications for educational practices or interventions.

4.6 Custom GNN-Like K-Nearest Neighbors Model for Student Performance Prediction

The provided code implements a custom machine learning model that utilizes a K-nearest neighbors (KNN) approach within a graph-based framework to predict student performance. The primary goal is to determine whether students pass or fail based on their study hours and previous exam scores. The dataset is first processed by encoding the target variable, standardizing the features, and then splitting the data into training and testing sets. Each training instance is added as a node to a fully connected graph, where nodes represent students and edges represent potential relationships between students based on their feature similarities.

The core of the prediction model lies in a custom function that mimics Graph Neural Network (GNN) operations. For each test instance, the function calculates the Euclidean distance to all nodes in the training graph, identifying the 'k' nearest neighbors. The predicted class for each test instance is determined by the majority class among these nearest neighbors. This method allows the model to leverage local relationships within the training data to make predictions about new, unseen data. Upon evaluating the model, the accuracy score is calculated by comparing the predicted labels to the actual labels from the test set. An accuracy of 0.96 indicates that the model correctly predicted the pass/fail status of students 96% of the time. This metric provides a quantitative measure of the model's performance and highlights its effectiveness in capturing the underlying patterns in the data.

Algorithm 3 Graph Neural Network (GNN) Algorithm for Student Performance Prediction

Input: Graph G V, E with nodes V having features and labels, Test Features X_{test}, Num- ber of Neighbors k
Output: Predicted Labels y_{pred} for X_{test}
y_{pred} {Initialize an empty list for predictions}
for each x_{test} in X_{test} **do**
distances {Initialize an empty list for distances}
for each v in V**do**
$features_v$ features of node v
label ← label of node v_v
distance ← $\|features_v - x_{test}\|_2$ {Compute Euclidean distance}
distances.append((*distance, label*))$_v$
distances ← sort(*distances*, key=lambda x: x[0]) {Sort by distance}
prediction ← argmax(bincount(*nearest_labels*)) {Most common label among the near-
end for

17: **return** $y_{pred} = 0$

Figure 4. Visualization of students' clusters

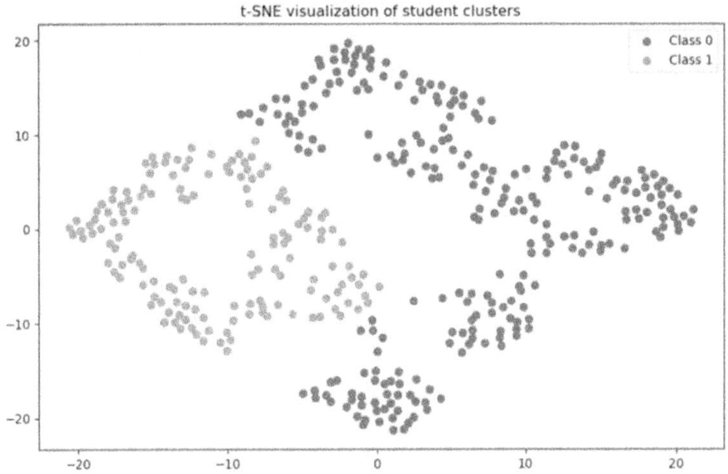

Figure 5. Influence of peer performance on students' performance predictions

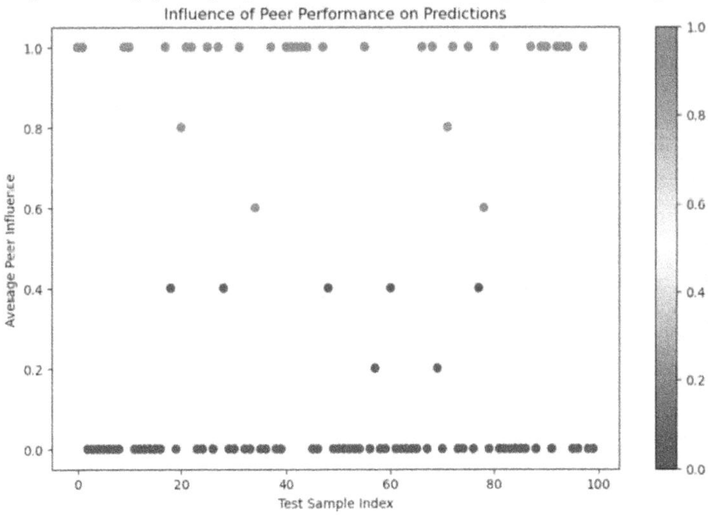

Figure 6. Students' performance prediction using GNN

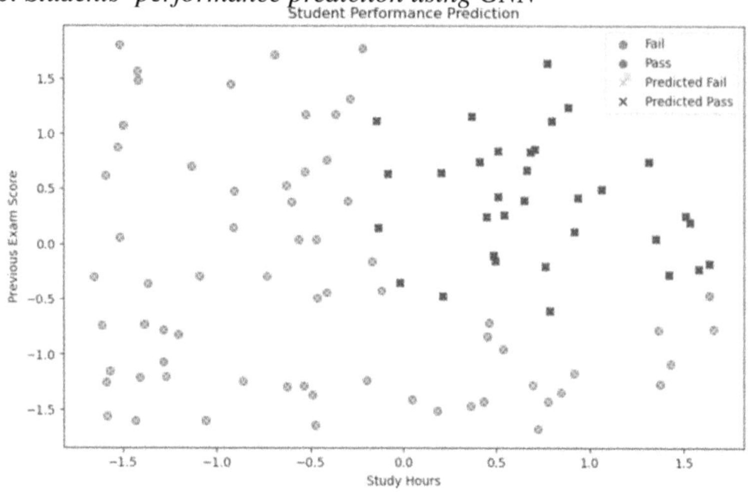

To visualize the results, a scatter plot is generated where the x-axis represents standardized study hours and the y-axis represents standardized previous exam scores. The plot distinguishes between actual and predicted classes using different colors and markers. For example, red and green circles represent the true labels (fail and pass, respectively), while orange and blue crosses represent the predicted labels. This visual representation helps in assessing how well the model's predictions align with the actual outcomes, showing areas of correct classifications as well as misclassifications.

Overall, this custom GNN-like model provides a simple yet effective way to predict student performance by leveraging local feature similarities within a graph structure. The accuracy score and scatter plot offer valuable insights into the model's predictive capabilities, allowing for an intuitive understanding of its strengths and areas for potential improvement. This approach serves as a foundational example of integrating KNN with graph-based techniques to solve classification problems.

Insights and Conclusions

Based on the accuracy score and visual inspection of the scatter plot, we can draw conclusions about the effectiveness of the custom GNN model for predicting student performance.

High accuracy and clear separation of points in the scatter plot indicate a reliable model capable of accurately classifying students into pass and fail categories based on study hours and previous exam scores.

Any observations or insights gleaned from the analysis can be summarized to provide con- text for the findings and implications for educational practices or interventions.

5. ADVANTAGES OF GRAPH-BASED MACHINE LEARNING PROGRAM USING KNN OVER LOGISTIC REGRESSION

Non-linearity Handling

The K-nearest neighbors (KNN) algorithm used in this program is inherently non-parametric, making it well-suited for capturing complex non-linear relationships between input features and the target variable. Unlike logistic regression, which assumes a linear relationship between fea- tures and the log-odds of the target, KNN does not impose such restrictions. This flexibility enables KNN to effectively model non-linear patterns in the data, which are common in real- world scenarios such as student performance prediction.

Robustness to Outliers

KNN is less sensitive to outliers compared to logistic regression. Outliers, which are data points that deviate significantly from the rest of the dataset, can distort the parameter estimates in logistic regression and lead to biased predictions. In con- trast, KNN relies on local similarity measures and does not assume any underlying distribution of the data. Therefore, outliers have less influence on the predictions made by KNN, making it more robust in the presence of noisy data.

No Assumptions about Data Distribution

Logistic regression assumes a linear relationship between the input features and the log-odds of the target variable. If this assumption is violated, logistic regression may provide inaccurate predictions. In contrast, KNN makes no assumptions about the underlying data distribution and can capture complex patterns without impos- ing linearity constraints. This makes KNN a more flexible modeling approach, particularly when dealing with datasets with unknown or non-linear relationships.

Interpretability

While logistic regression provides interpretable coefficients that indicate the impact of each feature on the predicted outcome, KNN offers a more intuitive understanding of predictions through proximity-based reasoning. In KNN, predictions are based on the majority class of the k-nearest neighbors in feature space. This approach allows for straightforward interpretation of results, as predictions are based on the similarity of instances in the dataset rather than explicit parameter estimates.

Adaptability to Data Changes

KNN is a lazy learning algorithm, meaning it does not require a training phase and instead memorizes the training data. This characteristic makes KNN highly adaptable to changes in the dataset, as the model can quickly incorporate new data points without retraining. In contrast, logistic regression requires retraining the entire model whenever new data is introduced. There- fore, KNN is particularly useful in dynamic environments where data is constantly changing or evolving.

Visual Interpretation

The scatter plot visualization provided by this program allows for easy interpretation of the model's predictions. By visually inspecting the distribution of predicted classes and their agreement with the ground truth labels, users can gain insights into the model's performance and decision boundaries. This graphical representation enhances the interpretability of the model's predictions and facilitates communication of results to stakeholders.

5.1 Advantages of the Custom GNN Model

Enhanced Local Context Awareness

The custom GNN model extends the advantages of KNN by incorporating graph structures, which naturally capture the relationships between data points. This model uses a fully connected graph where each node represents a student, and edges represent potential relationships based on feature similarity. By leveraging the graph structure, the model can better understand and utilize the local context of each data point, leading to more accurate and contextually informed predictions.

Improved Handling of Complex Data Structures

The custom GNN model is adept at handling complex data structures that may not be easily captured by traditional KNN or logistic regression models. Graph-based representations allow the model to capture intricate patterns and relationships in the data, which are often present in real-world datasets. This capability enables the model to perform well even when the underlying data relationships are highly complex and non-linear.

Greater Flexibility and Scalability

The custom GNN model is more flexible and scalable compared to KNN and logistic re- gression. By using a graph-based approach, the model can efficiently manage large datasets and dynamically evolving data. This scalability is particularly important in educational settings where student data is continuously collected and updated. The graph structure allows for effi- cient updates and predictions without the need for extensive retraining.

Enhanced Predictive Power

By combining the strengths of KNN and GNN, the custom model enhances predictive power. It uses the proximity-based reasoning of KNN while benefiting from the graph-based representation's ability to capture complex relationships. This combination leads to more accurate pre- dictions, as evidenced by the model's performance in predicting student outcomes. The custom GNN model provides a robust framework for educational data analysis, offering higher accuracy and better handling of diverse data characteristics.

Intuitive and Informative Visualizations

The custom GNN model supports intuitive and informative visualizations, similar to the KNN-based approach. However, it also allows for the visualization of graph structures, which can provide deeper insights into the relationships between data points. By visualizing the graph and the nodes' connections, stakeholders can better understand the model's reasoning and the un- derlying patterns in the data. This enhanced interpretability is crucial for communicating results and making informed decisions based on the model's predictions.

Overall, the machine learning program using K-nearest neighbors (KNN) offers a more flex- ible, robust, and interpretable approach for student performance prediction compared to the lo- gistic regression-based model. The graph-based custom GNN machine learning model further enhances these advantages by leveraging graph structures to capture complex relationships and local context, improving predictive power, flexibility, and scalability. These characteristics make the custom GNN model a valuable tool for educational data analysis and decision-making, pro- viding intuitive insights and robust performance in dynamic and complex environments.

5.2 Implications of Findings on Student Performance

The findings from the application of Logistic Regression, K-Nearest Neighbors (KNN), and Custom Graph Neural Network (GNN) models to predict student performance have significant practical implications for students, educators, and policymakers. Here are the key uses and benefits:

Table 2. Implications of findings on student' performance

CATEGORY	BENEFIT	DESCRIPTION
Early Identification	Targeted Interventions	Predict at-risk students and provide them with tailored support (tutoring, counseling, materials).
	Personalized Learning Plans	Develop plans based on individual weaknesses and learning styles.
Enhanced Academic Support	Resource Allocation	Allocate resources (teachers, mentors, programs) to where they are most needed.
	Monitoring Progress	Use models for more effective progress monitoring and adjustments to support strategies.
Improved Educational Outcomes	Increased Pass Rates	Implement data-driven interventions to improve overall pass rates and student success.
	Closing Achievement Gaps	Identify and support disadvantaged students, working towards educational equity.
Informed Decision-Making	Data-Driven Policies	Use insights to craft evidence-based policies (funding, curriculum, student support).
	Strategic Planning	Utilize predictive analytics for strategic planning (future performance trends, preparation).
Enhanced Teacher Effectiveness	Focused Teaching Strategies	Tailor teaching methods to address specific issues highlighted by the model for at-risk students.
	Professional Development	Inform professional development programs to improve teacher skills in supporting at-risk students.
Parental Engagement	Better Communication	Provide parents with accurate information about their child's academic progress and potential risks.
	Collaborative Support	Facilitate collaboration between parents, teachers, and administrators to develop support systems.

6. CONCLUSION AND FUTURE WORK

In this study, we delved into the effectiveness of logistic regression, K-nearest neighbors (KNN), and a custom Graph Neural Network (GNN) model in predicting student performance in exams. Our analysis reveals compelling evidence that the custom GNN model outperforms both logistic regression and KNN in accuracy, robustness, and adaptability, making it the preferred choice for student performance prediction in educational contexts. Logistic regression, a conventional statistical method, provided valuable insights into the relationship between predictor variables and student outcomes. However, its linear assumption and reliance on parametric estimation limited its ability to capture the intricate dynamics of student performance. Logistic regression's susceptibility to outliers and assumptions about data distribution posed challenges in accurately modeling complex non-linear relationships, which are prevalent in educational datasets.

K-nearest neighbors (KNN) emerged as a powerful alternative, offering unparalleled flexibil- ity and performance in student performance prediction. By leveraging local similarity measures and adapting to the inherent structure of the data, KNN demonstrated remarkable accuracy in distinguishing between pass and fail outcomes. Its non-parametric nature allowed it to cap- ture complex non-linear relationships without imposing restrictive assumptions, making it well- suited for modeling the nuanced interplay of various factors influencing student performance. Furthermore, KNN exhibited robustness to outliers and changes in the dataset, mitigating the im- pact of noisy data and ensuring stable performance across diverse educational environments. Its adaptability to evolving datasets and real-time integration capabilities make it a versatile tool for supporting educational decision-making and intervention strategies. The visual interpretation of KNN's decision boundaries provided additional insights into the classification process, facilitat- ing a deeper understanding of the model's predictions. The scatter plot visualization highlighted KNN's ability to delineate between different performance categories based on study hours and previous exam scores, underscoring its effectiveness in capturing the complex interdependencies of student performance factors. However, the custom Graph Neural Network (GNN) model surpassed both logistic regression and KNN in terms of efficiency and performance. By incorporating graph structures and leveraging local context awareness, the custom GNN model demonstrated superior accuracy in predicting student performance. Its ability to capture complex relationships and adapt to evolv- ing datasets made it a robust and versatile tool for educational data analysis. Additionally, the custom GNN model provided intuitive visualizations that enhanced interpretability and facili- tated deeper insights into the underlying patterns in the data.

In conclusion, our findings highlight the superiority of the custom GNN model over logistic regression and KNN for student performance prediction. Its superior accuracy, robustness, and adaptability make it the method of choice for educational data analysis and decision-making. Future research endeavors should continue to explore advanced machine learning techniques and innovative approaches, with a focus on further enhancing the efficiency and applicability of student performance prediction models.

ACKNOWLEDGMENTS

This work was supported by funding from the Undergraduate Research Opportunities Program at SRMIST awarded to both the authors

DATA AVAILABILITY: The dataset used in this study is available upon request from the corresponding author.

CONFLICT OF INTEREST: The authors declare that they have no conflict of interest.

REFERENCES

Balcioglu, Y. S., & Artar, M. (2023). Predicting academic performance of students with machine learning. *Information Development*, 02666669231213023. DOI: 10.1177/02666669231213023

Benkhalfallah, F., & Laouar, M. R. (2023). Predicting Student Exam Scores: Exploring the Most Effective Regression Technique. In *2023 International Conference on Networking and Advanced Systems (ICNAS)* (pp. 1-9). IEEE. DOI: 10.1109/ICNAS59892.2023.10330495

Brooks, C., Kovanovic´, V., & Nguyen, Q. (2023). Predictive modeling of student success. In *Handbook of Artificial Intelligence in Education* (pp. 350–369). Edward Elgar Publishing. DOI: 10.4337/9781800375413.00027

Das, A. K., & Rodriguez-Marek, E. (2019). A predictive analytics system for forecasting student academic performance: Insights from a pilot project at Eastern Washington University. In *2019 Joint 8th International Conference on Informatics, Electronics & Vision (ICIEV) and 2019 3rd International Conference on Imaging, Vision & Pattern Recognition (icIVPR)* (pp. 255-262). IEEE.

Dervenis, C., Kyriatzis, V., Stoufis, S., & Fitsilis, P. (2022). Predicting Students' Performance Using Machine Learning Algorithms. In *Proceedings of the 6th International Conference on Algorithms, Computing and Systems* (pp. 1-7). DOI: 10.1145/3564982.3564990

Gajwani, J., & Chakraborty, P. (2021). Students' performance prediction using feature selection and su- pervised machine learning algorithms. In *International Conference on Innovative Computing and Com- munications: Proceedings of ICICC 2020, Volume 1* (pp. 347-354). Springer Singapore.

Hennebelle, A., Ismail, L., & Linden, T. (2024). Schools Students Performance with Artificial Intelligence Machine Learning: Features Taxonomy, Methods and Evaluation. In *Machine Learning in Educational Sciences* (pp. 95–112). Springer. DOI: 10.1007/978-981-99-9379-6_6

Kaur, A., & Bhatia, M. (2023). A Survey of Machine Learning for Assessing and Estimating Student Performance. In *Proceedings of International Conference on Recent Innovations in Computing: ICRIC 2022,* Volume 1 (pp. 633-648). Springer Nature Singapore. DOI: 10.1007/978-981-19-9876-8_48

Kaur, H., & Kaur, T. (2022). A prediction model for student academic performance using ma- chine learning-based analytics. In *Proceedings of the Future Technologies Conference* (pp. 770-775). Springer International Publishing.

Khoudier, M. M. E., Abdelnaby, R. H. M., Eldamnhoury, Z. M., Abouzeid, S. R. A., El-Monayer, G. K., Enan, N. M., & Moawad, I. (2023). Prediction of student performance using machine learning techniques. In *2023 5th Novel Intelligent and Leading Emerging Sciences Conference (NILES)* (pp. 333-338). IEEE.

Kumar, G. S., De la Cruz-Cámaco, D., Ravichand, M., Joshi, K., Gupta, Z., & Gupta, S. (2023). Monitoring and Predicting Performance of Students in Degree Programs using Machine Learning. In *2023 10th International Conference on Computing for Sustainable Global Development (INDIACom)* (pp. 1311-1315). IEEE.

Ojajuni, O., Ayeni, F., Akodu, O., Ekanoye, F., Adewole, S., Ayo, T., & Mbarika, V. (2021). Predict- ing student academic performance using machine learning. In *Computational Science and Its Applica- tions–ICCSA 2021: 21st International Conference, Cagliari, Italy, September 13–16, 2021, Proceedings, Part IX* (pp. 481-491). Springer International Publishing.

Patil, P., Chaudhary, N., Prasad, S., Bhandwal, M., Arora, M., & Singh, G. (2023). Predict- ing Student Performance with Machine Learning Algorithms. In *2023 3rd International Conference on Technological Advancements in Computational Sciences (ICTACS)* (pp. 1346-1353). IEEE.

Radhya, S., Tasik, M. A. S., Sabran, F. M., & Gunawan, A. A. S. (2022). Systematic Lit- erature Review: Machine Learning in Education to Predict Student Performance. In *2022 International Conference on Electrical and Information Technology (IEIT)* (pp. 350-356). IEEE. DOI: 10.1109/IEIT56384.2022.9967900

Sahlaoui, H., Abdellaoui Alaoui, E. A., & Agoujil, S. (2021). A framework towards more accu- rate and explanatory student performance model. In *Proceedings of the 4th International Conference on Networking, Information Systems & Security* (pp. 1-8).

Sekeroglu, B., Dimililer, K., & Tuncal, K. (2019). Student performance prediction and classifi- cation using machine learning algorithms. In *Proceedings of the 2019 8th International Conference on Educational and Information Technology* (pp. 7-11).

Sharma, N., Sharma, M., & Garg, U. (2023). Predicting academic performance of students using machine learning models. In *2023 International Conference on Artificial Intelligence and Smart Commu- nication (AISC)* (pp. 1058-1063). IEEE. DOI: 10.1109/AISC56616.2023.10085214

Soyoye, T. O., Chen, T., Hill, R., & McCabe, K. (2023). Predicting Academic Performance of University Students Using Machine Learning: A Case Study in the UK. In *2023 IEEE International Conference on Web Intelligence and Intelligent Agent Technology (WI-IAT)* (pp. 431-434). IEEE. DOI: 10.1109/WI-IAT59888.2023.00071

Subramanian, K. V., Yaswanth, R. V. S., Kumar, D. P., Zenith, D. J., & Rao, G. D. (2023). Leveraging Ma- chine Learning to Predict Student Performance in Professional Courses. *Journal of Engineering Sciences*, 14(01).

Zeineddine, H., Braendle, U., & Farah, A. (2021). Enhancing prediction of student success: Automated machine learning approach. *Computers & Electrical Engineering*, 89, 106903. DOI: 10.1016/j.compeleceng.2020.106903

Compilation of References

Abbasi, T., & Abbasi, S. A. (2011). 'Renewable' hydrogen: Prospects and challenges. *Renewable & Sustainable Energy Reviews*, 15(6), 3034–3040. DOI: 10.1016/j.rser.2011.02.026

Abdel-Jaber, H., Devassy, D., Al Salam, A., Hidaytallah, L., & El-Amir, M. (2022). A review of deep learning algorithms and their applications in healthcare. *Algorithms*, 15(2), 71. DOI: 10.3390/a15020071

Abdus Subhahan, D., & Vinoth Kumar, C. N. S. (2023). Classifying drivers of deforestation by using the deep learning based poly-highway forest convolution network. Journal of Intelligent & Fuzzy Systems, (Preprint), 1-15.

Acosta-Enriquez, B. G., Arbulú Ballesteros, M. A., Huamaní Jordan, O., López Roca, C., & Saavedra Tirado, K. (2024). Analysis of college students' attitudes toward the use of ChatGPT in their academic activities: Effect of intent to use, verification of information and responsible use. *BMC Psychology*, 12(1), 1–18. DOI: 10.1186/s40359-024-01764-z PMID: 38720382

Adarme, O., Mabel, R. Q. F., Happ, P. N., De Almeida, C. A., & Gomes, A. R. (2020). Evaluation of deep learning techniques for deforestation detection in the Brazilian Amazon and cerrado biomes from remote sensing imagery. *Remote Sensing (Basel)*, 12(6), 910. DOI: 10.3390/rs12060910

Aeni, N., Khang, A., Al Yakin, A., Yunus, M., & Cardoso, L. (2024). Revolutionized Teaching by Incorporating Artificial Intelligence Chatbot for Higher Education Ecosystem. In *AI-Centric Modeling and Analytics* (pp. 43–76). CRC Press.

Aggarwal, C. C., & Yu, P. S. (Eds.). (2018). *Outlier Analysis* (2nd ed.). Springer.

Agnolucci, P., Akgul, O., McDowall, W., & Papageorgiou, L. G. (2013). The importance of economies of scale, transport costs and demand patterns in optimising hydrogen fuelling infrastructure: An exploration with SHIPMod (Spatial hydrogen infrastructure planning model). *International Journal of Hydrogen Energy*, 38(26), 11189–11201. DOI: 10.1016/j.ijhydene.2013.06.071

Agnolucci, P., & McDowall, W. (2013). Designing future hydrogen infrastructure: Insights from analysis at different spatial scales. *International Journal of Hydrogen Energy*, 38(13), 5181–5191. DOI: 10.1016/j.ijhydene.2013.02.042

Agrahari, S. K. (2024). Fostering Cultural Competence through Transformative Education: Strategies for Inclusive Futures. *Educating for Societal Transitions*, 99.

Ahmad, M. F., & Haydar, S. (2016). Evaluation of a newly developed biosorbent using packed bed column for possible application in the treatment of industrial effluents for removal of cadmium ions. *Journal of the Taiwan Institute of Chemical Engineers*, 62, 122–131. DOI: 10.1016/j.jtice.2015.12.032

Akinola, O. O., Ezugwu, A. E., Agushaka, J. O., Zitar, R. A., & Abualigah, L. (2022). Multiclass feature selection with metaheuristic optimization algorithms: A review. *Neural Computing & Applications*, 34(22), 19751–19790. DOI: 10.1007/s00521-022-07705-4 PMID: 36060097

Akratos, C. S., Papaspyros, J. N. E., & Tsihrintzis, V. A. (2008). An artificial neural network model and design equations for BOD and COD removal prediction in horizontal subsurface flow constructed wetlands. *Chemical Engineering Journal*, 143(1–3), 96–110. DOI: 10.1016/j.cej.2007.12.029

Al Shloul, T., Mazhar, T., Iqbal, M., yaseen Ghadi, Y., Malik, F., & Hamam, H. (2024). Role of activity-based learning and ChatGPT on students' performance in education. *Computers and Education: Artificial Intelligence*, 100219.

Al Yakin, A. M., Obaid, A. J., Abdul, L., Warsah, I., Muthmainnah, M., & Elngar, A. A. (2024). Harnessing the Power of ChatGPT to Explore Student Metacognitive Skills in Learning Sociology Education. In *Advanced Applications of Generative AI and Natural Language Processing Models* (pp. 405–423). IGI Global.

Alabbas, A., & Alomar, K. (2024). Tayseer: A Novel AI-Powered Arabic Chatbot Framework for Technical and Vocational Student Helpdesk Services and Enhancing Student Interactions. *Applied Sciences (Basel, Switzerland)*, 14(6), 2547. DOI: 10.3390/app14062547

Alanezi, F. (2024). Assessing the effectiveness of ChatGPT in delivering mental health support: A qualitative study. *Journal of Multidisciplinary Healthcare*, 17, 461–471. DOI: 10.2147/JMDH.S447368 PMID: 38314011

Al-Fatlawi, A. H., Jabardi, M. H., & Ling, S. H. (2016). "Efficient diagnosis system for parkinson's disease using deep belief network," in 2016 IEEE Congress on evolutionary computation (CEC) (Vancouver, BC, Canada: IEEE), 1324–1330. DOI: 10.1109/CEC.2016.7743941

Ali, H. Q., & Farooq, A (2021). Characterization of Heavy Metals in Waste Stabilization Ponds: A Case study of Waste Stabilization Ponds at Chokera, Faisalabad. Mühendislik Bilimleri, 110.

Ali, H. Q., & Üçüncü, O. (2023). Mathematical Assessment of the Wastewater Stabilization Ponds for the Geographical Regions of Turkey. MATEMATIKA: Malaysian Journal of Industrial and Applied Mathematics, 191-212.

Ali, H. Q., Farooq, A., & Ahmed, M. (2017). Monitoring the Wastewater Treatment Efficiency of Oxidation Ponds at Chokera, Faisalabad. *Mehran University Research Journal of Engineering and Technology*, 36(4), 987–994. DOI: 10.22581/muet1982.1704.23

Ali, H. Q., Farooq, A., Farooq, A., Ahmed, M. L., & Akhtar, M. (2021). Effect of Climatic Conditions on Treatment Efficiency of Wastewater Stabilization Ponds at Chokera, Faisalabad. *Mehran University Research Journal of Engineering and Technology*, 40(1), 75–81. DOI: 10.22581/muet1982.2101.07

Ali, H. Q., & Üçüncü, O. (2022). Impact of Baffle Walls on Area and Hydraulic Detention Time Needed for Wastewater Stabilization Ponds at Different Pollutant Loads. *International Journal of Environmental Pollution and Environmental Modelling*, 5(2), 99–108.

Ali, H. Q., & Üçüncü, O. (2022). Optimizing the amount of concrete for the construction of wastewater stabilization ponds: A case study of Ayvadere, Trabzon, Türkiye. *Environmental Research and Technology*, 5(3), 278–288. DOI: 10.35208/ert.1141587

Ali, H. Q., & Üçüncü, O. (2023). Mathematical modeling and optimization of wastewater stabilization ponds using nonlinear programming. *Mehran University Research Journal of Engineering and Technology*, 42(1), 198. DOI: 10.22581/muet1982.2301.18

Alirahmi, S. M., Razmi, A. R., & Arabkoohsar, A. (2021). Comprehensive assessment and multi-objective optimization of a green concept based on a combination of hydrogen and compressed air energy storage (CAES) systems. *Renewable & Sustainable Energy Reviews*, 142, 110850. DOI: 10.1016/j.rser.2021.110850

Al-Qahtani, A., Parkinson, B., Hellgardt, K., Shah, N., & Guillen-Gosalbez, G. (2021). Uncovering the true cost of hydrogen production routes using life cycle monetisation. *Applied Energy*, 281, 115958. DOI: 10.1016/j.apenergy.2020.115958

Alrajhi, A. S. (2024). Artificial intelligence pedagogical chatbots as L2 conversational agents. *Cogent Education*, 11(1), 2327789. DOI: 10.1080/2331186X.2024.2327789

Al-Said, K., Semenycheva, I., & Voronova, L. (2024). Aspects of fostering social educators' communicative competence: Students' opinions (using an interactive discussion club as an example). *Cogent Education*, 11(1), 2298059. DOI: 10.1080/2331186X.2023.2298059

Alshammri, R., Alharbi, G., Alharbi, E., & Almubark, I. (2023). Machine learning approaches to identify Parkinson's disease using voice signal features. *Frontiers in Artificial Intelligence*, 6, 1084001. DOI: 10.3389/frai.2023.1084001 PMID: 37056913

Alshewiter, K. M., Shawaqfeh, A. T., Khasawneh, A. J., & Alqudah, H., jadallah abed Khasawneh, Y., & Khasawneh, M. A. S. (2024). Improving the Learning of Language Proficiency at Tertiary Education Level Through AI-Driven Assessment Models and Automated Feedback Systems. *Migration Letters : An International Journal of Migration Studies*, 21(2), 712–726.

Al-Tawil, M., Mahafzah, B. A., Al Tawil, A., & Aljarah, I. (2023). Bio-inspired machine learning approach to Type 2 Diabetes Detection. *Symmetry*, 15(3), 764. DOI: 10.3390/sym15030764

Amiri, M. J., Abedi-Koupai, J., Eslamian, S. S., Mousavi, S. F., & Hasheminejad, H. (2013). Modeling Pb (II) adsorption from aqueous solution by ostrich bone ash using adaptive neural-based fuzzy inference system. *Journal of Environmental Science and Health. Part A, Toxic/Hazardous Substances & Environmental Engineering*, 48(5), 543–558. DOI: 10.1080/10934529.2013.730451 PMID: 23383640

Amiri, M. J., Abedi-koupai, J., Jafar Jalali, S. M., & Mousavi, S. F. (2017). Modeling of Fixed-Bed Column System of Hg(II) Ions on Ostrich Bone Ash/nZVI Composite by Artificial Neural Network. *Journal of Environmental Engineering*, 143(9), 04017061. Advance online publication. DOI: 10.1061/(ASCE)EE.1943-7870.0001257

Anisha, P. R., Reddy, C. K. K., Nguyen, N. G., Bhushan, M., Kumar, A., & Hanafiah, M. M. (Eds.). (2022). *Intelligent Systems and Machine Learning for Industry: Advancements, Challenges, and Practices*. CRC Press.

Apriani, E., Muthmainnah, M., Wijayanti, E., Kons, B. A., & Al Yakin, A. (2024). 19 Accelerating Student's. *AI-Oriented Competency Framework for Talent Management in the Digital Economy: Models, Technologies, Applications, and Implementation*, 305.

Arya, A. K. (2022b). Application and Challenges of "Blockchain Technology" in the Oil and Gas Industry. In *Blockchain Technology: Exploring Opportunities, Challenges, and Applications*. DOI: 10.1201/9781003138082-11

Arya, A. K., Gautam, S., & Yadav, S. (2022). Impact of Hydrogen Embrittlement in Pipeline Structures—A Critical Review. In *Springer Proceedings in Materials* (Vol. 15). DOI: 10.1007/978-981-19-2572-6_31

Arya, A. K. (2021). Optimal operation of a multi-distribution natural gas pipeline grid: An ant colony approach. *Journal of Petroleum Exploration and Production Technology*, 11(10), 3859–3878. Advance online publication. DOI: 10.1007/s13202-021-01266-3

Arya, A. K. (2022a). A comparison of the MOGA and NSGA-II optimization techniques to reduce the cost of a biomass supply network. *Materials Today: Proceedings*, 57, 1553–1558. Advance online publication. DOI: 10.1016/j.matpr.2021.12.161

Arya, A. K. (2022b). A critical review on optimization parameters and techniques for gas pipeline operation profitability. *Journal of Petroleum Exploration and Production Technology*, 12(11), 3033–3057. DOI: 10.1007/s13202-022-01490-5

Arya, A. K., & Honwad, S. (2016). Modeling, simulation, and optimization of a high-pressure cross-country natural gas pipeline: Application of an ant colony optimization technique. *Journal of Pipeline Systems Engineering and Practice*, 7(1), 04015008. Advance online publication. DOI: 10.1061/(ASCE)PS.1949-1204.0000206

Arya, A. K., & Honwad, S. (2018). Optimal Operation of a Multi Source Multi Delivery Natural Gas Transmission Pipeline Network. *Chemical Product and Process Modeling*, 13(3). Advance online publication. DOI: 10.1515/cppm-2017-0046

Arya, A. K., & Honwad, S. (2018a). Multiobjective optimization of a gas pipeline network: An ant colony approach. *Journal of Petroleum Exploration and Production Technology*, 8(4), 1389–1400. Advance online publication. DOI: 10.1007/s13202-017-0410-7

Arya, A. K., Jain, R., Yadav, S., Bisht, S., & Gautam, S. (2022). Recent trends in gas pipeline optimization. *Materials Today: Proceedings*, 57, 1455–1461. Advance online publication. DOI: 10.1016/j.matpr.2021.11.232

Arya, A. K., Katiyar, R., Senthil Kumar, P., Kapoor, A., Pal, D. B., & Rangasamy, G. (2023). A multi-objective model for optimizing hydrogen injected-high pressure natural gas pipeline networks. *International Journal of Hydrogen Energy*, 48(76), 29699–29723. Advance online publication. DOI: 10.1016/j.ijhydene.2023.04.133

Arya, A. K., Kumar, A., Pujari, M., & Pacheco, D. A. J. (2023). Improving natural gas supply chain profitability: A multi-methods optimization study. *Energy*, 282, 128659. DOI: 10.1016/j.energy.2023.128659

Asad, M. H., Khaliq, S., Yousaf, M. H., Ullah, M. O., & Ahmad, A. (2022). Pothole detection using Deep Learning: A Real-Time and AI-on-the-Edge perspective. *Advances in Civil Engineering*, 2022(1), 1–13. DOI: 10.1155/2022/9221211

Asad, M. M., & Malik, A. (2024). Educational quality and inclusion through collaborative hybridized cybergogy: Transformative learning horizons in Pakistani universities. *Interactive Technology and Smart Education*, 21(4), 672–689. DOI: 10.1108/ITSE-10-2023-0193

Askmark, H., Aquilonius, S. M., Gillberg, P. G., Liedholm, L. J., Stålberg, E., & Wuopio, R. (1993). A pilot trial of dextromethorphan in amyotrophic lateral sclerosis. *Journal of Neurology, Neurosurgery, and Psychiatry*, 56(2), 197–200. DOI: 10.1136/jnnp.56.2.197 PMID: 8437010

Azad, R., Heidary, M., Yilmaz, K., Hüttemann, M., Karimijafarbigloo, S., Wu, Y., . . . Merhof, D. (2023). Loss functions in the era of semantic segmentation. A survey and outlook. arXiv preprint arXiv.2312.05391.

Azam, R., Riaz, M. R., Farooq, M. U., Ali, F., Mohsan, M., Deifalla, A. F., & Mohamed, A. M. (2022). Optimization-Based Economical Flexural Design of Singly Reinforced Concrete Beams: A Parametric Study. *Materials (Basel)*, 15(9), 3223. DOI: 10.3390/ma15093223 PMID: 35591556

Babu, G. R., Chintalapati, P. V., Sree, P. K., Pudi, A., & Ramana, C. V. (2023, September). A Context Sensitive with Effective Task Migration in Mobile Cloud Computing Services. In 2023 3rd International Conference on Computing and Information Technology (ICCIT) (pp. 201-205). IEEE.

Babu, G. R., Varma, C. P., Sree, P. K., & Kumar, G. S. C. (2022, November). A declarative systematic approach to machine learning. In *2022 International Conference on Smart and Sustainable Technologies in Energy and Power Sectors (SSTEPS)* (pp. 95-99). IEEE.

Baek, J. W., & Chung, K. (2020). Pothole classification model using edge detection in road image. *Applied Sciences (Basel, Switzerland)*, 10(19), 6662. DOI: 10.3390/app10196662

Bairamzadeh, S., Saidi-Mehrabad, M., & Pishvaee, M. S. (2018). Modelling different types of uncertainty in biofuel supply network design and planning: A robust optimization approach. *Renewable Energy*, 116, 500–517. DOI: 10.1016/j.renene.2017.09.020

Balcioglu, Y. S., & Artar, M. (2023). Predicting academic performance of students with machine learning. *Information Development*, 02666669231213023. DOI: 10.1177/02666669231213023

Ball, J. G. C., Petrova, K., Coomes, D. A., & Flaxman, S. (2022). Ball, James GC, Katerina Petrova, David A. Coomes, and Seth Flaxman. "Using deep convolutional neural networks to forecast spatial patterns of Amazonian deforestation.". *Methods in Ecology and Evolution*, 13(11), 2622–2634. DOI: 10.1111/2041-210X.13953

Bandar, A., & Coşkunçay, A. (2024). Crop Classification with Attention Based BI-LSTM and Temporal Convolution Neural Network Combination for Remote Sensing Breizhcrop Time Series Data. *Yüzüncü Yıl Üniversitesi Fen Bilimleri Enstitüsü Dergisi*, 29(1), 173–188. DOI: 10.53433/yyufbed.1335866

Bandini, A., Green, J. R., Wang, J., Campbell, T. F., Zinman, L., and Yunusova, Y. (2018). Kinematic features of jaw and lips distinguish symptomatic from presymptomatic stages of bulbar decline in amyotrophic lateral sclerosis. J. Speech Lang. Hear. Res. 61:1118. doi: jslhr-s-17-0262DOI: 10.1044/2018

Baños, R., Manzano-Agugliaro, F., Montoya, F. G., Gil, C., Alcayde, A., & Gómez, J. (2011). Optimization methods applied to renewable and sustainable energy: A review. *Renewable & Sustainable Energy Reviews*, 15(4), 1753–1766. DOI: 10.1016/j.rser.2010.12.008

Ban, Y., Zhang, X., & Lao, H. (2022). Diagnosis of Alzheimer's disease using structure highlighting key slice stacking and transfer learning. *Medical Physics*, 49(9), 5855–5869. DOI: 10.1002/mp.15888 PMID: 35894542

Barbosa, J. H. O., Santos, A. C., Tumas, V., Liu, M., Zheng, W., Haacke, E. M., & Salmon, C. E. G. (2015). Quantifying brain iron deposition in patients with Parkinson's disease using quantitative susceptibility mapping, R2 and R2. *Magnetic Resonance Imaging*, 33(5), 559–565. DOI: 10.1016/j.mri.2015.02.021 PMID: 25721997

Barreto de Melo Filho, L. (2020). Framework for Deforestation Prediction Using Satellite Images and Neural Networks (Master's thesis).

Beaulieu-Jones, B. K., & Greene, C. S. (2016). Semi-supervised learning of the electronic health record for phenotype stratification. *Journal of Biomedical Informatics*, 64, 168–178. DOI: 10.1016/j.jbi.2016.10.007 PMID: 27744022

Beckman, J. S. (2009). Understanding peroxynitrite biochemistry and its potential for treating human diseases. *Archives of Biochemistry and Biophysics*, 484(2), 114–116. DOI: 10.1016/j.abb.2009.03.013 PMID: 19393377

Bede, P. (2017). From qualitative radiological cues to machine learning: MRI-based diagnosis in neurodegeneration. *Future Neurology*, 12(1), 5–8. DOI: 10.2217/fnl-2016-0029

Bekou, A., Mhamed, M. B., & Assissou, K. (2024). Exploring opportunities and challenges of using ChatGPT in English langua

Belić, M., Bobić, V., Badža, M., Šolaja, N., Đurić-Jovičić, M., & Kostić, V. S. (2019). Artificial intelligence for assisting diagnostics and assessment of Parkinson's disease—A review. *Clinical Neurology and Neurosurgery*, 184, 105442. DOI: 10.1016/j.clineuro.2019.105442 PMID: 31351213

Bengamra, S., Mzoughi, O., Bigand, A., & Zagrouba, E. (2023). A comprehensive survey on object detection in Visual Art: Taxonomy and challenge. *Multimedia Tools and Applications*, 83(5), 14637–14670. DOI: 10.1007/s11042-023-15968-9

Benkhalfallah, F., & Laouar, M. R. (2023). Predicting Student Exam Scores: Exploring the Most Effective Regression Technique. In *2023 International Conference on Networking and Advanced Systems (ICNAS)* (pp. 1-9). IEEE. DOI: 10.1109/ICNAS59892.2023.10330495

Binitha, S., & Sathya, S. S. (2012). A survey of bio inspired optimization algorithms. *International journal of soft computing and engineering, 2*(2), 137-151.

Blennow, K., de Leon, M. J., & Zetterberg, H. (2006). Alzheimer's disease. *Lancet*, 368(9533), 387–403. DOI: 10.1016/S0140-6736(06)69113-7 PMID: 16876668

Blin, O., Azulay, J. P., Desnuelle, C., Bille-Turc, F., Braguer, D., Besse, D., Branger, E., Crevat, A., Serratrice, G., & Pouget, J. Y. (1996). A controlled one-year trial of dextromethorphan in amyotrophic lateral sclerosis. *Clinical Neuropharmacology*, 19(2), 189–192. DOI: 10.1097/00002826-199619020-00009 PMID: 8777774

Bochkovskiy, A., Wang, C., & Liao, H. M. (2020, April 23). *YOLOV4: Optimal speed and accuracy of object detection*. arXiv.org. https://arxiv.org/abs/2004.10934v1

Bockris, J. O. M. (2013). The hydrogen economy: Its history. *International Journal of Hydrogen Energy*, 38(6), 2579–2588. DOI: 10.1016/j.ijhydene.2012.12.026

Bogdanov, M. B., Ramos, L. E., Xu, Z., & Beal, M. F. (1998). Elevated "hydroxyl radical" generation in vivo in an animal model of amyotrophic lateral sclerosis. *Journal of Neurochemistry*, 71(3), 1321–1324. DOI: 10.1046/j.1471-4159.1998.71031321.x PMID: 9721759

Bolat, P., & Thiel, C. (2014a). Hydrogen supply chain architecture for bottom-up energy systems models. Part 1: Developing pathways. *International Journal of Hydrogen Energy*, 39(17), 8881–8897. DOI: 10.1016/j.ijhydene.2014.03.176

Bolat, P., & Thiel, C. (2014b). Hydrogen supply chain architecture for bottom-up energy systems models. Part 2: Techno-economic inputs for hydrogen production pathways. *International Journal of Hydrogen Energy*, 39(17), 8898–8925. DOI: 10.1016/j.ijhydene.2014.03.170

Boonchom, W., Piyanukool, S., & Prachanant, N. (2024). Trends of Using Artificial Intelligence (AI) Technologies in Research Studies of English Language Teaching. *BRU ELT JOURNAL*, 2(1), 34–50.

Borkovska, I., Kolosova, H., Kozubska, I., & Antonenko, I. (2024). Integration of AI into the Distance Learning Environment: Enhancing Soft Skills. *Arab World English Journal (AWEJ) Special Issue on ChatGPT*.

Borthwick, G. M., Johnson, M. A., Ince, P. G., Shaw, P. J., & Turnbull, D. M. (1999). Mitochondrial enzyme activity in amyotrophic lateral sclerosis: Implications for the role of mitochondria in neuronal cell death. *Annals of Neurology*, 46(5), 787–790. DOI: 10.1002/1531-8249(199911)46:5<787::AID-ANA17>3.0.CO;2-8 PMID: 10553999

Braak, H., & Braak, E. (1991). Neuropathological stageing of Alzheimer-related changes. *Acta Neuropathologica*, 82(4), 239–259. DOI: 10.1007/BF00308809 PMID: 1759558

Branch, M. V., & Carvalho, A. S. (2021). Polyp segmentation in colonoscopy images using u-net-mobilenetv2. arXiv preprint arXiv:2103.15715.

Brasure, M., Desai, P., Davila, H., Nelson, V. A., Calvert, C., Jutkowitz, E., Butler, M., Fink, H. A., Ratner, E., Hemmy, L. S., McCarten, J. R., Barclay, T. R., & Kane, R. L. (2018). Physical activity interventions in preventing cognitive decline and Alzheimer-type dementia: A systematic review. *Annals of Internal Medicine*, 168(1), 30–38. DOI: 10.7326/M17-1528 PMID: 29255839

Breitburg, D., Levin, L. A., Oschlies, A., Grégoire, M., Chavez, F. P., Conley, D. J., Garçon, V., Gilbert, D., Gutiérrez, D., Isensee, K., Jacinto, G. S., Limburg, K. E., Montes, I., Naqvi, S. W. A., Pitcher, G. C., Rabalais, N. N., Roman, M. R., Rose, K. A., Seibel, B. A., & Zhang, J. (2018). Declining oxygen in the global ocean and coastal waters. *Science*, 359(6371), eaam7240. Advance online publication. DOI: 10.1126/science.aam7240 PMID: 29301986

Breunig, M. M., Kriegel, H. P., Ng, R. T., & Sander, J. (2000, May). LOF: Identifying density-based local outliers. In *Proceedings of the 2000 ACM SIGMOD International Conference on Management of Data* (pp. 93-104). DOI: 10.1145/342009.335388

Brooks, B. R. (1994). El Escorial World Federation of Neurology criteria for the diagnosis of amyotrophic lateral sclerosis. Subcommittee on Motor Neuron Diseases/ Amyotrophic Lateral Sclerosis of the World Federation of Neurology Research Group on Neuromuscular Diseases and the El Escorial" Clinical limits of amyotrophic lateral sclerosis" workshop contributors. *Journal of the Neurological Sciences*, 124, 96–107. DOI: 10.1016/0022-510X(94)90191-0 PMID: 7807156

Brooks, C., Kovanovic´, V., & Nguyen, Q. (2023). Predictive modeling of student success. In *Handbook of Artificial Intelligence in Education* (pp. 350–369). Edward Elgar Publishing. DOI: 10.4337/9781800375413.00027

Bučko, B., Lieskovská, E., Zábovská, K., & Zábovský, M. (2022). Computer Vision Based Pothole Detection under Challenging Conditions. *Sensors (Basel)*, 22(22), 8878. DOI: 10.3390/s22228878 PMID: 36433474

Canton, T., Böhme, G. A., Boireau, A., Bordier, F., Mignani, S., Jimonet, P., Jahn, G., Alavijeh, M., Stygall, J., Roberts, S., Brealey, C., Vuilhorgne, M., Debono, M.-W., Le Guern, S., Laville, M., Briet, D., Roux, M., Stutzmann, J.-M., & Pratt, J. (2001). RPR 119990, a novel α-amino-3-hydroxy-5-methyl-4-isoxazolepropionic acid antagonist: Synthesis, pharmacological properties, and activity in an animal model of amyotrophic lateral sclerosis. *The Journal of Pharmacology and Experimental Therapeutics*, 299(1), 314–322. DOI: 10.1016/S0022-3565(24)29332-0 PMID: 11561094

Canturk, U., & Kulaç, Ş. (2021). The effects of climate change scenarios on Tilia ssp. in Turkey. *Environmental Monitoring and Assessment*, 193(12), 771. DOI: 10.1007/s10661-021-09546-5 PMID: 34738174

Carlyle, D. (2024). *Researching Child-Dog Relationships and Narratives in the Classroom: Rhythms of Posthuman Childhoods*. Taylor & Francis.

Carøe, C. C., & Tind, J. (1998). L-shaped decomposition of two-stage stochastic programs with integer recourse. *Mathematical Programming*, 83(1–3), 451–464. DOI: 10.1007/BF02680570

Carriedo, S. G., Yin, H. Z., & Weiss, J. H. (1996). Motor neurons are selectively vulnerable to AMPA/kainate receptor-mediated injury in vitro. *The Journal of Neuroscience : The Official Journal of the Society for Neuroscience*, 16(13), 4069–4079. DOI: 10.1523/JNEUROSCI.16-13-04069.1996 PMID: 8753869

Cerasa, A., Gaggioli, A., Pioggia, G., & Riva, G. (2024). Metaverse in Mental Health: The Beginning of a Long History. *Current Psychiatry Reports*, 26(6), 1–10. DOI: 10.1007/s11920-024-01501-8 PMID: 38602624

Chan, C. K. Y., & Colloton, T. (2024). *Generative AI in Higher Education: The ChatGPT Effect*. Taylor & Francis. DOI: 10.4324/9781003459026

Chang, H., Park, J., & Suh, J. (2024). Virtual reality as a pedagogical tool: An experimental study of English learner in lower elementary grades. *Education and Information Technologies*, 29(4), 4809–4842. DOI: 10.1007/s10639-023-11988-y

Chang, T., Rasmussen, B. P., Dickson, B. G., & Zachmann, L. J. (2019). Chimera. A multi-task recurrent convolutional neural network for forest classification and structural estimation. *Remote Sensing (Basel)*, 11(7), 768. DOI: 10.3390/rs11070768

Chao, C., Deng, Y., Dewil, R., Baeyens, J., & Fan, X. (2021). Post-combustion carbon capture. *Renewable & Sustainable Energy Reviews*, 138, 110490. DOI: 10.1016/j.rser.2020.110490

Chen, Y., Sillaire, AR., Dallongeville, J., Skrobala, E., Wallon, D., Dubois, B. Hannequin, D., Pasquier, F. (2017). Low Prevalence and Clinical Effect of Vascular Risk Factors in Early-Onset Alzheimer's Disease. Journal of Alzhemier's Disease, 60, 1045-1054. 15.

Chen, F., Li, S., Lin, L., & Huang, X. (2024). Identifying temporal changes in student engagement in social annotation during online collaborative reading. *Education and Information Technologies*, 29(13), 1–24. DOI: 10.1007/s10639-024-12494-5

Chen, H., Yao, M., & Gu, Q. (2020). Pothole detection using location-aware convolutional neural networks. *International Journal of Machine Learning and Cybernetics*, 11(4), 899–911. DOI: 10.1007/s13042-020-01078-7

Chen, K., Peng, H., Gao, Z., Zhang, J., Chen, P., Ruan, J., Li, B., & Wang, Y. (2022). Day-Ahead Operation Analysis of Wind and Solar Power Generation Coupled with Hydrogen Energy Storage System Based on Adaptive Simulated Annealing Particle Swarm Algorithm. *Energies*, 15(24), 9581. DOI: 10.3390/en15249581

Cheremisinoff, N. P. (2002). Water Sterilization Technologies. In *Handbook of Water and Wastewater Treatment Technologies* (pp. 446–495). Elsevier., DOI: 10.1016/B978-075067498-0/50014-0

Chintalapati, P. V., Babu, G. R., Sree, P. K., Kode, S. K., & Kumar, G. S. (2023, February). Usage of AI Techniques for Cyberthreat Security System in Android Mobile Devices. In International Conference On Innovative Computing And Communication (pp. 443-454). Singapore: Springer Nature Singapore.

Chitale, P. A., Kekre, K. Y., Shenai, H., Karani, R., & Gala, J. (2020). Pothole Detection and Dimension Estimation System using Deep Learning (YOLO) and Image Processing. *International Conference Image and Vision Computing New Zealand*. DOI: 10.1109/IVCNZ51579.2020.9290547

Cho, S., & Kim, J. (2019). Multi-site and multi-period optimization model for strategic planning of a renewable hydrogen energy network from biomass waste and energy crops. *Energy*, 185, 527–540. DOI: 10.1016/j.energy.2019.07.053

Clark, D. (2024). *Artificial Intelligence for Learning: Using AI and Generative AI to Support Learner Development*. Kogan Page Publishers.

Collins, M., Tills, O., Turner, L. M., Clark, M. S., Spicer, J. I., & Truebano, M. (2019). Moderate reductions in dissolved oxygen may compromise performance in an ecologically-important estuarine invertebrate. *The Science of the Total Environment*, 693, 133444. DOI: 10.1016/j.scitotenv.2019.07.250 PMID: 31362229

Cook, S., Price, O., King, A., Finnegan, C., van Egmond, R., Schäfer, H., Pearson, J. M., Abolfathi, S., & Bending, G. D. (2020). Bedform characteristics and biofilm community development interact to modify hyporheic exchange. *The Science of the Total Environment*, 749, 141397. DOI: 10.1016/j.scitotenv.2020.141397 PMID: 32841855

Cooray, S., Hettiarachchi, C., Nanayakkara, V., Matthies, D., Samaradivakara, Y., & Nanayakkara, S. (2024, April). Kavy: Fostering Language Speaking Skills and Self-Confidence Through Conversational AI. In *Proceedings of the Augmented Humans International Conference 2024* (pp. 226-236). DOI: 10.1145/3652920.3652944

Copado-Méndez, P. J., Blum, C., Guillén-Gosálbez, G., & Jiménez, L. (2013). Large neighbourhood search applied to the efficient solution of spatially explicit strategic supply chain management problems. *Computers &amp. Computers & Chemical Engineering*, 49, 114–126. DOI: 10.1016/j.compchemeng.2012.09.006

Coraggio, L., & Coretto, P. (2023). Selecting the number of clusters, clustering models, and algorithms. A unifying approach based on the quadratic discriminant score. *Journal of Multivariate Analysis*, 196, 105181. DOI: 10.1016/j.jmva.2023.105181

Cri an, G., Moldovean-Cioroianu, N. S., Timaru, D. G., Andrie , G., Căinap, C., & Chi , V. (2022). Radiopharmaceuticals for PET and SPECT imaging: A literature review over the last decade. *International Journal of Molecular Sciences*, 23(9), 5023. DOI: 10.3390/ijms23095023 PMID: 35563414

Crow, J. P., Ye, Y. Z., Strong, M., Kirk, M., Barnes, S., & Beckman, J. S. (1997). Superoxide dismutase catalyzes nitration of tyrosines by peroxynitrite in the rod and head domains of neurofilament-L. *Journal of Neurochemistry*, 69(5), 1945–1953. DOI: 10.1046/j.1471-4159.1997.69051945.x PMID: 9349539

Currell, S., & Filho, C. N. (2000). The environmental cost of urban sprawl. *Ecological Economics*, 30(2), 101–123.

D'hulst, L., Weehaeghe, D. V., Chiò, A., Calvo, A., Moglia, C., & Canosa, A.. (2018). Multicenter validation of [18f]-FDG PET and support-vector machine discriminant analysis in automatically classifying patients with amyotrophic lateral sclerosis versus controls. *Amyotrophic Lateral Sclerosis & Frontotemporal Degeneration*, 19(7-8), 570–577. Advance online publication. DOI: 10.1080/21678421.2018.1476548 PMID: 29862846

Dagdougui, H. (2012). Models, methods and approaches for the planning and design of the future hydrogen supply chain. *International Journal of Hydrogen Energy*, 37(6), 5318–5327. DOI: 10.1016/j.ijhydene.2011.08.041

Dagdougui, H., Ouammi, A., & Sacile, R. (2012). Modelling and control of hydrogen and energy flows in a network of green hydrogen refuelling stations powered by mixed renewable energy systems. *International Journal of Hydrogen Energy*, 37(6), 5360–5371. DOI: 10.1016/j.ijhydene.2011.07.096

Dara, O. A., Lopez-Guede, J. M., Raheem, H. I., Rahebi, J., Zulueta, E., & Fernandez-Gamiz, U. (2023). Alzheimer's Disease Diagnosis Using Machine Learning: A Survey. *Applied Sciences (Basel, Switzerland)*, 13(14), 8298. DOI: 10.3390/app13148298

Darwish, O., Al-shorman, A., AlSobeh, A., & Tashtoush, Y. (2024, April). A Survey Analysis of Internet of Things (IoT) Education Across the Top 25 Universities in the United States. In *International Conference on Advanced Information Networking and Applications* (pp. 497-510). Cham: Springer Nature Switzerland. DOI: 10.1007/978-3-031-57931-8_47

Das, A. K., & Rodriguez-Marek, E. (2019). A predictive analytics system for forecasting student academic performance: Insights from a pilot project at Eastern Washington University. In *2019 Joint 8th International Conference on Informatics, Electronics & Vision (ICIEV) and 2019 3rd International Conference on Imaging, Vision & Pattern Recognition (icIVPR)* (pp. 255-262). IEEE.

Das, R. (2010). A comparison of multiple classification methods for diagnosis of parkinson disease. *Expert Systems with Applications*, 37(2), 1568–1572. DOI: 10.1016/j.eswa.2009.06.040

de Albuquerque, V. H. C., Gupta, D., De Falco, I., Sannino, G., & Bouguila, N. (2020). Special issue on Bio-inspired optimization techniques for Biomedical Data Analysis: Methods and applications. *Applied Soft Computing*, 95, 106672.

de Andrade, R. B., Guilherme, L. A. M., & da Costa, G. A. O. P. (2022). Deforestation detection in the amazon using DeepLabv3+ semantic segmentation model variants. *Remote Sensing (Basel)*, 14(19), 4694. DOI: 10.3390/rs14194694

De-León Almaraz, S., Azzaro-Pantel, C., Montastruc, L., & Boix, M. (2015). Deployment of a hydrogen supply chain by multi-objective/multi-period optimisation at regional and national scales. *Chemical Engineering Research & Design*, 104, 11–31. DOI: 10.1016/j.cherd.2015.07.005

De-León Almaraz, S., Azzaro-Pantel, C., Montastruc, L., & Domenech, S. (2014). Hydrogen supply chain optimization for deployment scenarios in the Midi-Pyrénées region, France. *International Journal of Hydrogen Energy*, 39(23), 11831–11845. DOI: 10.1016/j.ijhydene.2014.05.165

De-León Almaraz, S., Azzaro-Pantel, C., Montastruc, L., Pibouleau, L., & Senties, O. B. (2013). Assessment of mono and multi-objective optimization to design a hydrogen supply chain. *International Journal of Hydrogen Energy*, 38(33), 14121–14145. DOI: 10.1016/j.ijhydene.2013.07.059

Delucchi, M. A., & Lipman, T. E. (2010). Lifetime Cost of Battery, Fuel-Cell, and Plug-in Hybrid Electric Vehicles. In *Electric and Hybrid Vehicles* (pp. 19–60). Elsevier., DOI: 10.1016/B978-0-444-53565-8.00002-6

DeMaagd, G., & Philip, A. (2015). Parkinson's disease and its management: part 1: disease entity, risk factors, pathophysiology, clinical presentation, and diagnosis. *P&T*, 40, 504–532. PMID: 26236139

Demirel, D. F., Gönül-Sezer, E. D., & Pehlivan, S. A. (2022). Analyzing the waste-water treatment facility location/network design problem via system dynamics: Antalya, Turkey case. *Journal of Environmental Management*, 320, 115814. DOI: 10.1016/j.jenvman.2022.115814 PMID: 35952561

Dempere, J., Modugu, K., Hesham, A., & Ramasamy, L. K. (2023, September). The impact of ChatGPT on higher education. [). Frontiers Media SA.]. *Frontiers in Education*, 8, 1206936. DOI: 10.3389/feduc.2023.1206936

Deng, Y., Dewil, R., Appels, L., Van Tulden, F., Li, S., Yang, M., & Baeyens, J. (2022). Hydrogen-enriched natural gas in a decarbonization perspective. *Fuel*, 318, 123680. DOI: 10.1016/j.fuel.2022.123680

Dervenis, C., Kyriatzis, V., Stoufis, S., & Fitsilis, P. (2022). Predicting Students' Performance Using Machine Learning Algorithms. In *Proceedings of the 6th International Conference on Algorithms, Computing and Systems* (pp. 1-7). DOI: 10.1145/3564982.3564990

Dewangan, D. K., & Sahu, S. P. (2021). PotNet: Pothole detection for autonomous vehicle system using convolutional neural network. *Electronics Letters*, 57(2), 53–56. DOI: 10.1049/ell2.12062

Dhivvya, J. P., & Karnati, S. B. (2024, March). BuddyBot: AI Powered Chatbot for Enhancing English Language Learning. In *2024 IEEE International Conference on Interdisciplinary Approaches in Technology and Management for Social Innovation (IATMSI)* (Vol. 2, pp. 1-6). IEEE.

Diaz-Elsayed, N., Rezaei, N., Ndiaye, A., & Zhang, Q. (2020). Trends in the environmental and economic sustainability of wastewater-based resource recovery: A review. *Journal of Cleaner Production*, 265, 121598. DOI: 10.1016/j.jclepro.2020.121598

Dib, J., Sirlantzis, K., & Howells, G. (2023). An annotated water-filled, and dry potholes dataset for deep learning applications. *Data in Brief*, 48, 109206. DOI: 10.1016/j.dib.2023.109206 PMID: 37213553

Dincer, I., & AlZahrani, A. A. (2018). 4.25 Electrolyzers. In *Comprehensive Energy Systems* (pp. 985–1025). Elsevier., DOI: 10.1016/B978-0-12-809597-3.00442-9

Ding, H., Li, Z., Zhao, S., Li, S., & Wang, C. (2020). A graph convolutional neural network based traffic recommendation model with spatial-temporal attention. *IEEE Access : Practical Innovations, Open Solutions*, 8, 129234–129244.

Dolgen, D., & Alpaslan, M. N. (2023). *Reuse of Wastewater from the Circular Economy (CE) Perspective. A Sustainable Green Future*. Springer., DOI: 10.1007/978-3-031-24942-6_18

Doniavi, E., Babazadeh, R., & Hasanzadeh, R. (2023). Optimization of Renewable Energy Supply Chain for Sustainable Hydrogen Energy Production from Plastic Waste. *Sustainability (Basel)*, 15(24), 16813. DOI: 10.3390/su152416813

Dorigo, M., & Blum, C. (2005) Ant colony optimization theory. A survey,Theoretical Computer Science,Vol 344, Issues 2–3,Pages 243-278,ISSN 0304-3975.

Dost, G. (2024). Students' perspectives on the 'STEM belonging'concept at A-level, undergraduate, and postgraduate levels: An examination of gender and ethnicity in student descriptions. *International Journal of STEM Education*, 11(1), 1–33. DOI: 10.1186/s40594-024-00472-9

Dragnić-Cindrić, D., Lobczowski, N. G., Greene, J. A., & Murphy, P. K. (2024). Exploring the Teacher's Role in Discourse and Social Regulation of Learning: Insights from Collaborative Sessions in High-School Physics Classrooms. *Cognition and Instruction*, 42(1), 92–123. DOI: 10.1080/07370008.2023.2266847

Dubey, R., Bajpai, J., & Bajpai, A. K. (2015). Green synthesis of graphene sand composite (GSC) as novel adsorbent for efficient removal of Cr (VI) ions from aqueous solution. *Journal of Water Process Engineering*, 5, 83–94. DOI: 10.1016/j.jwpe.2015.01.004

Du, G., Lewis, M. M., Styner, M., Shaffer, M. L., Sen, S., Yang, Q. X., & Huang, X. (2011). Combined R2* and diffusion tensor imaging changes in the substantia nigra in Parkinson's disease. *Movement Disorders*, 26(9), 1627–1632. DOI: 10.1002/mds.23643 PMID: 21618607

Dulfer, N., Gowing, A., & Mitchell, J. (2024). Building belonging in online classrooms: Relationships at the core. *Teaching in Higher Education*, •••, 1–17. DOI: 10.1080/13562517.2024.2349993

Dumitru, R. G., Peteleaza, D., & Craciun, C. (2023). Using DUCK-Net for polyp image segmentation. *Scientific Reports*, 13(1), 9803.

Duong, T., & Suppasetseree, S. (2024). The Effects of an Artificial Intelligence Voice Chatbot on Improving Vietnamese Undergraduate Students' English Speaking Skills. *International Journal of Learning. Teaching and Educational Research*, 23(3), 293–321.

Durach, C. F., & Gutierrez, L. (2024). "Hello, this is your AI co-pilot"–operational implications of artificial intelligence chatbots. *International Journal of Physical Distribution & Logistics Management*, 54(3), 229–246. DOI: 10.1108/IJPDLM-01-2024-0031

Ebadi, S., & Amini, A. (2024). Examining the roles of social presence and human-likeness on Iranian EFL learners' motivation using artificial intelligence technology: A case of CSIEC chatbot. *Interactive Learning Environments*, 32(2), 655–673. DOI: 10.1080/10494820.2022.2096638

El Hanandeh, A., Mahdi, Z., & Imtiaz, M. S. (2021). Modelling of the adsorption of Pb, Cu and Ni ions from single and multi-component aqueous solutions by date seed derived biochar: Comparison of six machine learning approaches. *Environmental Research*, 192, 110338. DOI: 10.1016/j.envres.2020.110338 PMID: 33075354

Elbanna, S., & Armstrong, L. (2024). Exploring the integration of ChatGPT in education: Adapting for the future. *Management & Sustainability: An Arab Review*, 3(1), 16–29. DOI: 10.1108/MSAR-03-2023-0016

Elkiran, G., Nourani, V., & Abba, S. I. (2019). Multi-step ahead modelling of river water quality parameters using ensemble artificial intelligence-based approach. *Journal of Hydrology (Amsterdam)*, 577, 123962. DOI: 10.1016/j.jhydrol.2019.123962

Ellis, K. A., Bush, A. I., Darby, D., De Fazio, D., Foster, J., Hudson, P., Lautenschlager, N. T., Lenzo, N., Martins, R. N., Maruff, P., Masters, C., Milner, A., Pike, K., Rowe, C., Savage, G., Szoeke, C., Taddei, K., Villemagne, V., Woodward, M., & Ames, D. (2009). The Australian Imaging, Biomarkers and Lifestyle (AIBL) study of aging: Methodology and baseline characteristics of 1112 individuals recruited for a longitudinal study of Alzheimer's disease. *International Psychogeriatrics*, 21(4), 672–687. DOI: 10.1017/S1041610209009405 PMID: 19470201

Elshamy, R., Abu-Elnasr, O., Elhoseny, M., & Elmougy, S. (2023). Improving the efficiency of RMSProp optimizer by utilizing Nestrove in deep learning. *Scientific Reports*, 13(1), 8814. DOI: 10.1038/s41598-023-35663-x PMID: 37258633

Eppen, G. D., Martin, R. K., & Schrage, L. (1989). OR Practice—A Scenario Approach to Capacity Planning. *Operations Research*, 37(4), 517–527. DOI: 10.1287/opre.37.4.517

Ersoy Tonyaloğlu, E., & Nurlu, E. (2023). The Challenges to Sustainable Urban Future in a Rapidly Changing World: A Case Study of Izmir, Turkey. Sustainable Urban Transitions: Research. *Policy & Practice*, •••, 73–86. DOI: 10.1007/978-981-99-2695-4_5

Eskildsen, S. F., Coupé, P., Fonov, V., & Collins, D. L. (2014, September). Detecting Alzheimer's disease by morphological MRI using hippocampal grading and cortical thickness. In *Proc MICCAI Workshop Challenge on Computer-Aided Diagnosis of Dementia Based on Structural MRI Data* (pp. 38-47).

Essel, H. B., Vlachopoulos, D., Essuman, A. B., & Amankwa, J. O. (2024). ChatGPT effects on cognitive skills of undergraduate students: Receiving instant responses from AI-based conversational large language models (LLMs). *Computers and Education: Artificial Intelligence*, 6, 100198. DOI: 10.1016/j.caeai.2023.100198

Everitt, B. S., & Hothorn, T. (2011). *An Introduction to Applied Multivariate Analysis with R*. Springer Science & Business Media. DOI: 10.1007/978-1-4419-9650-3

Fahad, S. A., Salloum, S. A., & Shaalan, K. (2024). The Role of ChatGpt in Knowledge Sharing and Collaboration Within Digital Workplaces: A Systematic Review. *Artificial Intelligence in Education: The Power and Dangers of ChatGPT in the Classroom*, 259-282.

Fan, D. P., Ji, G. P., Zhou, T., Chen, G., Fu, H., Shen, J., & Shao, L. (2020, September). Pranet: Parallel reverse attention network for polyp segmentation. In International conference on medical image computing and computer-assisted intervention (pp. 263-273). Cham: Springer International Publishing.

Farazouli, A., Cerratto-Pargman, T., Bolander-Laksov, K., & McGrath, C. (2024). Hello GPT! Goodbye home examination? An exploratory study of AI chatbots impact on university teachers' assessment practices. *Assessment & Evaluation in Higher Education*, 49(3), 363–375. DOI: 10.1080/02602938.2023.2241676

Farooq, U., Kozinski, J. A., Khan, M. A., & Athar, M. (2010). Biosorption of heavy metal ions using wheat based biosorbents – A review of the recent literature. *Bioresource Technology*, 101(14), 5043–5053. DOI: 10.1016/j.biortech.2010.02.030 PMID: 20223652

Fasting, R. B., & Breilid, N. (2024). Cross-professional collaboration to improve inclusive education. *Scandinavian Journal of Educational Research*, 68(4), 661–676. DOI: 10.1080/00313831.2023.2175248

Fathi, J., Rahimi, M., & Derakhshan, A. (2024). Improving EFL learners' speaking skills and willingness to communicate via artificial intelligence-mediated interactions. *System*, 121, 103254. DOI: 10.1016/j.system.2024.103254

Fernández-Batanero, J. M., Montenegro-Rueda, M., Fernández-Cerero, J., & López Meneses, E. (2024). Adoption of the Internet of Things in higher education: Opportunities and challenges. *Interactive Technology and Smart Education*, 21(2), 292–307. DOI: 10.1108/ITSE-01-2023-0025

Fiche, A., Hardy, C., & Mora, B. (2022). Towards a Tree Cover Change Early Warning System Based on SENTINEL-1 Data and a Neural Network Architecture. *The International Archives of the Photogrammetry, Remote Sensing and Spatial Information Sciences*, 43, 603–609. DOI: 10.5194/isprs-archives-XLIII-B3-2022-603-2022

Field, A., Miles, J., & Field, Z. (2012). *Discovering Statistics Using R*. SAGE Publications Ltd.

Flores-Piñero, M. D. C., Valdivia-Moral, P., Ramos-Mondejar, L., & González-Hernández, J. (2024). Motivational Climate, Physical Self-Concept, and Social Relationships in Adolescents in Physical Education Classes: A Systematic Review. *Education Sciences*, 14(2), 199. DOI: 10.3390/educsci14020199

Gajwani, J., & Chakraborty, P. (2021). Students' performance prediction using feature selection and su- pervised machine learning algorithms. In *International Conference on Innovative Computing and Com- munications: Proceedings of ICICC 2020, Volume 1* (pp. 347-354). Springer Singapore.

Gandomi, A. H., Gun, J. Y., Yang, X.-S., & Talatahari, S. (2013),Chaos-enhanced accelerated particle swarm optimization,Communications in Nonlincar Scicncc and Numerical Simulation,Volume 18, Issue 2,Pages 327-340,ISSN 1007-5704.

Garg, P., Maass, F., Sundaram, S. M., Mollenhauer, B., Mahajani, S., van Riesen, C., Kügler, S., & Bähr, M. (2022). The relevance of synuclein autoantibodies as a biomarker for Parkinson's disease. *Molecular and Cellular Neurosciences*, 121, 103746. DOI: 10.1016/j.mcn.2022.103746 PMID: 35660088

Garrido-Cardenas, J. A., Esteban-García, B., Agüera, A., Sánchez-Pérez, J. A., & Manzano-Agugliaro, F. (2019). Wastewater Treatment by Advanced Oxidation Process and Their Worldwide Research Trends. *International Journal of Environmental Research and Public Health*, 17(1), 170. DOI: 10.3390/ijerph17010170 PMID: 31881722

Geetha Varma, V., Jha, S., Himesh Karthik Raju, L., Lalith Kishore, R., & Ranjith, V. (2022). A review on decentralized wastewater treatment systems in India. *Chemosphere*, 300, 134462. DOI: 10.1016/j.chemosphere.2022.134462 PMID: 35390407

Gichamo, T., Gökçekuş, H., Ozsahin, D. U., Gelete, G., & Uzun, B. (2020). Evaluation of different natural wastewater treatment alternatives by fuzzy promethee method. *Desalination and Water Treatment*, 177, 400–407. DOI: 10.5004/dwt.2020.25049

Gil, D., & Manuel, D. J. (2009). Diagnosing parkinson by using artificial neural networks and support vector machines. *Glob. J. Comput. Sci. Technol.*, 9, 63–71.

Gill, S. S., & Buyya, R. (2019). Bio-inspired algorithms for big data analytics. a survey, taxonomy, and open challenges. In *Big data analytics for intelligent healthcare management* (pp. 1–17). Academic Press. DOI: 10.1016/B978-0-12-818146-1.00001-5

Global Hydrogen Review 2022. (2022). OECD. DOI: 10.1787/a15b8442-en

Gomeni, R., & Fava, M. (2013). Amyotrophic lateral sclerosis disease progression model. *Amyotrophic Lateral Sclerosis & Frontotemporal Degeneration*, 15(1-2), 119–129. DOI: 10.3109/21678421.2013.838970 PMID: 24070404

Gomez-Gonzalez, R., Cerino-Córdova, F. J., Garcia-León, A. M., Soto-Regalado, E., Davila-Guzman, N. E., & Salazar-Rabago, J. J. (2016). Lead biosorption onto coffee grounds: Comparative analysis of several optimization techniques using equilibrium adsorption models and ANN. *Journal of the Taiwan Institute of Chemical Engineers*, 68, 201–210. DOI: 10.1016/j.jtice.2016.08.038

Gómez-Ramírez, M., & Tenorio-Sánchez, A. S. (2021). Treatment of Solid Waste Containing Metals by Biological Methods. In *Natural Resources Management and Biological Sciences*. IntechOpen. DOI: 10.5772/intechopen.92211

Goodarzi, D., Mohammadian, A., Pearson, J., & Abolfathi, S. (2022). Numerical modelling of hydraulic efficiency and pollution transport in waste stabilization ponds. *Ecological Engineering*, 182, 106702. DOI: 10.1016/j.ecoleng.2022.106702

Goodfellow, I., Pouget-Abadie, J., Mirza, M., Xu, B., Warde-Farley, D., Ozair, S., Courville, A., & Bengio, Y. (2020). Generative adversarial networks. *Communications of the ACM*, 63(11), 139–144. DOI: 10.1145/3422622

Gordon, P. H., Cheng, B., Katz, I. B., Mitsumoto, H., & Rowland, L. P. (2009). Clinical features that distinguish PLS, upper motor neuron-dominant ALS, and typical ALS. *Neurology*, 72(22), 1948–1952. DOI: 10.1212/WNL.0b013e3181a8269b PMID: 19487653

Govindan, K., Fattahi, M., & Keyvanshokooh, E. (2017). Supply chain network design under uncertainty: A comprehensive review and future research directions. *European Journal of Operational Research*, 263(1), 108–141. DOI: 10.1016/j.ejor.2017.04.009

Gray, L. E., & Dunn, S. D. (Eds.). (2024). *Incorporating the Human Element in Online Teaching and Learning*. IGI Global. DOI: 10.4018/979-8-3693-4131-5

Greenlund, L. J., Deckwerth, T. L., & Johnson, E. M.Jr. (1995). Superoxide dismutase delays neuronal apoptosis: A role for reactive oxygen species in programmed neuronal death. *Neuron*, 14(2), 303–315. DOI: 10.1016/0896-6273(95)90287-2 PMID: 7857640

Grings, F., Roitberg, E., & Barraza, V. (2019). EVI time-series breakpoint detection using convolutional networks for online deforestation monitoring in Chaco forest. *IEEE Transactions on Geoscience and Remote Sensing*, 58(2), 1303–1312. DOI: 10.1109/TGRS.2019.2945719

Groth, K. M., & Al-Douri, A. (2023). Hydrogen safety, risk, and reliability analysis. In *Hydrogen Economy* (pp. 487–510). Elsevier., DOI: 10.1016/B978-0-323-99514-6.00012-1

Guerrero Elecalde, R., Contreras García, J., Bonilla Martos, A. L., & Serrano Arnáez, B. (2024). Digital and Social-Civic Skills in Future Primary Education Teachers: A Study from the Didactics of Social Sciences for the Improvement of Teacher Training in Competences. *Education Sciences*, 14(2), 211. DOI: 10.3390/educsci14020211

Guillén-Gosálbez, G., Mele, F. D., & Grossmann, I. E. (2009). A bi-criterion optimization approach for the design and planning of hydrogen supply chains for vehicle use. *AIChE Journal. American Institute of Chemical Engineers*, 56(3), 650–667. DOI: 10.1002/aic.12024

Gu, J., Wang, Z., Kuen, J., Ma, L., Shahroudy, A., Shuai, B., Liu, T., Wang, X., Wang, G., Cai, J., & Chen, T. (2018). Recent advances in convolutional neural networks. *Pattern Recognition*, 77, 354–377. DOI: 10.1016/j.patcog.2017.10.013

Güler, M. G., Geçici, E., & Erdoğan, A. (2021). Design of a future hydrogen supply chain: A multi period model for Turkey. *International Journal of Hydrogen Energy*, 46(30), 16279–16298. DOI: 10.1016/j.ijhydene.2020.09.018

Gulsen, H., Turan, M., & Altay, A. (2000). The application of an empirical design model in the development of facultative pond design criteria for turkey. *Environmental Technology*, 21(12), 1363–1369. DOI: 10.1080/09593332208618174

Guo, X., Li, Y., Chen, J., & Li, H. (2019). Deep learning for real-time traffic prediction using big data. *Transportation Research Part C, Emerging Technologies*, 111, 88–103.

Gupta, N., Joshi, S., Arora, K., & Negi, N. (2024, February). Zone Labelling Model to Prevent Deforestation Using Image Classification. In 2024 2nd International Conference on Computer, Communication and Control (IC4) (pp. 1-6). IEEE.

Gupta, S. S., Arya, A. K., & Vijay, P. (2019). Designing a model for optimization of maintenance and inspection efforts against third party damage to cross country Pipelines in India. *International Journal of Innovative Technology and Exploring Engineering*, 8(12), 4529–4540. Advance online publication. DOI: 10.35940/ijitee.L3209.1081219

Haass, C., & Selkoe, D. J. (2007). Soluble protein oligomers in neurodegeneration: Lessons from the Alzheimer's amyloid β-peptide. *Nature Reviews. Molecular Cell Biology*, 8(2), 101–112. DOI: 10.1038/nrm2101 PMID: 17245412

Hafeez, A., Shamair, Z., Shezad, N., Javed, F., Fazal, T., Rehman, S., Bazmi, A. A., & Rehman, F. (2021). Solar powered decentralized water systems: A cleaner solution of the industrial wastewater treatment and clean drinking water supply challenges. *Journal of Cleaner Production*, 289, 125717. DOI: 10.1016/j.jclepro.2020.125717

Hair, J. F.Jr, Black, W. C., Babin, B. J., & Anderson, R. E. (2019). *Multivariate Data Analysis* (8th ed.). Cengage Learning.

Han, J., & Lee, D. (2024). Research on the development of principles for designing elementary English-speaking lessons using artificial intelligence chatbots. *Humanities & Social Sciences Communications*, 11(1), 1–15. DOI: 10.1057/s41599-024-02646-w

Hawkins, D. M. (1980). *Identification of Outliers*. Chapman and Hall. DOI: 10.1007/978-94-015-3994-4

Hayes, M. T. (2019). Parkinson's disease and parkinsonism. *The American Journal of Medicine*, 132(7), 802–807. DOI: 10.1016/j.amjmed.2019.03.001 PMID: 30890425

Heldman, D. A., Harris, D. A., Felong, T., Andrzejewski, K. L., Dorsey, E. R., Giuffrida, J. P., Goldberg, B., & Burack, M. A. (2017). Telehealth management of Parkinson's disease using wearable sensors: An exploratory study. *Digital Biomarkers*, 1(1), 43–51. DOI: 10.1159/000475801 PMID: 29725667

Hennebelle, A., Ismail, L., & Linden, T. (2024). Schools Students Performance with Artificial Intelligence Machine Learning: Features Taxonomy, Methods and Evaluation. In *Machine Learning in Educational Sciences* (pp. 95–112). Springer. DOI: 10.1007/978-981-99-9379-6_6

Herschan, J., Tsinda, A., Okurut, K., Malcolm, R., Lapworth, D. J., & Pond, K. (2023). Progress of Using Risk Assessment to Manage Small Drinking-Water Supplies in Rwanda: A Preliminary Study. *Processes (Basel, Switzerland)*, 11(3), 748. DOI: 10.3390/pr11030748

Heuser, P.-M., Ryberg, D. S., Grube, T., Robinius, M., & Stolten, D. (2019). Techno-economic analysis of a potential energy trading link between Patagonia and Japan based on CO2 free hydrogen. *International Journal of Hydrogen Energy*, 44(25), 12733–12747. DOI: 10.1016/j.ijhydene.2018.12.156

Heuvel, M. P. (2017). Deep learning predictions of survival based on MRI in amyotrophic lateral sclerosis. *NeuroImage. Clinical*, 13, 361–369. DOI: 10.1016/j.nicl.2016.10.008 PMID: 28070484

Hong, X., Thaore, V. B., Karimi, I. A., Farooq, S., Wang, X., Usadi, A. K., Chapman, B. R., & Johnson, R. A. (2021). Techno-enviro-economic analyses of hydrogen supply chains with an ASEAN case study. *International Journal of Hydrogen Energy*, 46(65), 32914–32928. DOI: 10.1016/j.ijhydene.2021.07.138

Hoq, M., Uddin, M. N., & Park, S. B. (2021). Vocal feature extraction-based artificial intelligent model for Parkinson's disease detection. *Diagnostics (Basel)*, 11(6), 1076. DOI: 10.3390/diagnostics11061076 PMID: 34208330

Hothorn, T., & Jung, H. H. (2014). RandomForest4life: A random forest for predicting ALS disease progression. *Amyotrophic Lateral Sclerosis & Frontotemporal Degeneration*, 15(5-6), 444–452. DOI: 10.3109/21678421.2014.893361 PMID: 25141076

http://ljournal.ru/wp-content/uploads/2017/03/a-2017-023.pdf. (2017). LJournal. DOI: 10.18411/a-2017-023

Huang, G. (2009). Erratum. In *Relict Species* (p. 450). Springer Berlin Heidelberg., DOI: 10.1007/978-3-540-92160-8_27

Huang, W., Song, G., Hong, H., & Xie, K. (2017). Deep learning for intelligent transportation systems: A survey. *Transport Reviews*, 37(8), 1083–1104.

Huang, Z., Zhang, H., Boss, J., Goutman, S. A., Mukherjee, B., Dinov, I. D., & Guan, Y. (2017). Complete hazard ranking to analyze rightcensored data: An ALS survival study. *PLoS Computational Biology*, 13(12), e1005887. DOI: 10.1371/journal.pcbi.1005887 PMID: 29253881

Hughes, S. (2024). Teaching assistants as facilitators of learning and engagement: Promoting social justice for students with special educational needs and disabilities (SEND). In *Social Justice in Practice in Education* (pp. 156-174). Routledge.

Hunot, S., & Hirsch, E. C. (2003). Neuroinflammatory processes in Parkinson's disease. *Annals of Neurology*, 53(S3), S49–S60. DOI: 10.1002/ana.10481 PMID: 12666098

Hurtig, H. (2000). What is Parkinson's disease? Neuropathology, neurochemistry, and pathophysiology. In *Parkinson's disease and movement disorders: Diagnosis and treatment guidelines for the practicing physician* (pp. 57–70). Humana Press.

Hwangbo, S., Heo, S., & Yoo, C. (2018). Network modeling of future hydrogen production by combining conventional steam methane reforming and a cascade of waste biogas treatment processes under uncertain demand conditions. *Energy Conversion and Management*, 165, 316–333. DOI: 10.1016/j.enconman.2018.03.069

Hydrogen and Fuel Cells. (2015). In *IEA Technology Roadmaps*. OECD. DOI: 10.1787/9789264239760-en

Hyndman, R. J., & Fan, Y. (1996). Sample quantiles in statistical packages. *The American Statistician*, 50(4), 361–365. DOI: 10.1080/00031305.1996.10473566

Ikonomidou, C., Qin, Y. Q., Labruyere, J., & Olney, J. W. (1996). Motor neuron degeneration induced by excitotoxin agonists has features in common with those seen in the SOD-1 transgenic mouse model of amyotrophic lateral sclerosis. *Journal of Neuropathology and Experimental Neurology*, 55(2), 211–224. DOI: 10.1097/00005072-199602000-00010 PMID: 8786380

Illarionova, S., Shadrin, D., Tregubova, P., Ignatiev, V., Efimov, A., Oseledets, I., & Burnaev, E. (2022). A survey of computer vision techniques for forest characterization and carbon monitoring tasks. *Remote Sensing (Basel)*, 14(22), 5861. DOI: 10.3390/rs14225861

Ingason, H., Pallingolfsson, H., & Jensson, P.INGASON. (2008). Optimizing site selection for hydrogen production in Iceland. *International Journal of Hydrogen Energy*, 33(14), 3632–3643. DOI: 10.1016/j.ijhydene.2008.04.046

International Energy Agency (IEA). (2008). *Encyclopedia of Global Warming and Climate Change*. SAGE Publications, Inc., DOI: 10.4135/9781412963893.n340

Jack, C. R.Jr, Barnes, J., Bernstein, M. A., Borowski, B. J., Brewer, J., Clegg, S., Dale, A. M., Carmichael, O., Ching, C., DeCarli, C., Desikan, R. S., Fennema-Notestine, C., Fjell, A. M., Fletcher, E., Fox, N. C., Gunter, J., Gutman, B. A., Holland, D., Hua, X., & Weiner, M. (2015). Magnetic resonance imaging in Alzheimer's disease neuroimaging initiative 2. *Alzheimer's & Dementia*, 11(7), 740–756. DOI: 10.1016/j.jalz.2015.05.002 PMID: 26194310

Jahandideh, S., Taylor, A. A., Beaulieu, D., Keymer, M., Meng, L., Bian, A., Atassi, N., Andrews, J., & Ennist, D. L. (2017). Longitudinal modeling to predict vital capacity in amyotrophic lateral sclerosis. *Amyotrophic Lateral Sclerosis & Frontotemporal Degeneration*, 19(3-4), 294–302. DOI: 10.1080/21678421.2017.1418003 PMID: 29260584

Jamil, A., Bokhari, T. H., Javed, T., Mustafa, R., Sajid, M., Noreen, S., Zuber, M., Nazir, A., Iqbal, M., & Jilani, M. I. (2020). Photocatalytic degradation of disperse dye Violet-26 using TiO_2 and ZnO nanomaterials and process variable optimization. *Journal of Materials Research and Technology*, 9(1), 1119–1128. DOI: 10.1016/j.jmrt.2019.11.035

Jamshed, M. A., Theodorou, C., Kalsoom, T., Anjum, N., Abbasi, Q. H., & Ur-Rehman, M. (2022). Intelligent computing based forecasting of deforestation using fire alerts. A deep learning approach. *Physical Communication*, 55, 101941. DOI: 10.1016/j.phycom.2022.101941

Jellinger, K. A. (2014). Neuropathology of Parkinson's disease. In *Inflammation in Parkinson's disease: scientific and clinical aspects* (pp. 25–74). Springer International Publishing. DOI: 10.1007/978-3-319-08046-8_2

Jeon, J. (2024). Exploring AI chatbot affordances in the EFL classroom: Young learners' experiences and perspectives. *Computer Assisted Language Learning*, 37(1-2), 1–26. DOI: 10.1080/09588221.2021.2021241

Jha, D., Smedsrud, P. H., Johansen, D., de Lange, T., Johansen, H. D., Halvorsen, P., & Riegler, M. A. (2021). A comprehensive study on colorectal polyp segmentation with ResUNet++, conditional random field and test-time augmentation. *IEEE Journal of Biomedical and Health Informatics*, 25(6), 2029–2040. DOI: 10.1109/JBHI.2021.3049304 PMID: 33400658

Jia, B., Tsau, J.-S., & Barati, R. (2019). A review of the current progress of CO_2 injection EOR and carbon storage in shale oil reservoirs. *Fuel*, 236, 404–427. DOI: 10.1016/j.fuel.2018.08.103

Jiang, F., Jiang, Y., Zhi, H., Dong, Y., Li, H., Ma, S., Wang, Y., Dong, Q., Shen, H., & Wang, Y. (2017). Artificial intelligence in healthcare. past, present and future. *Stroke and Vascular Neurology*, 2(4), 230–243. Advance online publication. DOI: 10.1136/svn-2017-000101 PMID: 29507784

Jinming, D. U., & Daniel, B. K. (2024). A Systematic Review of AI-Powered Chatbots in EFL Speaking Practice: Transforming Language Education. *Computers and Education: Artificial Intelligence*, 100230.

Jo, H., & Park, D. H. (2024). Effects of ChatGPT's AI capabilities and human-like traits on spreading information in work environments. *Scientific Reports*, 14(1), 7806. DOI: 10.1038/s41598-024-57977-0 PMID: 38565880

Johnson, N., & Ogden, J. (2012). A spatially-explicit optimization model for long-term hydrogen pipeline planning. *International Journal of Hydrogen Energy*, 37(6), 5421–5433. DOI: 10.1016/j.ijhydene.2011.08.109

Kadam, V. J., & Jadhav, S. M. (2019). In Iyer, B., Nalbalwar, S., & Pathak, N. (Eds.), *"Feature ensemble learning based on sparse autoencoders for diagnosis of parkinson's disease," in Computing, Communication and Signal Processing. Advances in Intelligent Systems and Computing* (Vol. 810, pp. 567–581). Springer., DOI: 10.1007/978-981-13-1513-8_58

Kadri, S. U. T., Tavanappanavar, A. N., Nagesh Babu, R., Bilal, M., Singh, B., Gupta, S. K., Bharagava, R. N., Govarthanan, M., Savanur, M. A., & Mulla, S. I. (2021). *Overview of Waste Stabilization Ponds in Developing Countries. Cost-efficient Wastewater Treatment Technologies. The Handbook of Environmental Chemistry* (Vol. 117). Springer., DOI: 10.1007/698_2021_790

Kamalakannan, S., Navaneethan, S., Deshmukh, Y. S., Sujay, V., Ramalingam, V., Jagadeesan, D., & Venkatesh, C. (2024). A novel pothole detection model based on YOLO algorithm for VANET. *International Journal of Intelligent Systems and Applications in Engineering*, 12(11s), 56–61.

Kansok-Dusche, J., Wettstein, A., Castellanos, M., Schwab, C., Subramaniam, A., Wachs, S., & Bilz, L. (2024). Associations between social competence, perceived parents' prosocial educational goals and adolescents' hate speech perpetration in school. *Victims & Offenders*, 19(3), 419–446. DOI: 10.1080/15564886.2023.2189191

Kar, A. K. (2016). Bio inspired computing–a review of algorithms and scope of applications. *Expert Systems with Applications*, 59, 20–32. DOI: 10.1016/j.eswa.2016.04.018

Karthikha, R., Najumnissa Jamal, D., & Syed Rafiammal, S. (2024). An approach of polyp segmentation from colonoscopy images using Dilated-U-Net-Seg–A deep learning network. *Biomedical Signal Processing and Control*, 93, 106197. DOI: 10.1016/j.bspc.2024.106197

Kaselimi, M., Voulodimos, A., Daskalopoulos, I., Doulamis, N., & Doulamis, A. (2022). A vision transformer model for convolution-free multilabel classification of satellite imagery in deforestation monitoring. *IEEE Transactions on Neural Networks and Learning Systems*. PMID: 35108212

Katoch, S., Chauhan, S. S., & Kumar, V. (2021). A review on genetic algorithm: Past, present, and future. *Multimedia Tools and Applications*, 80(5), 8091–8126. DOI: 10.1007/s11042-020-10139-6 PMID: 33162782

Kaur, A., & Bhatia, M. (2023). A Survey of Machine Learning for Assessing and Estimating Student Performance. In *Proceedings of International Conference on Recent Innovations in Computing: ICRIC 2022,* Volume 1 (pp. 633-648). Springer Nature Singapore. DOI: 10.1007/978-981-19-9876-8_48

Kaur, H., & Kaur, T. (2022). A prediction model for student academic performance using ma- chine learning-based analytics. In *Proceedings of the Future Technologies Conference* (pp. 770-775). Springer International Publishing.

Kavas-Torris, O., Lackey, N., & Güvenç, L. (2021). Simulating the effect of autonomous vehicles on roadway mobility in a microscopic traffic simulator. *International Journal of Automotive Technology*, 22(3), 713–733. DOI: 10.1007/s12239-021-0066-7

Kelly, M. (2024). Re-imagining pedagogical approaches to support student engagement in secondary schools. In *Re-imagining Teaching Improvement: From Early Childhood to University* (pp. 47–70). Springer Nature Singapore.

Khalifeh Soltani, S. R., Mostafaeipour, A., Almutairi, K., Hosseini Dehshiri, S. J., Hosseini Dehshiri, S. S., & Techato, K. (2022). Predicting effect of floating photovoltaic power plant on water loss through surface evaporation for wastewater pond using artificial intelligence: A case study. *Sustainable Energy Technologies and Assessments*, 50, 101849. DOI: 10.1016/j.seta.2021.101849

Khalil, S. (2024). Understanding Society: An Introduction to Social Science. *Pakistan Journal of Social Science Review*, 2(01), 57–68.

Khang, A., Rath, K. C., Panda, S., Sree, P. K., & Panda, S. K. (2023). Revolutionizing agriculture. exploring advanced technologies for plant protection in the agriculture sector. In *Handbook of Research on AI-Equipped IoT Applications in High-Tech Agriculture* (pp. 1–22). IGI Global. DOI: 10.4018/978-1-6684-9231-4.ch001

Khan, M., Raza, M. A., Abbas, G., Othmen, S., Yousef, A., & Jumani, T. A. (2024). Pothole detection for autonomous vehicles using deep learning: A robust and efficient solution. *Frontiers in Built Environment*, 9, 1323792. Advance online publication. DOI: 10.3389/fbuil.2023.1323792

Khlaisang, J., & Teo, T. (2024). An innovation-based virtual flipped learning system in a ubiquitous learning environment the 21st century skills of higher education learners. *Journal of Educational Technology & Society*, 27(1), 100–116.

Khoudier, M. M. E., Abdelnaby, R. H. M., Eldamnhoury, Z. M., Abouzeid, S. R. A., El-Monayer, G. K., Enan, N. M., & Moawad, I. (2023). Prediction of student performance using machine learning techniques. In *2023 5th Novel Intelligent and Leading Emerging Sciences Conference (NILES)* (pp. 333-338). IEEE.

Kiemel, S., Smolinka, T., Lehner, F., Full, J., Sauer, A., & Miehe, R. (2021). Critical materials for water electrolysers at the example of the energy transition in Germany. *International Journal of Energy Research*, 45(7), 9914–9935. DOI: 10.1002/er.6487

Kim, A., & Su, Y. (2024). How implementing an AI chatbot impacts Korean as a foreign language learners' willingness to communicate in Korean. *System*, 122, 103256. DOI: 10.1016/j.system.2024.103256

Kim, J., & Moon, I. (2008). Strategic design of hydrogen infrastructure considering cost and safety using multiobjective optimization. *International Journal of Hydrogen Energy*, 33(21), 5887–5896. DOI: 10.1016/j.ijhydene.2008.07.028

Kingma, D. P., & Ba, J. (2014). Adam. A method for stochastic optimization. arXiv preprint arXiv.1412.6980.

Kipf, T. N., & Welling, M. (2017). Semi-supervised classification with graph convolutional networks. *arXiv preprint arXiv:1609.02907.*

Kisi, O., Alizamir, M., & Docheshmeh Gorgij, A. (2020). Dissolved oxygen prediction using a new ensemble method. *Environmental Science and Pollution Research International*, 27(9), 9589–9603. DOI: 10.1007/s11356-019-07574-w PMID: 31925684

Klibi, W., Martel, A., & Guitouni, A. (2010). The design of robust value-creating supply chain networks: A critical review. *European Journal of Operational Research*, 203(2), 283–293. DOI: 10.1016/j.ejor.2009.06.011

Koç, F. Ş. (2024). The development of listening and speaking skills in EFL via an artificially intelligent chatbot application: A quasi-experimental design study.

Ko, K. D., El-Ghazawi, T., Kim, D., & Morizono, H. (2014). "Predicting theseverity of motor neuron disease progression using electronic health record data with a cloud computing big data approach," in *2014 IEEE Conference on Computational Intelligence in Bioinformatics and Computational Biology* (Honolulu, HI: IEEE), 1–6.

Kolade, O., Owoseni, A., & Egbetokun, A. (2024). Is AI changing learning and assessment as we know it? Evidence from a ChatGPT experiment and a conceptual framework. *Heliyon*, 10(4), e25953. DOI: 10.1016/j.heliyon.2024.e25953 PMID: 38379960

Konak, A., Coit, D. W., & Smith, A. E. (2006). Multi-objective optimization using genetic algorithms: A tutorial. *Reliability Engineering &amp. Reliability Engineering & System Safety*, 91(9), 992–1007. DOI: 10.1016/j.ress.2005.11.018

Kooh, M. R. R., Dahri, M. K., Lim, L. B. L., Lim, L. H., & Malik, O. A. (2016). Batch adsorption studies of the removal of methyl violet 2B by soya bean waste: Isotherm, kinetics and artificial neural network modelling. *Environmental Earth Sciences*, 75(9), 783. Advance online publication. DOI: 10.1007/s12665-016-5582-9

Kookana, R. S., Drechsel, P., Jamwal, P., & Vanderzalm, J. (2020). Urbanisation and emerging economies: Issues and potential solutions for water and food security. *The Science of the Total Environment*, 732, 139057. DOI: 10.1016/j.scitotenv.2020.139057 PMID: 32438167

Kosko, B., & Isaka, S. (1993). Fuzzy Logic. *Scientific American*, 269(1), 76–81. DOI: 10.1038/scientificamerican0793-76

Kostikova, I., Holubnycha, L., Besarab, T., Moshynska, O., Moroz, T., & Shamaieva, I. (2024). ChatGPT for Professional English Course Development. *International Journal of Interactive Mobile Technologies*, 18(2).

Koteswaramma, S., Sri, M. D., Manohar, M., Srihari, K., & Yabbeju, M. (2024). Advancing Road Safety: Pothole Detection Using Yolov8 And Wandb Deep Learning. Journal of Advanced Zoology, 45.

Koyama, C. N., Watanabe, M., Shimada, M., & Tadono, T. "Long-Term Pantropical Forest Monitoring with ALOS-2 L-Band SAR." In 2021 7th Asia-Pacific Conference on Synthetic Aperture Radar (APSAR), pp. 1-5. IEEE, 2021. DOI: 10.1109/APSAR52370.2021.9688516

Kumar, G. S., De la Cruz-Cámaco, D., Ravichand, M., Joshi, K., Gupta, Z., & Gupta, S. (2023). Monitoring and Predicting Performance of Students in Degree Programs using Machine Learning. In *2023 10th International Conference on Computing for Sustainable Global Development (INDIACom)* (pp. 1311-1315). IEEE.

Kumari, J., Kumar, E., & Kumar, D. (2023). A structured analysis to study the role of machine learning and deep learning in the healthcare sector with big data analytics. *Archives of Computational Methods in Engineering*, 30(6), 3673–3701. DOI: 10.1007/s11831-023-09915-y PMID: 37359744

Kundu, K. G. C., & Kundu, K. G. C. (2023, January 1). *A DEEP LEARNING BASED APPROACH TO DETECT POTHOLES USING YOLO VERSION 7*. www.ijcrt.org. https://www.ijcrt.org/viewfull.php?&p_id=IJCRT2301187

Kundu, S., & Gupta, A. K. (2006). Arsenic adsorption onto iron oxide-coated cement (IOCC): Regression analysis of equilibrium data with several isotherm models and their optimization. *Chemical Engineering Journal*, 122(1–2), 93–106. DOI: 10.1016/j.cej.2006.06.002

Kurniawan, T. A., Chan, G. Y. S., Lo, W.-H., & Babel, S. (2006). Physico–chemical treatment techniques for wastewater laden with heavy metals. *Chemical Engineering Journal*, 118(1–2), 83–98. DOI: 10.1016/j.cej.2006.01.015

Kwok, J. (2021). Towards a hydrogen economy – a sustainable pathway for global energy transition. *HKIE Transactions*, 28(2), 102–107. DOI: 10.33430/V28N2THIE-2020-0046

Latourelle, J. C., Beste, M. T., Hadzi, T. C., Miller, R. E., Oppenheim, J. N., Valko, M. P., Wuest, D. M., Church, B. W., Khalil, I. G., Hayete, B., & Venuto, C. S. (2017). Large-scale identification of clinical and genetic predictors of motor progression in patients with newly diagnosed Parkinson's disease: A longitudinal cohort study and validation. *Lancet Neurology*, 16(11), 908–916. DOI: 10.1016/S1474-4422(17)30328-9 PMID: 28958801

Lee, K. A., & Lim, S. B. (2023). Designing a leveled conversational teachable agent for English language learners. *Applied Sciences (Basel, Switzerland)*, 13(11), 6541. DOI: 10.3390/app13116541

Lehman, W. (2024, May). Chat (bot) ting Up a Storm: How AI Is Changing the Game in Foreign Language Teaching. In *World Languages and Cultures in the Public Sphere: Selected Proceedings of the 25th Southeast Conference on Languages, Literatures, and Film* (p. 165). BrownWalker Press.

Lemus, R. G., & Martínez Duart, J. M. (2010). Updated hydrogen production costs and parities for conventional and renewable technologies. *International Journal of Hydrogen Energy*, 35(9), 3929–3936. DOI: 10.1016/j.ijhydene.2010.02.034

Leng, L. (2024). Challenge, integration, and change: ChatGPT and future anatomical education. *Medical Education Online*, 29(1), 2304973. DOI: 10.1080/10872981.2024.2304973 PMID: 38217884

Lester, D. B., Rogers, T. D., & Blaha, C. D. (2010). Acetylcholine–dopamine interactions in the pathophysiology and treatment of CNS disorders. *CNS Neuroscience & Therapeutics*, 16(3), 137–162. DOI: 10.1111/j.1755-5949.2010.00142.x PMID: 20370804

Leys, C., Ley, C., Klein, O., Bernard, P., & Licata, L. (2013). Detecting outliers: Do not use standard deviation around the mean, use absolute deviation around the median. *Journal of Experimental Social Psychology*, 49(4), 764–766. DOI: 10.1016/j.jesp.2013.03.013

Lhommée, E., Batir, A., Quesada, J. L., Ardouin, C., Fraix, V., Seigneuret, E., & Krack, P. (2014). Dopamine and the biology of creativity: Lessons from Parkinson's disease. *Frontiers in Neurology*, 5, 55. DOI: 10.3389/fneur.2014.00055 PMID: 24795692

Lian, Y., Coggins, L. X., Hay, J., van de Ven, A., & Ghadouani, A. (2022). Effect of Attached Growth on Treatment Performance in Waste Stabilization Ponds. *Water (Basel)*, 14(20), 3245. DOI: 10.3390/w14203245

Li, B., Bonk, C. J., Wang, C., & Kou, X. (2024). Reconceptualizing self-directed learning in the era of generative AI: An exploratory analysis of language learning. *IEEE Transactions on Learning Technologies*, 17, 1489–1503. DOI: 10.1109/TLT.2024.3386098

Lieskovská, E., Jakubec, M., Bučko, B., & Zábovská, K. (2023). Automatic pothole detection. *Transportation Research Procedia*, 74, 1164–1170. DOI: 10.1016/j.trpro.2023.11.257

Li, J., Wang, J., Lin, F., Heidari, A. A., Chen, Y., Chen, H., & Wu, W. (2024). PRCNet. A parallel reverse convolutional attention network for colorectal polyp segmentation. *Biomedical Signal Processing and Control*, 95, 106336. DOI: 10.1016/j.bspc.2024.106336

Li, L., Rong, S., Wang, R., & Yu, S. (2021). Recent advances in artificial intelligence and machine learning for nonlinear relationship analysis and process control in drinking water treatment: A review. *Chemical Engineering Journal*, 405, 126673. DOI: 10.1016/j.cej.2020.126673

Lin, J., & Wang, Y. (2024). Unpacking the mediating role of classroom interaction between student satisfaction and perceived online learning among Chinese EFL tertiary learners in the new normal of post-COVID-19. *Acta Psychologica*, 245, 104233. DOI: 10.1016/j.actpsy.2024.104233 PMID: 38522350

Lin, T. Y., Goyal, P., Girshick, R., He, K., & Dollár, P. (2017). Focal loss for dense object detection. In *Proceedings of the IEEE international conference on computer vision* (pp. 2980-2988).

Lin, Z., Chen, C.-W., Ogden, J., & Fan, Y. (2008). The least-cost hydrogen for Southern California. *International Journal of Hydrogen Energy*, 33(12), 3009–3014. DOI: 10.1016/j.ijhydene.2008.01.039

Li, T., Howells, J., Lin, C., Garg, N., Kiernan, M., & Park, S. (2018). 8. Predicting motor disorders from nerve excitability studies. *Clinical Neurophysiology*, 129(4), e4. DOI: 10.1016/j.clinph.2017.12.021

Litjens, G., Kooi, T., Bejnordi, B. E., Setio, A. A. A., Ciompi, F., Ghafoorian, M., van der Laak, J. A. W. M., van Ginneken, B., & Sánchez, C. I. (2017). A survey on deep learning in medical image analysis. *Medical Image Analysis*, 42, 60–88. DOI: 10.1016/j.media.2017.07.005 PMID: 28778026

Liu H, M. J. (2020). A review of models and methods for hydrogen supply chain system planning. *CSEE Journal of Power and Energy Systems*. DOI: 10.17775/CSEEJPES.2020.02280

Liu, G., & Ma, C. (2024). Measuring EFL learners' use of ChatGPT in informal digital learning of English based on the technology acceptance model. *Innovation in Language Learning and Teaching*, 18(2), 125–138. DOI: 10.1080/17501229.2023.2240316

Liu, Y., Wu, Z., Wang, Y., Zhang, H., & Yu, J. (2020). A survey of deep learning for intelligent transporta- tion systems. *IEEE Transactions on Intelligent Transportation Systems*, 22(1), 70–88.

Liu, Z., Zhu, Z., Zhu, L., Jiang, E., Hu, X., Peppler, K. A., & Ramani, K. (2024, May). ClassMeta: Designing Interactive Virtual Classmate to Promote VR Classroom Participation. In *Proceedings of the CHI Conference on Human Factors in Computing Systems* (pp. 1-17). DOI: 10.1145/3613904.3642947

Li, W., Shen, Z., & Li, P. (2019). Crack Detection of Track Plate Based on YOLO. *12th International Symposium on Computational Intelligence and Design*. DOI: 10.1109/ISCID.2019.10086

Li, Y., Chen, X., Zhao, W., Xu, Z., & Zhou, S. (2020). Real-time traffic anomaly detection based on long short-term memory network with gated recurrent units. *Journal of Advanced Transportation*, •••, 2020.

Li, Y., Gong, H., Yu, B., Yu, C., & Mao, Y. (2020). Spatial-temporal graph convolutional networks for traffic flow forecasting. *Applied Intelligence*, 50(1), 186–201. DOI: 10.1007/s10489-022-03420-8

Li, Y., Sun, L., Li, S., He, Z., & Sun, C. (2020). A survey of deep learning for traffic light signal timing optimization. *IEEE Transactions on Intelligent Transportation Systems*, 21(5), 2040–2053.

Li, Z., Song, Y., Zhao, C., & Liu, X. (2019). A survey of machine learning in intelligent transportation systems. *IEEE Transactions on Intelligent Transportation Systems*, 20(5), 3820–3838.

Lou, A.. (2022). *"CaraNet. context axial reverse attention network for segmentation of small medical objects." Medical Imaging 2022. Image Processing* (Vol. 12032). SPIE.

Lovell, J., & Webber, M. (2024). Adaptation of Connecting People to address loneliness and social isolation in university students: A feasibility study. *Journal of Higher Education Policy and Management*, 46(1), 32–47. DOI: 10.1080/1360080X.2023.2242079

Lu, F. M., & Yuan, Z. (2015). PET/SPECT molecular imaging in clinical neuroscience: Recent advances in the investigation of CNS diseases. *Quantitative Imaging in Medicine and Surgery*, 5(3), 433. PMID: 26029646

Lv, Y., Duan, Y., Cao, W., Wang, Z., & Zhou, F. (2020). T-GCN: A temporal graph convolutional net- work for real-time traffic flow prediction. *IEEE Transactions on Intelligent Transportation Systems*, 21(8), 3848–3858.

Magwaza, S. T., Magwaza, L. S., Odindo, A. O., & Mditshwa, A. (2020). Hydroponic technology as decentralised system for domestic wastewater treatment and vegetable production in urban agriculture: A review. *The Science of the Total Environment*, 698, 134154. DOI: 10.1016/j.scitotenv.2019.134154 PMID: 31505342

Mahajani, S., Bähr, M., & Kügler, S. (2021). Patterning inconsistencies restrict the true potential of dopaminergic neurons derived from human induced pluripotent stem cells. *Neural Regeneration Research*, 16(4), 692–693. DOI: 10.4103/1673-5374.295316 PMID: 33063729

Mahapatra, S., Samal, K., & Dash, R. R. (2022). Waste Stabilization Pond (WSP) for wastewater treatment: A review on factors, modelling and cost analysis. *Journal of Environmental Management*, 308, 114668. DOI: 10.1016/j.jenvman.2022.114668 PMID: 35152038

Mahyari, F. G., Boegman, L., Rey, A., Mulligan, R., Champagne, P., Filion, Y., & da Silva, A. M. F. (2023). Evaluation of a Three-Dimensional Hydrodynamic and Water Quality Model for Design of Wastewater Stabilization Ponds. *Journal of Environmental Engineering*, 149(4), 5023003. DOI: 10.1061/JOEEDU.EEENG-6987

Malta, A., Mendes, M., & Farinha, J. T. (2021). Augmented Reality Maintenance Assistant using YOLOV5. *Applied Sciences (Basel, Switzerland)*, 11(11), 4758. DOI: 10.3390/app11114758

Mamatnabiyev, Z., Chronis, C., Varlamis, I., & Zhaparov, M. (2024). Development of an Educational Robot for Exploring the Internet of Things. *International Journal of Advanced Computer Science and Applications*, 15(4). Advance online publication. DOI: 10.14569/IJACSA.2024.01504120

Mandave, D. D., & Patil, L. V. (2023). Bio-inspired computing algorithms in dementia diagnosis–a application-oriented review. *Results in Control and Optimization*, 12, 100276.

Mangalampalli, S., Sree, P. K., Swain, S. K., & Karri, G. R. (2023). *"Cloud computing and virtualization." Convergence of Cloud with AI for Big Data Analytics*. Foundations and Innovation.

Mansouri, M., & Taylor, H. (2024). Does Cultural Robotics Need Culture? Conceptual Fragmentation and the Problems of Merging Culture with Robot Design. *International Journal of Social Robotics*, 16(2), 385–401. DOI: 10.1007/s12369-023-01085-y

Marcus, D. S., Wang, T. H., Parker, J., Csernansky, J. G., Morris, J. C., & Buckner, R. L. (2007). Open Access Series of Imaging Studies (OASIS): Cross-sectional MRI data in young, middle aged, nondemented, and demented older adults. *Journal of Cognitive Neuroscience*, 19(9), 1498–1507. DOI: 10.1162/jocn.2007.19.9.1498 PMID: 17714011

Marginson, S. (2024). Student self-formation: An emerging paradigm in higher education. *Studies in Higher Education*, 49(4), 748–762. DOI: 10.1080/03075079.2023.2252826

Markert, F., Marangon, A., Carcassi, M., & Duijm, N. J. (2017). Risk and sustainability analysis of complex hydrogen infrastructures. *International Journal of Hydrogen Energy*, 42(11), 7698–7706. DOI: 10.1016/j.ijhydene.2016.06.058

Marler, R. T., & Arora, J. S. (2004). Survey of multi-objective optimization methods for engineering. *Structural and Multidisciplinary Optimization*, 26(6), 369–395. DOI: 10.1007/s00158-003-0368-6

Maroušek, J., Strunecký, O., Bartoš, V., & Vochozka, M. (2022). Revisiting competitiveness of hydrogen and algae biodiesel. *Fuel*, 328, 125317.

Martínez, F. C., Cansino, A. T., García, M. A. A., Kalashnikov, V., & Rojas, R. L. (2014). Mathematical analysis for the optimization of a design in a facultative pond: Indicator organism and organic matter. *Mathematical Problems in Engineering*, 2014(1), 1–12. DOI: 10.1155/2014/652509

Martini, S., Afroze, S., & Ahmad Roni, K. (2020). Modified eucalyptus bark as a sorbent for simultaneous removal of COD, oil, and Cr(III) from industrial wastewater. *Alexandria Engineering Journal*, 59(3), 1637–1648. DOI: 10.1016/j.aej.2020.04.010

Maryam, S. (2017). Review of modelling approaches used in the HSC context for the UK. *International Journal of Hydrogen Energy*, 42(39), 24927–24938. DOI: 10.1016/j.ijhydene.2017.04.303

Masolele, R. N., De Sy, V., Herold, M., Marcos, D., Verbesselt, J., Gieseke, F., Mullissa, A. G., & Martius, C. (2021). Spatial and temporal deep learning methods for deriving land-use following deforestation. A pan-tropical case study using Landsat time series. *Remote Sensing of Environment*, 264, 112600. DOI: 10.1016/j.rse.2021.112600

Mateos-Pérez, J. M., Dadar, M., Lacalle-Aurioles, M., Iturria-Medina, Y., Zeighami, Y., & Evans, A. C. (2018). Structural neuroimaging as clinical predictor: A review of machine learning applications. *NeuroImage. Clinical*, 20, 506–522. DOI: 10.1016/j.nicl.2018.08.019 PMID: 30167371

Mattson, K. (2024). *Digital citizenship in action: empowering students to engage in online communities.* International Society for Technology in Education.

Melo, M. T., Nickel, S., & Saldanha-da-Gama, F. (2009). Facility location and supply chain management – A review. *European Journal of Operational Research*, 196(2), 401–412. DOI: 10.1016/j.ejor.2008.05.007

Meraz-Ríos, M. A., Lira-De León, K. I., Campos-Peña, V., De Anda-Hernández, M. A., & Mena-López, R. (2010). Tau oligomers and aggregation in Alzheimer's disease. *Journal of Neurochemistry*, 112(6), 1353–1367. DOI: 10.1111/j.1471-4159.2009.06511.x PMID: 19943854

Meti, V., & Dabburu, K. (2017). Anti-anxiety activity of ethanolic extract of Spermacoce hispida leaves in swiss albino mice. *International Journal of Basic &amp. International Journal of Basic and Clinical Pharmacology*, 6(9), 2224. DOI: 10.18203/2319-2003.ijbcp20173749

Mittal, M., Goyal, L. M., Kaur, S., Kaur, I., Verma, A., & Hemanth, D. J. (2019), Deep learning based enhanced tumor segmentation approach for MR brain images,Applied Soft Computing,Volume 78,Pages 346-354,ISSN 1568-4946.

Mohamed, A. M. (2024). Exploring the potential of an AI-based Chatbot (ChatGPT) in enhancing English as a Foreign Language (EFL) teaching: Perceptions of EFL Faculty Members. *Education and Information Technologies*, 29(3), 3195–3217. DOI: 10.1007/s10639-023-11917-z

Mohapatra, S., Pati, G. K., Mishra, M., & Swarnkar, T. (2022). Upolyseg. A u-net-based polyp segmentation network using colonoscopy images. *Gastroenterology Insights*, 13(3), 264–274. DOI: 10.3390/gastroent13030027

Monika, M., & Suganthan, C. (2024). A Study on Analyzing the Role of Chatgpt in English Acquisition among ESL Learners during English Language Classroom. *language, 8*(2).

Moradi, A., Gorji, S. A., Hakemi, A., Ganjavi, A., & Sera, D. (2022). Study of a DC Micro-Gird Configuration to Produce Hydrogen (DCMG-H2). In *2022 IEEE 7th Southern Power Electronics Conference (SPEC)* (Vol. 24, pp. 1–5). IEEE. DOI: 10.1109/SPEC55080.2022.10058435

Moradi, E., Pepe, A., Gaser, C., Huttunen, H., & Tohka, J. (2015). Machine learning framework for early MRI-based Alzheimer's conversion prediction in MCI subjects. *NeuroImage*, 104, 398–412. DOI: 10.1016/j.neuroimage.2014.10.002 PMID: 25312773

Mosavi, A., Salimi, M., Ardabili, S. F., Rabczuk, T., Shamshirband, S., & Varkonyi-Koczy, A. R. (2019). State of the Art of Machine Learning Models in Energy Systems, a Systematic Review. *Energies*, 12(7), 1301. DOI: 10.3390/en12071301

Moussa, M. N., Steen, M. R., Laurienti, P. J., & Hayasaka, S. (2012). Consistency of network modules in resting-state FMRI connectome data, 7(8):e44428. DOI: 10.1371/journal.pone.0044428

Mulvey, J. M., Vanderbei, R. J., & Zenios, S. A. (1995). Robust Optimization of Large-Scale Systems. *Operations Research*, 43(2), 264–281. DOI: 10.1287/opre.43.2.264

Murthy Konda, N. V. S. N., Shah, N., & Brandon, N. P. (2011). Optimal transition towards a large-scale hydrogen infrastructure for the transport sector: The case for the Netherlands. *International Journal of Hydrogen Energy*, 36(8), 4619–4635. DOI: 10.1016/j.ijhydene.2011.01.104

Muthmainnah, P. M. I. S., & Oteir, I. "Playing with AI to Investigate Human-Computer Interaction Technology and Improve Critical Thinking Skills to Catch Up with the 21st Century Era", Education Research International, vol. 2022, Article ID 6468995, 17 pages, 2022. DOI: 10.1155/2022/6468995

Muthmainnah, M., Darmawati, B., Khang, A., Al Yakin, A., Obaid, A. J., & Elngar, A. A. Shaping Artificial Intelligence-Perceived Hybrid Learning Environment at University Toward the Global Talent Development Strategy. In *AI-Oriented Competency Framework for Talent Management in the Digital Economy* (pp. 219–243). CRC Press. DOI: 10.1201/9781003440901-14

Nair, P., Trisno, R., Baghini, M. S., Pendharkar, G., & Chung, H. (2020). Predicting early stage drug induced Parkinsonism using unsupervised and supervised machine learning. *Annual International Conference of the IEEE Engineering in Medicine and Biology Society*, 2020, 776–779. DOI: 10.1109/EMBC44109.2020.9175343 PMID: 33018101

Najah Ahmed, A., Binti Othman, F., Abdulmohsin Afan, H., Khaleel Ibrahim, R., Ming Fai, C., Shabbir Hossain, M., Ehteram, M., & Elshafie, A. (2019). Machine learning methods for better water quality prediction. *Journal of Hydrology (Amsterdam)*, 578, 124084. DOI: 10.1016/j.jhydrol.2019.124084

Nakane, T., Nihashi, T., Kawai, H., & Naganawa, S. (2008). Visualization of neuromelanin in the Substantia nigra and locus ceruleus at 1.5 T using a 3D-gradient echo sequence with magnetization transfer contrast. *Magnetic Resonance in Medical Sciences*, 7(4), 205–210. DOI: 10.2463/mrms.7.205 PMID: 19110515

Nasr, M., & Shahawy, E. L. (2016). Artificial Intelligence for Electrocoagulation Treatment of Olive Mill Wastewater. *Journal of Bioremediation &amp. Journal of Bioremediation & Biodegradation*, 07(03). Advance online publication. DOI: 10.4172/2155-6199.1000345

Nassef, A. M., Abdelkareem, M. A., Maghrabie, H. M., & Baroutaji, A. (2023). Metaheuristic-Based Algorithms for Optimizing Fractional-Order Controllers—A Recent, Systematic, and Comprehensive Review. *Fractal and Fractional*, 7(7), 553. DOI: 10.3390/fractalfract7070553

Newell, P., & Ilgen, A. G. (2019). Overview of Geological Carbon Storage (GCS). In *Science of Carbon Storage in Deep Saline Formations* (pp. 1–13). Elsevier. DOI: 10.1016/B978-0-12-812752-0.00001-0

Newman, S. E., & Kenworthy, J. R. (1989). *Cities and the automobile*. University of California Press.

Ng, D. T. K., Tan, C. W., & Leung, J. K. L. (2024). Empowering student self-regulated learning and science education through ChatGPT: A pioneering pilot study. *British Journal of Educational Technology*, 55(4), 1328–1353. DOI: 10.1111/bjet.13454

Ng, K. C., Thu, K., Oh, S. J., Ang, L., Shahzad, M. W., & Ismail, A. (2015). Recent developments in thermally-driven seawater desalination: Energy efficiency improvement by hybridization of the MED and AD cycles. *Desalination*, 356, 255–270. DOI: 10.1016/j.desal.2014.10.025

Nguyen, A., Kremantzis, M., Essien, A., Petrounias, I., & Hosseini, S. (2024). Enhancing Student Engagement Through Artificial Intelligence (AI): Understanding the Basics, Opportunities, and Challenges. *Journal of University Teaching & Learning Practice*, 21(06). Advance online publication. DOI: 10.53761/caraaq92

Nguyen, T.-A., Rußwurm, M., Lenczner, G., & Tuia, D. (2024). Multi-temporal forest monitoring in the Swiss Alps with knowledge-guided deep learning. *Remote Sensing of Environment*, 305, 114109. DOI: 10.1016/j.rse.2024.114109

Nickel, S., Steinhardt, C., Schlenker, H., Burkart, W., & Reuter-Oppermann, M. (2020). IBM ILOG CPLEX Optimization Studio. In *Angewandte Optimierung mit IBM ILOG CPLEX Optimization Studio* (pp. 9–23). Springer Berlin Heidelberg., DOI: 10.1007/978-3-662-62185-1_2

Niso, G., Botvinik-Nezer, R., Appelhoff, S., De La Vega, A., Esteban, O., Etzel, J. A., Finc, K., Ganz, M., Gau, R., Halchenko, Y. O., Herholz, P., Karakuzu, A., Keator, D. B., Markiewicz, C. J., Maumet, C., Pernet, C. R., Pestilli, F., Queder, N., Schmitt, T., & Rieger, J. W. (2022). Open and reproducible neuroimaging: From study inception to publication. *NeuroImage*, 263, 119623. DOI: 10.1016/j.neuroimage.2022.119623 PMID: 36100172

Nourani, V., Elkiran, G., & Abba, S. I. (2018). Wastewater treatment plant performance analysis using artificial intelligence – an ensemble approach. *Water Science and Technology*, 78(10), 2064–2076. DOI: 10.2166/wst.2018.477 PMID: 30629534

NYC. The Official Website of the City of New York, www.nyc.gov/html/tlc/html/about/triv_record_data.shtml. (up to date).

Oba, D., & Berger, J. (2024). How communication mediums shape the message. *Journal of Consumer Psychology*, 34(3), 406–424. DOI: 10.1002/jcpy.1372

Ochoa Bique, A., Maia, L. K. K., La Mantia, F., Manca, D., & Zondervan, E. (2019). Balancing costs, safety and CO2 emissions in the design of hydrogen supply chains. *Computers &amp. Computers & Chemical Engineering*, 129, 106493. DOI: 10.1016/j.compchemeng.2019.06.018

Oh, K., Chung, Y. C., Kim, K. W., Kim, W. S., & Oh, I. S. (2019). Classification and visualization of Alzheimer's disease using volumetric convolutional neural network and transfer learning. *Scientific Reports*, 9(1), 18150. DOI: 10.1038/s41598-019-54548-6 PMID: 31796817

Ohtsuka, C., Sasaki, M., Konno, K., Koide, M., Kato, K., Takahashi, J., Takahashi, S., Kudo, K., Yamashita, F., & Terayama, Y. (2013). Changes in substantia nigra and locus coeruleus in patients with early-stage Parkinson's disease using neuromelanin-sensitive MR imaging. *Neuroscience Letters*, 541, 93–98. DOI: 10.1016/j.neulet.2013.02.012 PMID: 23428505

Ojajuni, O., Ayeni, F., Akodu, O., Ekanoye, F., Adewole, S., Ayo, T., & Mbarika, V. (2021). Predict- ing student academic performance using machine learning. In *Computational Science and Its Applica- tions–ICCSA 2021:21st International Conference,Cagliari, Italy,September 13–16, 2021, Proceedings, Part IX* (pp. 481-491). Springer International Publishing.

Okada, A., Panselinas, G., Bizoi, M., Malagrida, R., & Torres, P. L. (2024). Fostering Transversal Skills through Open Schooling with the CARE-KNOW-DO Framework for Sustainable Education. *Sustainability (Basel)*, 16(7), 2794. DOI: 10.3390/su16072794

Okur, B., & Örçen, N. (2020). Soil salinization and climate change. Climate Change and Soil Interactions, 331–350. .DOI: 10.1016/B978-0-12-818032-7.00012-6

Oladipo, A. A., & Gazi, M. (2015). Nickel removal from aqueous solutions by alginate-based composite beads: Central composite design and artificial neural network modeling. *Journal of Water Process Engineering*, 8, e81–e91. DOI: 10.1016/j.jwpe.2014.12.002

Oliveiraesquerre, K., Seborg, D., Mori, M., & Bruns, R.OLIVEIRAESQUERRE. (2004). Application of steady-state and dynamic modeling for the prediction of the BOD of an aerated lagoon at a pulp and paper millPart II. Nonlinear approaches. *Chemical Engineering Journal*, 105(1–2), 61–69. DOI: 10.1016/j.cej.2004.06.012

Oliver, J. M., Esteban, M. D., López-Gutiérrez, J.-S., Negro, V., & Neves, M. G. (2021). Optimizing Wave Overtopping Energy Converters by ANN Modelling: Evaluating the Overtopping Rate Forecasting as the First Step. *Sustainability (Basel)*, 13(3), 1483. DOI: 10.3390/su13031483

Onesi-Ozigagun, O., Ololade, Y. J., Eyo-Udo, N. L., & Ogundipe, D. O. (2024). Revolutionizing education through AI: A comprehensive review of enhancing learning experiences. *International Journal of Applied Research in Social Sciences*, 6(4), 589–607. DOI: 10.51594/ijarss.v6i4.1011

Ong, M.-L., Tan, P. F., & Holbrook, J. D. (2017). Predicting functional decline and survival amyotrophic lateral sclerosis. *PLoS One*, 12(4), e0174925. DOI: 10.1371/journal.pone.0174925 PMID: 28406915

Ordidge, R. J., Gorell, J. M., Deniau, J. C., Knight, R. A., & Helpern, J. A. (1994). Assessment of relative brain iron concentrations using T2-weighted and T2*-weighted MRI at 3 Tesla. *Magnetic Resonance in Medicine*, 32(3), 335–341. DOI: 10.1002/mrm.1910320309 PMID: 7984066

Osman, A. I., Mehta, N., Elgarahy, A. M., Hefny, M., Al-Hinai, A., Al-Muhtaseb, A. H., & Rooney, D. W. (2021). Hydrogen production, storage, utilisation and environmental impacts: A review. *Environmental Chemistry Letters*, 20(1), 153–188. DOI: 10.1007/s10311-021-01322-8

Othman, R. M., Abushunar, M. S., & Al-Shboul, S. S. (2024). Investigating the Effectiveness of Synchronous E-Learning Tools on Developing EFL Learners' Language Communication Skills Among the Hashemite University Students. *Theory and Practice in Language Studies*, 14(2), 597–605. DOI: 10.17507/tpls.1402.33

Ouchama, K. (2022). *Efficiency of Constructed Wetlands and Wastewater Stabilization Ponds for Wastewater Treatment in Northern Algerian Sahara. Constructed Wetlands for Wastewater Treatment in Hot and Arid Climates. Wetlands: Ecology, Conservation and Management, 7.* Springer., DOI: 10.1007/978-3-031-03600-2_2

Ouyang, F., Zhang, L., Wu, M., & Jiao, P. (2024). Empowering collaborative knowledge construction through the implementation of a collaborative argument map tool. *The Internet and Higher Education*, 62, 100946. DOI: 10.1016/j.iheduc.2024.100946

Ozkay, U. D., Oztürk, Y., & Can, O. D. (2011). The disease of the aging world: Alzheimer's disease. S.D.U. *Faculty of Medicine Journal*, 18(1), 35–42.

Ozpak, L., Pazarbası, A., & Keser, N. (2017). Genetics and Epigenetics of Alzheimer's Disease. *Journal of Archive Literature Review*, 26(1), 34–49.

Parekh, D., Poddar, N., Rajpurkar, A., Chahal, M., Kumar, N., Joshi, G. P., & Cho, W. (2022). A review on Autonomous Vehicles: Progress, methods and challenges. *Electronics (Basel)*, 11(14), 2162. DOI: 10.3390/electronics11142162

Parker, N., Fan, Y., & Ogden, J. (2010). From waste to hydrogen: An optimal design of energy production and distribution network. *Transportation Research Part E, Logistics and Transportation Review*, 46(4), 534–545. DOI: 10.1016/j.tre.2009.04.002

Park, H., & Ahn, D. (2024, May). The Promise and Peril of ChatGPT in Higher Education: Opportunities, Challenges, and Design Implications. In *Proceedings of the CHI Conference on Human Factors in Computing Systems* (pp. 1-21). DOI: 10.1145/3613904.3642785

Park, S., Tran, V., & Lee, D. (2021). Application of various YOLO models for computer Vision-Based Real-Time pothole detection. *Applied Sciences (Basel, Switzerland)*, 11(23), 11229. DOI: 10.3390/app112311229

Park, W., & Kwon, H. (2024). Implementing artificial intelligence education for middle school technology education in Republic of Korea. *International Journal of Technology and Design Education*, 34(1), 109–135. DOI: 10.1007/s10798-023-09812-2 PMID: 36844448

Parsazadeh, N., Yousefi, F., Ghaedi, M., Dashtian, K., & Borousan, F. (2018). Preparation and characterization of monoliths HKUST-1 MOF *via* straightforward conversion of $Cu(OH)_2$-based monoliths and its application for wastewater treatment: Artificial neural network and central composite design modeling. *New Journal of Chemistry*, 42(12), 10327–10336. DOI: 10.1039/C8NJ01067F

Patel, H. V., Gorji, S. A., Shahi, S. S. M., & Love, J. G. (2021). Implementation of a Lab-Scale Green Hydrogen Production System with Solar PV Emulator and Energy Storage System. In *2021 11th International Conference on Power and Energy Systems (ICPES)* (Vol. 1, pp. 201–208). IEEE. DOI: 10.1109/ICPES53652.2021.9683797

Patel, K. P., Wymer, D. T., Bhatia, V. K., Duara, R., & Rajadhyaksha, C. D. (2020). Multimodality imaging of dementia: Clinical importance and role of integrated anatomic and molecular imaging. *Radiographics*, 40(1), 200–222. DOI: 10.1148/rg.2020190070 PMID: 31917652

Patil, P., Chaudhary, N., Prasad, S., Bhandwal, M., Arora, M., & Singh, G. (2023). Predict- ing Student Performance with Machine Learning Algorithms. In *2023 3rd International Conference on Technological Advancements in Computational Sciences (ICTACS)* (pp. 1346-1353). IEEE.

Pellegrini, E., Ballerini, L., Hernandez, M. D. C. V., Chappell, F. M., González-Castro, V., Anblagan, D., Danso, S., Muñoz-Maniega, S., Job, D., Pernet, C., Mair, G., MacGillivray, T. J., Trucco, E., & Wardlaw, J. M. (2018). Machine learning of neuroimaging for assisted diagnosis of cognitive impairment and dementia: A systematic review. *Alzheimer's & Dementia: Diagnosis, Assessment & Disease Monitoring*, 10(1), 519–535. DOI: 10.1016/j.dadm.2018.07.004 PMID: 30364671

Peng, M., Liu, Y., Khan, A., Ahmed, B., Sarker, S. K., Ghadi, Y. Y., Bhatti, U. A., Al-Razgan, M., & Ali, Y. A. (2024). Crop monitoring using remote sensing land use and land change data. Comparative analysis of deep learning methods using pre-trained CNN models. *Big Data Research*, 36, 100448. DOI: 10.1016/j.bdr.2024.100448

Péran, P., Cherubini, A., Assogna, F., Piras, F., Quattrocchi, C., Peppe, A., Celsis, P., Rascol, O., Démonet, J.-F., Stefani, A., Pierantozzi, M., Pontieri, F. E., Caltagirone, C., Spalletta, G., & Sabatini, U. (2010). Magnetic resonance imaging markers of Parkinson's disease nigrostriatal signature. *Brain*, 133(11), 3423–3433. DOI: 10.1093/brain/awq212 PMID: 20736190

Pfohl, S. R., Kim, R. B., Coan, G. S., & Mitchell, C. S. (2018). Unraveling the complexity of amyotrophic lateral sclerosis survival prediction. *Frontiers in Neuroinformatics*, 12, 36. DOI: 10.3389/fninf.2018.00036 PMID: 29962944

Pham, D. T., Ho, L., Espinoza-Palacios, J., Arevalo-Durazno, M., Van Echelpoel, W., & Goethals, P. (2020). Generalised linear models for prediction of dissolved oxygen in a waste stabilisation pond. *Water (Basel)*, 12(7), 1930. DOI: 10.3390/w12071930

Podder, M. S., & Majumder, C. B. (2016). The use of artificial neural network for modelling of phycoremediation of toxic elements As(III) and As(V) from wastewater using Botryococcus braunii. *Spectrochimica Acta. Part A: Molecular and Biomolecular Spectroscopy*, 155, 130–145. DOI: 10.1016/j.saa.2015.11.011 PMID: 26615452

Pollanen, M. S., Dickson, D. W., & Bergeron, C. (1993). Pathology and biology of the Lewy body. Journal of Neuropathology & Experimental Neurology, 52(3), 183-191. 74.

Pothole Detection Dataset. (2022, December 25). Kaggle. https://www.kaggle.com/datasets/rajdalsaniya/pothole-detection-dataset?source=post_page-----f605f6eb4d9c--------------------------------

Pothole detection. (2020, June 9). Kaggle. https://www.kaggle.com/datasets/andrewmvd/pothole-detection

Pothole Image Data-Set. (2019, August 20). Kaggle. https://www.kaggle.com/datasets/sachinpatel21/pothole-image-dataset

Powers, R., Etezadi-Amoli, M., Arnold, E. M., Kianian, S., Mance, I., Gibiansky, M., Trietsch, D., Alvarado, A. S., Kretlow, J. D., Herrington, T. M., Brillman, S., Huang, N. C., Lin, P., Pham, H. A., & Ullal, A. (2021). Smartwatch inertial sensors continuously monitor real-world motor fluctuations in Parkinson's disease. *Science Translational Medicine*, 13(579), eabd7865. DOI: 10.1126/scitranslmed.abd7865 PMID: 33536284

Prasantha, H. S., Shashidhara, H. L., Murthy, K. N. B., & Madhavi, L. G. (2010). Medical image segmentation. *International Journal on Computer Science and Engineering*, 2(4), 1209–1218.

Psol, M., Darvas, S. G., Leite, K., Mahajani, S. U., Bähr, M., & Kügler, S. (2021). Dementia with Lewy bodies-associated β-synuclein mutations V70M and P123H cause mutation-specific neuropathological lesions. *Human Molecular Genetics*, 30(3-4), 247–264. DOI: 10.1093/hmg/ddab036 PMID: 33760043

Querin, G., Mendili, M.-M. E., Bede, P., Delphine, S., & Lenglet, T. (2018). Multimodal spinal cord MRI offers accurate diagnostic classification in ALS. *Journal of Neurology, Neurosurgery, and Psychiatry*, 89(11), 1220–1221. DOI: 10.1136/jnnp-2017-317214 PMID: 29353238

Radhya, S., Tasik, M. A. S., Sabran, F. M., & Gunawan, A. A. S. (2022). Systematic Lit- erature Review: Machine Learning in Education to Predict Student Performance. In *2022 International Conference on Electrical and Information Technology (IEIT)* (pp. 350-356). IEEE. DOI: 10.1109/IEIT56384.2022.9967900

Rafique, H., Nazeer, I., & Rehman, J. (2024). The Impact of ChatGPT on Language Evolution: A Linguistic Analysis. *Journal of Education and Social Studies*, 5(1), 56–68. DOI: 10.52223/jess.2024.5106

Rahmaniani, R., Crainic, T. G., Gendreau, M., & Rei, W. (2017). The Benders decomposition algorithm: A literature review. *European Journal of Operational Research*, 259(3), 801–817. DOI: 10.1016/j.ejor.2016.12.005

Raina, A., Leite, K., Guerin, S., Mahajani, S. U., Chakrabarti, K. S., Voll, D., Becker, S., Griesinger, C., Bähr, M., & Kügler, S. (2021). Dopamine promotes the neurodegenerative potential of β-synuclein. *Journal of Neurochemistry*, 156(5), 674–691. DOI: 10.1111/jnc.15134 PMID: 32730640

Raina, A., Mahajani, S., Bähr, M., & Kügler, S. (2020). Neuronal trans-differentiation by transcription factors Ascl1 and Nurr1: Induction of a dopaminergic neurotransmitter phenotype in cortical GABAergic neurons. *Molecular Neurobiology*, 57(1), 249–260. DOI: 10.1007/s12035-019-01701-x PMID: 31317490

Rajaei, A. (2024). Teaching in the Age of AI/ChatGPT in Mental-Health-Related Fields. *The Family Journal (Alexandria, Va.)*, 32(1), 6–10. DOI: 10.1177/10664807231209721

Rajamohan, K., Rangasamy, S., Manoj, C. N., Mary, L., Sabu, R., & Jose, S. S. (2024). Internet of Things (IoT) as a Game Changer to the Education Sector. In *Advanced IoT Technologies and Applications in the Industry 4.0 Digital Economy* (pp. 251–274). CRC Press. DOI: 10.1201/9781003434269-14

Rajesh, M., Wadhwa, T., Cherukuri, A. K., Kamalov, F., Jonnalagadda, A., & Ray, S. (2024). Preventing Illegal Deforestation using Acoustic Surveillance. *International Journal of Computer Information Systems and Industrial Management Applications*, 16(1), 17–17.

Raju, P. J. R., Kiran, K. V. D., & Sree, P. K. (2023). Shalem, K. V. D. Kiran, and Pokkuluri Kiran Sree. "Digital image watermarking based on hybrid FRT-HD-DWT domain and flamingo search optimisation.". *International Journal of Computational Vision and Robotics*, 13(6), 573–598. DOI: 10.1504/IJCVR.2023.134319

Rani, A., & Chang, C.-T. (2021). *Modeling and optimization of wastewater treatment processes. Current Developments in Biotechnology and Bioengineering*. Elsevier., DOI: 10.1016/B978-0-12-821009-3.00006-3

Rao, N. R., Kiran, S. P., Amena, T., Senthilkumar, A., Sivakumar, R., Kumar, M. A., & Velusamy, S. (2024). Enhancing rainwater harvesting and groundwater recharge efficiency with multi-dimensional LSTM and clonal selection algorithm. *Groundwater for Sustainable Development*, 25, 101167.

Rao, S. D., & Weiss, J. H. (2004). Excitotoxic and oxidative cross-talk between motor neurons and glia in ALS pathogenesis. *Trends in Neurosciences*, 27(1), 17–23. DOI: 10.1016/j.tins.2003.11.001 PMID: 14698606

Rasheed, J., Hameed, A. A., Ajlouni, N., Jamil, A., Ozyavas, A., & Orman, Z. (2020). "Application of adaptive back-propagation neural networks for parkinson's disease prediction," in *2020 International Conference on Data Analytics for Business and Industry: Way Towards a Sustainable Economy (ICDABI)* (IEEE), 1–5. DOI: 10.1109/ICDABI51230.2020.9325709

Rathore, S., Habes, M., Iftikhar, M. A., Shacklett, A., & Davatzikos, C. (2017). A review on neuroimaging-based classification studies and associated feature extraction methods for Alzheimer's disease and its prodromal stages. *NeuroImage*, 155, 530–548. DOI: 10.1016/j.neuroimage.2017.03.057 PMID: 28414186

Reddy, C. K. K., Anisha, P. R., Khan, S., Hanafiah, M. M., Pamulaparty, L., & Mohana, R. M. (Eds.). (2024). *Sustainability in Industry 5.0: Theory and Applications*. CRC Press.

Reddy, P. Y., Reddy, C. K. K., & Sithole, N. T. (2024). A study and analysis on diagnosis of melanoma cancer with deep learning: A case study. In *Federated Learning and AI for Healthcare 5.0* (pp. 203–218). IGI Global Scientific Publishing.

Redmon, J., & Farhadi, A. (2017). YOLO9000: Better, Faster, Stronger. *Proceedings - 30th IEEE Conference on Computer Vision and Pattern Recognition*. DOI: 10.1109/CVPR.2017.690

Redmon, J., & Farhadi, A. (2018, April 8). *YOLOV3: an incremental improvement*. arXiv.org. https://arxiv.org/abs/1804.02767v1

Redmon, J., Divvala, S., Girshick, R., & Farhadi, A. (2016). You Only Look Once: Unified, Real-Time Object Detection. *Proceedings of the IEEE Computer Society Conference on Computer Vision and Pattern Recognition*. DOI: 10.1109/CVPR.2016.91

Rencher, A. C., & Christensen, W. F. (2012). *Methods of Multivariate Analysis* (3rd ed.). Wiley. DOI: 10.1002/9781118391686

Rey, A., Mulligan, R., da Silva, A. M. F., Filion, Y., Champagne, P., & Boegman, L. (2021). Three-Dimensional Hydrodynamic Behavior of an Operational Waste-Stabilization Pond. *Journal of Environmental Engineering*, 147(2), 5020009. DOI: 10.1061/(ASCE)EE.1943-7870.0001834

Reyes-Barquet, L. M., Rico-Contreras, J. O., Azzaro-Pantel, C., Moras-Sánchez, C. G., González-Huerta, M. A., Villanueva-Vásquez, D., & Aguilar-Lasserre, A. A. (2022). Multi-Objective Optimal Design of a Hydrogen Supply Chain Powered with Agro-Industrial Wastes from the Sugarcane Industry: A Mexican Case Study. *Mathematics*, 10(3), 437. DOI: 10.3390/math10030437

Rezaei, S., & Ansary, A. (Eds.). (2024). *Artificial Intelligence of Things (AIoT) for Productivity and Organizational Transition*. IGI Global. DOI: 10.4018/979-8-3693-0993-3

Riera, J. A., Lima, R. M., Hoteit, I., & Knio, O. (2022). Simulated co-optimization of renewable energy and desalination systems in Neom, Saudi Arabia. *Nature Communications*, 13(1), 3514. DOI: 10.1038/s41467-022-31233-3 PMID: 35717452

Robles, J. O., Azzaro-Pantel, C., Garcia, G. M., & Lasserre, A. A. (2020). Social cost-benefit assessment as a post-optimal analysis for hydrogen supply chain design and deployment: Application to Occitania (France). *Sustainable Production and Consumption*, 24, 105–120. DOI: 10.1016/j.spc.2020.06.010

Roboflow: Computer vision tools for developers and enterprises. (n.d.). https://roboflow.com/

Ronneberger, O., Fischer, P., & Brox, T. (2015). U-net: Convolutional networks for biomedical image segmentation. In Medical image computing and computer-assisted intervention–MICCAI 2015: 18th international conference, Munich, Germany, October 5-9, 2015, proceedings, part III 18 (pp. 234-241). Springer International Publishing.

Rothstein, J. D., Bristol, L. A., Hosler, B., Brown, R. H.Jr, & Kuncl, R. W. (1994). Chronic inhibition of superoxide dismutase produces apoptotic death of spinal neurons. *Proceedings of the National Academy of Sciences of the United States of America*, 91(10), 4155–4159. DOI: 10.1073/pnas.91.10.4155 PMID: 7910402

Rousseeuw, P. J., & Croux, C. (1993). Alternatives to the median absolute deviation. *Journal of the American Statistical Association*, 88(424), 1273–1283. DOI: 10.1080/01621459.1993.10476408

Rout, N. K., Dutta, G., Sinha, V., Dey, A., Mukherjee, S., & Gupta, G. (2023, November 10). *Improved pothole detection using YOLOV7 and ESRGAN*. arXiv.org. https://arxiv.org/abs/2401.08588v1

Rovini, E., Maremmani, C., & Cavallo, F. (2020). A wearable system to objectify assessment of motor tasks for supporting parkinson's disease diagnosis. *Sensors (Basel)*, 20(9), 2630. DOI: 10.3390/s20092630 PMID: 32380675

Rowley, P. A., Samsonov, A. A., Betthauser, T. J., Pirasteh, A., Johnson, S. C., & Eisenmenger, L. B. (2020). Amyloid and Tau PET imaging of Alzheimer disease and other neurodegenerative conditions. In *Seminars in Ultrasound, CT and MRI. 41(6): 572-583*. WB Saunders. DOI: 10.1053/j.sult.2020.08.011

Rubinger, L., Gazendam, A., Ekhtiari, S., & Bhandari, M. (2023),Machine learning and artificial intelligence in research and healthcare,Volume 54, Supplement 3,Pages S69-S73,ISSN 0020-1383,Injury.

Sabio, N., Gadalla, M., Guillén-Gosálbez, G., & Jiménez, L. (2010). Strategic planning with risk control of hydrogen supply chains for vehicle use under uncertainty in operating costs: A case study of Spain. *International Journal of Hydrogen Energy*, 35(13), 6836–6852. DOI: 10.1016/j.ijhydene.2010.04.010

Saboe, D., Ghasemi, H., Gao, M. M., Samardzic, M., Hristovski, K. D., Boscovic, D., Burge, S. R., Burge, R. G., & Hoffman, D. A. (2021). Real-time monitoring and prediction of water quality parameters and algae concentrations using microbial potentiometric sensor signals and machine learning tools. *The Science of the Total Environment*, 764, 142876. DOI: 10.1016/j.scitotenv.2020.142876 PMID: 33757235

Safarov, S., & Whangbo, T. K. (2021). A-DenseUNet: Adaptive densely connected UNet for polyp segmentation in colonoscopy images with atrous convolution. *Sensors (Basel)*, 21(4), 1441.

Sahin, H. A. (2009). Alzheimer hastalığını klinik belirtileri ve seyri. *Türkiye Klinikleri*, 2(1), 3135.

Sahlaoui, H., Abdellaoui Alaoui, E. A., & Agoujil, S. (2021). A framework towards more accu- rate and explanatory student performance model. In *Proceedings of the 4th International Conference on Networking, Information Systems & Security* (pp. 1-8).

Saleem, F., Chikhaoui, E., & Malik, M. I. (2024, February). Technostress in students and quality of online learning: Role of instructor and university support. [). Frontiers Media SA.]. *Frontiers in Education*, 9, 1309642. DOI: 10.3389/feduc.2024.1309642

Sanbornscott, D. (2004). Inside fuelcells. *International Journal of Hydrogen Energy*, 29(12), 1203–1211. DOI: 10.1016/j.ijhydene.2004.01.013

Sandler, M., Howard, A., Zhu, M., Zhmoginov, A., & Chen, L. C. (2018). Mobilenetv2: Inverted residuals and linear bottlenecks. In *Proceedings of the IEEE conference on computer vision and pattern recognition* (pp. 4510-4520).

Sarkar, A., & Arya, A. K. (2022). A Survey on Optimization Parameters and Techniques for Crude Oil Pipeline Transportation. In *Smart Innovation* (Vol. 292). Systems and Technologies., DOI: 10.1007/978-981-19-0836-1_43

Sasaki, M., Shibata, E., Tohyama, K., Takahashi, J., Otsuka, K., Tsuchiya, K., Takahashi, S., Ehara, S., Terayama, Y., & Sakai, A. (2006). Neuromelanin magnetic resonance imaging of locus ceruleus and substantia nigra in Parkinson's disease. *Neuroreport*, 17(11), 1215–1218. DOI: 10.1097/01.wnr.0000227984.84927.a7 PMID: 16837857

Saufi Sulaiman, M., Singh, B., & Mohamed, W. A. N. W. (2019). Experimental and theoretical study of thermoelectric generator waste heat recovery model for an ultra-low temperature PEM fuel cell powered vehicle. *Energy*, 179, 628–646. DOI: 10.1016/j.energy.2019.05.022

Saur, G., & Ramsden, T. (2011). *Wind Electrolysis: Hydrogen Cost Optimization*. Office of Scientific and Technical Information (OSTI). DOI: 10.2172/1015505

Saxer, K., Schnell, J., Mori, J., & Hascher, T. (2024). The role of teacher–student relationships and student–student relationships for secondary school students' well-being in Switzerland. *International Journal of Educational Research Open*, 6, 100318. DOI: 10.1016/j.ijedro.2023.100318

Schuster, C., Hardiman, O., & Bede, P. (2017). Survival prediction in amyotrophic lateral sclerosis based on MRI measures and clinical characteristics. *BMC Neurology*, 17(1), 73. DOI: 10.1186/s12883-017-0854-x PMID: 28412941

Schwarz, S. T., Abaei, M., Gontu, V., Morgan, P. S., Bajaj, N., & Auer, D. P. (2013). Diffusion tensor imaging of nigral degeneration in Parkinson's disease: A region-of-interest and voxel-based study at 3 T and systematic review with meta-analysis. *NeuroImage. Clinical*, 3, 481–488. DOI: 10.1016/j.nicl.2013.10.006 PMID: 24273730

Schwarz, S. T., Rittman, T., Gontu, V., Morgan, P. S., Bajaj, N., & Auer, D. P. (2011). T1-weighted MRI shows stage-dependent substantia nigra signal loss in Parkinson's disease. *Movement Disorders*, 26(9), 1633–1638. DOI: 10.1002/mds.23722 PMID: 21491489

Scm, D. S. S. (2015). *A comparative study on web scraping*. http://ir.kdu.ac.lk/handle/345/1051

Scuotto et al. (2024) stated despite the prevalence of digital communication, humans still have an innate desire for face-to-face interaction, making social interaction the most important item in the digital age. This helps people feel more connected, empathic, and at home in their community.

Scuotto, V., Manlio, D. G., Malhotra, A., & Pereira, V. (2024). The evolution of human-machine interaction, from Taylorism to human-centered approach. *Journal of Management History*, 30(2), 141–146. DOI: 10.1108/JMH-04-2024-305

Seibold, H., Zeileis, A., & Hothorn, T. (2017). Individual treatment effect prediction for amyotrophic lateral sclerosis patients. *Statistical Methods in Medical Research*, 27(10), 3104–3125. DOI: 10.1177/0962280217693034 PMID: 29298618

Sekeroglu, B., Dimililer, K., & Tuncal, K. (2019). Student performance prediction and classifi- cation using machine learning algorithms. In *Proceedings of the 2019 8th International Conference on Educational and Information Technology* (pp. 7-11).

Senter, M. S. (2024). The impact of social relationships on college student learning during the pandemic: Implications for sociologists. *Teaching Sociology*, 52(1), 39–54. DOI: 10.1177/0092055X231178505

Senturk, Z. K. (2020). Early diagnosis of parkinson's disease using machine learning algorithms. Medical Hypoth. 138, 109603. doi: . 109603DOI: 10.1016/j.mehy.2020

Shahbazi, M., Dehghan-Manshadi, A., Hossain, M. D., Ravishankar, R., & Mackinnon, I. (2021). Microstructure and Superconducting Properties of Copper Sheathed MgB_2 Wires With Ti Barrier Fabricated by in Situ Powder-in-Tube Method. *IEEE Transactions on Applied Superconductivity*, 31(5), 1–5. DOI: 10.1109/TASC.2021.3065634

Shamshirband, S., Fathi, M., Dehzangi, A., Chronopoulos, A. T., & Alinejad-Rokny, H. (2021), A review on deep learning approaches in healthcare systems. Taxonomies, challenges, and open issues,Journal of Biomedical Informatics,Volume 113,103627,ISSN 1532-0464.

Shannon, L. (2023). *Interconnected Realities: How the Metaverse Will Transform Our Relationship with Technology Forever*. John Wiley & Sons.

Sharafi, M., & ELMekkawy, T. Y. (2014). Multi-objective optimal design of hybrid renewable energy systems using PSO-simulation based approach. *Renewable Energy*, 68, 67–79. DOI: 10.1016/j.renene.2014.01.011

Sharma, N., Sharma, M., & Garg, U. (2023). Predicting academic performance of students using machine learning models. In *2023 International Conference on Artificial Intelligence and Smart Commu- nication (AISC)* (pp. 1058-1063). IEEE. DOI: 10.1109/AISC56616.2023.10085214

Sharmila, K. S., Revathi, S. T., & Sree, P. K. "Convolution Neural Networks based lungs disease detection and Severity classification." In *2023 International Conference on Computer Communication and Informatics (ICCCI)*, pp. 1-9. IEEE, 2023. DOI: 10.1109/ICCCI56745.2023.10128188

Shen, D., Wu, G., & Suk, H. I. (2017). Deep learning in medical image analysis. *Annual Review of Biomedical Engineering*, 19(1), 221–248. DOI: 10.1146/annurev-bioeng-071516-044442 PMID: 28301734

She, X., Hong-Wei, Z., Wang, Z., & Yan, J. (2021). Feasibility study of asphalt pavement pothole properties measurement using 3D line laser technology. *International Journal of Transportation Science and Technology*, 10(1), 83–92. DOI: 10.1016/j.ijtst.2020.07.004

Shin, D., Lee, J. H., & Noh, W. I. (2024). Realizing Corrective Feedback in Task-Based Chatbots Engineered for Second Language Learning. *RELC Journal*, •••, 00336882231221902. DOI: 10.1177/00336882231221902

Siebenmorgen, N., & Weber, M. (2003). A behavioral model for asset allocation. *Financial Markets and Portfolio Management*, 17(1), 15–42. DOI: 10.1007/s11408-003-0102-9

Siklós, L., Engelhardt, J., Harati, Y., Smith, R. G., Joó, F., & Appel, S. H. (1996). Ultrastructural evidence for altered calcium in motor nerve terminals in amyotrophc lateral sclerosis. *Annals of Neurology*, 39(2), 203–216. DOI: 10.1002/ana.410390210 PMID: 8967752

Skalski, P. (n.d.). *GitHub - SkalskiP/make-sense: Free to use online tool for labelling photos.* https://makesense.ai. GitHub. https://github.com/SkalskiP/make-sense/

Snell, S. A., Swart, J., Morris, S., & Boon, C. (2023). The HR ecosystem: Emerging trends and a future research agenda. *Human Resource Management*, 62(1), 5–14. DOI: 10.1002/hrm.22158

Solórzano, J. V., Mas, J. F., Gallardo-Cruz, J. A., Gao, Y., & Fernández-Montes de Oca, A. (2023). Deforestation detection using a spatio-temporal deep learning approach with synthetic aperture radar and multispectral images. *ISPRS Journal of Photogrammetry and Remote Sensing*, 199, 87–101. DOI: 10.1016/j.isprsjprs.2023.03.017

Soyoye, T. O., Chen, T., Hill, R., & McCabe, K. (2023). Predicting Academic Performance of University Students Using Machine Learning: A Case Study in the UK. In *2023 IEEE International Conference on Web Intelligence and Intelligent Agent Technology (WI-IAT)* (pp. 431-434). IEEE. DOI: 10.1109/WI-IAT59888.2023.00071

Sree, P. K., & Chintalapati, P. V. M. Prasad, Gurujukota Ramesh Babu, and PBV Raja Rao. "Waste Management Detection Using Deep Learning." In 2023 3rd International Conference on Computing and Information Technology (ICCIT), pp. 50-54. IEEE, 2023.

Sree, K., Babu, R., & Devi, N. U. (2010). Identification of Promoter Region in Genomic DNA Using Cellular Automata Based Text Clustering. *The International Arab Journal of Information Technology*, 7(1), 75–78.

Sree, P. (2008). Kiran, I. Ramesh Babu, J. V. R. Murty, R. Ramachandran, and NSSSN Usha Devi. "Power-aware hybrid intrusion detection system (PHIDS) using cellular automata in wireless ad hoc networks.". *WSEAS Transactions on Computers*, 7(11), 1848–1874.

Sree, P. K. (2008). Exploring a novel approach for providing software security using soft computing systems. *International Journal of Security and Its Applications*, 2(2), 51–58.

Sree, P. K., Babu, I. R., & Usha Devi, N. S. S. S. N. (2009). Investigating an Artificial Immune System to strengthen protein structure prediction and protein coding region identification using the Cellular Automata classifier. *International Journal of Bioinformatics Research and Applications*, 5(6), 647–662. DOI: 10.1504/IJBRA.2009.029044 PMID: 19887338

Sree, P. K., Chakrabarti, P., Margala, M., & Usha Devi, N. (2024). Auto encoders with Cellular Automata for Anomaly Detection. *Journal of Electrical Systems*, 20(no. 2s), 227–232. DOI: 10.52783/jes.1131

Srivastava, S., Shet, S. P., Shanmuga Priya, S., Sudhakar, K., & Tahir, M. (2022). Molecular simulation of copper based metal-organic framework (Cu-MOF) for hydrogen adsorption. *International Journal of Hydrogen Energy*, 47(35), 15820–15831. DOI: 10.1016/j.ijhydene.2022.03.089

Stahl, B. C., & Eke, D. (2024). The ethics of ChatGPT–Exploring the ethical issues of an emerging technology. *International Journal of Information Management*, 74, 102700. DOI: 10.1016/j.ijinfomgt.2023.102700

Steiss, J., Tate, T., Graham, S., Cruz, J., Hebert, M., Wang, J., & Olson, C. B. (2024). Comparing the quality of human and ChatGPT feedback of students' writing. *Learning and Instruction, 91*, 101894. ge teaching (ELT) in Morocco. *Focus on ELT Journal*, 6(1), 87–106.

Steward, D., Ramsden, T., & Zuboy, J. (2008). *H2A Production Model, Version 2 User Guide*. Office of Scientific and Technical Information (OSTI). DOI: 10.2172/939003

Strzelecki, A. (2024). Students' acceptance of ChatGPT in higher education: An extended unified theory of acceptance and use of technology. *Innovative Higher Education*, 49(2), 223–245. DOI: 10.1007/s10755-023-09686-1

Suárez-García, A., Álvarez-Hernández, M., Arce, E., & Ribas, J. R. (2024). Exploring the Efficacy of Binary Surveys versus Likert Scales in Assessing Student Perspectives Using Bayesian Analysis. *Applied Sciences (Basel, Switzerland)*, 14(10), 4189. DOI: 10.3390/app14104189

Subramanian, K. V., Yaswanth, R. V. S., Kumar, D. P., Zenith, D. J., & Rao, G. D. (2023). Leveraging Ma- chine Learning to Predict Student Performance in Professional Courses. *Journal of Engineering Sciences*, 14(01).

Sun, X., Zhang, P., Wang, D., Cao, Y., & Liu, B. (2019, December). Colorectal polyp segmentation by U-Net with dilation convolution. In 2019 18th IEEE international conference on machine learning and applications (ICMLA) (pp. 851-858). IEEE.

Sun, D., Boudouaia, A., Zhu, C., & Li, Y. (2024). Would ChatGPT-facilitated programming mode impact college students' programming behaviors, performances, and perceptions? An empirical study. *International Journal of Educational Technology in Higher Education*, 21(1), 14. DOI: 10.1186/s41239-024-00446-5

Suong, L. K., & Jangwoo, K. (2018). Detection of potholes using a deep convolutional neural network. *Journal of Universal Computer Science*, 24(9), 1244–1257.

Suresh, R., Rajendran, S., Kumar, P. S., Dutta, K., & Vo, D.-V. N. (2022). Current advances in microbial fuel cell technology toward removal of organic contaminants – A review. *Chemosphere*, 287, 132186. DOI: 10.1016/j.chemosphere.2021.132186 PMID: 34509759

Tabachnick, B. G., & Fidell, L. S. (2019). *Using Multivariate Statistics* (7th ed.). Pearson.

Tai, T. Y. (2024). Comparing the effects of intelligent personal assistant-human and human-human interactions on EFL learners' willingness to communicate beyond the classroom. *Computers & Education*, 210, 104965. DOI: 10.1016/j.compedu.2023.104965

Talebian, H., Herrera, O. E., & Mérida, W. (2019). Spatial and temporal optimization of hydrogen fuel supply chain for light duty passenger vehicles in British Columbia. *International Journal of Hydrogen Energy*, 44(47), 25939–25956. DOI: 10.1016/j.ijhydene.2019.07.218

Tang, F. (2024). Understanding the role of digital immersive technology in educating the students of english language: Does it promote critical thinking and self-directed learning for achieving sustainability in education with the help of teamwork? *BMC Psychology*, 12(1), 144. DOI: 10.1186/s40359-024-01636-6 PMID: 38481260

Tanhaei, B., Ayati, A., Lahtinen, M., Mahmoodzadeh Vaziri, B., & Sillanpää, M. (2016). A magnetic mesoporous chitosan based core-shells biopolymer for anionic dye adsorption: Kinetic and isothermal study and application of <scp>ANN</scp>. *Journal of Applied Polymer Science*, 133(22), app.43466. Advance online publication. DOI: 10.1002/app.43466

Tao, A. (2024). The Dynamics of Community Engagement in Distance Education: A Sociological Analysis Based on Online Learning Platforms. *International Journal of Interactive Mobile Technologies*, 18(7), 4–18. DOI: 10.3991/ijim.v18i07.48599

Tashk, A., Herp, J., & Nadimi, E. (2019, December). Fully automatic polyp detection based on a novel U-Net architecture and morphological post-process. In *2019 International Conference on Control, Artificial Intelligence, Robotics & Optimization (ICCAIRO)* (pp. 37-41). IEEE.

Taylor, J. P., Brown, R. H., and Cleveland, D. W. (2016). Decoding ALS: from genes to mechanism. Nature 539, 197–206. doi: 20413DOI: 10.1038/nature

Termenon, M., Grana, M., Besga, A., Echeveste, J., & Gonzalez-Pinto, A. (2013). Lattice independent component analysis feature selection on diffusion weighted imaging for Alzheimer's disease classification. *Neurocomputing*, 114, 132–141. DOI: 10.1016/j.neucom.2012.08.044

Thakur, A. K., Arya, A. K., & Sharma, P. (2022). Prediction and mitigation of AC interference on the pipeline system. *Corrosion Reviews*, 40(2), 149–157. Advance online publication. DOI: 10.1515/corrrev-2021-0061

Tlhagale, M., Liphadzi, S., Bhagwan, J., Naidoo, V., Jonas, K., van Vuuren, L., Medema, G., Andrews, L., Béen, F., Ferreira, M. L., Saatci, A. M., Alpaslan Kocamemi, B., Hassard, F., Singer, A. C., Bunce, J. T., Grimsley, J. M. S., Brown, M., & Jones, D. L. (2022). Establishment of local wastewater-based surveillance programmes in response to the spread and infection of COVID-19 – case studies from South Africa, the Netherlands, Turkey and England. *Journal of Water and Health*, 20(2), 287–299. DOI: 10.2166/wh.2022.185 PMID: 36366987

Tomar, N. K., Jha, D., & Bagci, U. (2023, January). Dilatedsegnet: A deep dilated segmentation network for polyp segmentation. In *International conference on multimedia modeling* (pp. 334-344). Cham: Springer International Publishing.

Tran, S.-T., Nguyen, M.-H., Dang, H.-P., & Nguyen, T.-T. (2022). Automatic polyp segmentation using modified recurrent residual Unet network. *IEEE Access : Practical Innovations, Open Solutions*, 10, 65951–65961. DOI: 10.1109/ACCESS.2022.3184773

Traver, A. E., Shin, H., & Bakalarz-Duverger, M. (2024). Introduction: Teaching and Learning a Humanistic Sociology. *Teaching Sociology*, 52(2), 107–114. DOI: 10.1177/0092055X241233882

Traynor, B. J., Codd, M. B., Corr, B., Forde, C., Frost, E., & Hardiman, O. M. (2000). Clinical features of amyotrophic lateral sclerosis according to the El Escorial and Airlie House diagnostic criteria: A population-based study. *Archives of Neurology*, 57(8), 1171–1176. DOI: 10.1001/archneur.57.8.1171 PMID: 10927797

Tuan, N. T., Moore, P., Thanh, D. H. V., & Pham, H. V. (2024). A Generative Artificial Intelligence Using Multilingual Large Language Models for ChatGPT Applications. *Applied Sciences (Basel, Switzerland)*, 14(7), 3036. DOI: 10.3390/app14073036

Tuan, P. T., Phuong, N. T. H., & Nam, N. T. (2024). Engaging Students in The Learning Process with Game-Based Learning. *International Journal of English Language Studies*, 6(1), 54–60. DOI: 10.32996/ijels.2024.6.1.5

Tukey, J. W. (1977). *Exploratory Data Analysis*. Addison-Wesley.

Tzimas, D. E., & Demetriadis, S. N. (2024). Impact of Learning Analytics Guidance on Student Self-Regulated Learning Skills, Performance, and Satisfaction: A Mixed Methods Study. *Education Sciences*, 14(1), 92. DOI: 10.3390/educsci14010092

Uhrig, R. E. (n.d.). Introduction to artificial neural networks. In *Proceedings of IECON '95 - 21st Annual Conference on IEEE Industrial Electronics*. IEEE. DOI: 10.1109/IECON.1995.483329

Ullah, I., Khan, I. U., Ouaissa, M., Ouaissa, M., & El Hajjami, S. (Eds.). (2024). *Future Communication Systems Using Artificial Intelligence, Internet of Things and Data Science*. CRC Press. DOI: 10.1201/9781032648309

Vagheei, R. (2021). Upgrading of waste stabilization ponds using a low-cost small-scale fine bubble diffused aeration system. *Water Science and Technology*, 84(10–11), 3104–3121. DOI: 10.2166/wst.2021.330 PMID: 34850715

Vaillancourt, D. E., Spraker, M. B., Prodoehl, J., Abraham, I., Corcos, D. M., Zhou, X. J., Comella, C. L., & Little, D. M. (2009). High-resolution diffusion tensor imaging in the substantia nigra of de novo Parkinson disease. *Neurology*, 72(16), 1378–1384. DOI: 10.1212/01.wnl.0000340982.01727.6e PMID: 19129507

Van Damme, P., Leyssen, M., Callewaert, G., Robberecht, W., & Van Den Bosch, L. (2003). The AMPA receptor antagonist NBQX prolongs survival in a transgenic mouse model of amyotrophic lateral sclerosis. *Neuroscience Letters*, 343(2), 81–84. DOI: 10.1016/S0304-3940(03)00314-8 PMID: 12759169

Van Damme, P., Van den Bosch, L., Van Houtte, E., Callewaert, G., & Robberecht, W. (2002). GluR2-dependent properties of AMPA receptors determine the selective vulnerability of motor neurons to excitotoxicity. *Journal of Neurophysiology*, 88(3), 1279–1287. DOI: 10.1152/jn.2002.88.3.1279 PMID: 12205149

Vandenberghe, W., Robberecht, W., & Brorson, J. R. (2000). AMPA receptor calcium permeability, GluR2 expression, and selective motoneuron vulnerability. *The Journal of Neuroscience : The Official Journal of the Society for Neuroscience*, 20(1), 123–132. DOI: 10.1523/JNEUROSCI.20-01-00123.2000 PMID: 10627588

Vandermaas-Peeler, M., Byrd, S. K., Reiness, C. G., Frederick, K. A., Holmes, B. E., Bates, S. C., & Wooley, C. A. (2024). Transformation through a Disciplinary Lens. In *Transforming Academic Culture and Curriculum* (pp. 72–93). Routledge. DOI: 10.4324/9781003448327-5

Veitch, D. P., Weiner, M. W., Aisen, P. S., Beckett, L. A., Cairns, N. J., Green, R. C., Harvey, D., Jack, C. R.Jr, Jagust, W., Morris, J. C., Petersen, R. C., Saykin, A. J., Shaw, L. M., Toga, A. W., & Trojanowski, J. Q. (2019). Understanding disease progression and improving Alzheimer's disease clinical trials: Recent highlights from the Alzheimer's Disease Neuroimaging Initiative. *Alzheimer's & Dementia*, 15(1), 106–152. DOI: 10.1016/j.jalz.2018.08.005 PMID: 30321505

vom Lehn, D., Gibson, W., & Ruiz-Junco, N. (2024). *People, technology, and social organization*. Routledge.

Wahshat, H., Khalifeh, A., Taha, A., Wahsheh, F., Amayreh, K., & Matalka, M. (2024). Individual, technological, organizational, and environmental factors impact of the internet of things on e-learning adoption in higher education institutions in Jordan. *International Journal of Data and Network Science*, 8(3), 1451–1462. DOI: 10.5267/j.ijdns.2024.3.020

Wakhloo, D., Oberhauser, J., Madira, A., & Mahajani, S. (2022). From cradle to grave: Neurogenesis, neuroregeneration and neurodegeneration in Alzheimer's and Parkinson's diseases. *Neural Regeneration Research*, 17(12), 2606–2614. DOI: 10.4103/1673-5374.336138 PMID: 35662189

Wang, C. Y., Bochkovskiy, A., & Liao, H. Y. M. (2023). YOLOv7: Trainable bag-of-freebies sets new state-of-the-art for real-time object detectors. In Proceedings of the IEEE/CVF conference on computer vision and pattern recognition (pp. 7464-7475).

Wang, H. W., Chen, C., Cheng, D. Y., Lin, C., & Lo, C. (2015). A Real-Time pothole detection approach for intelligent transportation system. *Mathematical Problems in Engineering*, 2015, 1–7. DOI: 10.1155/2015/813507

Wang, H., Lu, B., Li, J., Liu, T., Xing, Y., Lv, C., Cao, D., Li, J., Zhang, J., & Hashemi, E. (2022). Risk assessment and mitigation in local path planning for autonomous vehicles with LSTM based predictive model. *IEEE Transactions on Automation Science and Engineering*, 19(4), 2738–2749. DOI: 10.1109/TASE.2021.3075773

Wang, K. D., Burkholder, E., Wieman, C., Salehi, S., & Haber, N. (2024, January). Examining the potential and pitfalls of chatgpt in science and engineering problem-solving. [). Frontiers Media SA.]. *Frontiers in Education*, 8, 1330486. DOI: 10.3389/feduc.2023.1330486

Wang, R., Sun, Y., Li, Z., & Li, Y. (2021). Joint graph convolutional network and attention mechanism for personalized traffic recommendation. *IEEE Transactions on Intelligent Transportation Systems*, 22(1), 784–794.

Wang, Z., Li, Y., Wu, C., Zeng, Y., & Liu, Z. (2019). Gated spatial-temporal graph attention networks for traffic flow prediction. *Proceedings of the AAAI Conference on Artificial Intelligence*, 33(1), 4640–4647. PMID: 32219010

Wan, Y., & Moorhouse, B. L. (2024). Using Call Annie as a Generative Artificial Intelligence Speaking Partner for Language Learners. *RELC Journal*, •••, 00336882231224813. DOI: 10.1177/00336882231224813

Warlina, L. I., & Simanjuntak, W. (2014). Pengolahan Limbah Cair Hotel Dengan Kombinasi Metode Elektrokoagulasi Dan Adsorpsi Menggunakan Karbosil. *Jurnal Sylva Lestari*, 1(1), 85. DOI: 10.23960/jsl1185-92

Wei, D., Shi, X., Junge, H., Du, C., & Beller, M. (2023). Carbon neutral hydrogen storage and release cycles based on dual-functional roles of formamides. *Nature Communications*, 14(1), 3726. DOI: 10.1038/s41467-023-39309-4 PMID: 37349304

Weiner, M. W., Veitch, D. P., Aisen, P. S., Beckett, L. A., Cairns, N. J., Green, R. C., Harvey, D., Jack, C. R.Jr, Jagust, W., Morris, J. C., Petersen, R. C., Saykin, A. J., Shaw, L. M., Toga, A. W., & Trojanowski, J. Q. (2017). Recent publications from the Alzheimer's Disease Neuroimaging Initiative: Reviewing progress toward improved AD clinical trials. *Alzheimer's & Dementia*, 13(4), e1–e85. DOI: 10.1016/j.jalz.2016.11.007 PMID: 28342697

Weiss, J. H., & Sensi, S. L. (2000). Ca2+–Zn2+ permeable AMPA or kainate receptors: Possible key factors in selective neurodegeneration. *Trends in Neurosciences*, 23(8), 365–371. DOI: 10.1016/S0166-2236(00)01610-6 PMID: 10906800

Wei, Y., Jiao, Y., An, D., Li, D., Li, W., & Wei, Q. (2019). Review of Dissolved Oxygen Detection Technology: From Laboratory Analysis to Online Intelligent Detection. *Sensors (Basel)*, 19(18), 3995. DOI: 10.3390/s19183995 PMID: 31527482

Westeneng, H.-J., Debray, T. P. A., Visser, A. E., van Eijk, R. P. A., Rooney, J. P. K., Calvo, A., Martin, S., McDermott, C. J., Thompson, A. G., Pinto, S., Kobeleva, X., Rosenbohm, A., Stubendorff, B., Sommer, H., Middelkoop, B. M., Dekker, A. M., van Vugt, J. J. F. A., van Rheenen, W., Vajda, A., & van den Berg, L. H. (2018). Prognosis for patients with amyotrophic lateral sclerosis: Development and validation of a personalised prediction model. *Lancet Neurology*, 17(5), 423–433. DOI: 10.1016/S1474-4422(18)30089-9 PMID: 29598923

Wiedemann, F. R., Winkler, K., Kuznetsov, A. V., Bartels, C., Vielhaber, S., Feistner, H., & Kunz, W. S. (1998). Impairment of mitochondrial function in skeletal muscle of patients with amyotrophic lateral sclerosis. *Journal of the Neurological Sciences*, 156(1), 65–72. DOI: 10.1016/S0022-510X(98)00008-2 PMID: 9559989

Wikipedia. (2018). *The SAGE Encyclopedia of the Internet.* SAGE Publications, Inc., DOI: 10.4135/9781473960367.n287

Wodzinski, M., Skalski, A., Hemmerling, D., Orozco-Arroyave, J. R., & Nöth, E. (2019, July). Deep learning approach to Parkinson's disease detection using voice recordings and convolutional neural network dedicated to image classification. In 2019 41st annual international conference of the IEEE engineering in medicine and biology society (EMBC) (pp. 717-720). IEEE. DOI: 10.1109/EMBC.2019.8856972

Wolz, R., Julkunen, V., Koikkalainen, J., Niskanen, E., Zhang, D. P., Rueckert, D., Soininen, H., & Lötjönen, J. (2011). Multi-method analysis of MRI images in early diagnostics of Alzheimer's disease. *PLoS One*, 6(10), e25446. DOI: 10.1371/journal.pone.0025446 PMID: 22022397

Woo, S., Youtie, J., Ott, I., & Scheu, F. (2021). Understanding the long-term emergence of autonomous vehicles technologies. *Technological Forecasting & Social Change. Technological Forecasting and Social Change*, 170, 120852. DOI: 10.1016/j.techfore.2021.120852

Wu, F., Lv, Y., Zhao, X., & Fu, S. (2020). A survey of graph neural networks for traffic forecasting. *arXiv preprint arXiv:2001.07175.*

Wu, Z., Liu, Y., Wang, Y., & Liu, R. (2020). A survey on deep learning for urban traffic control. *IEEE Transactions on Intelligent Vehicles*, 5(3), 828–848.

Xiao, L., Liu, Z., Li, Y., Wang, F., & Wang, Y. (2021). Can graph neural networks help urban traffic management? A comprehensive survey. *IEEE Transactions on Intelligent Transportation Systems*, 99(1), 1–18.

Xu, C., Ke, Y., Li, Y., Chu, H., & Wu, Y. (2020). Data-driven configuration optimization of an off-grid wind/PV/hydrogen system based on modified NSGA-II and CRITIC-TOPSIS. *Energy Conversion and Management*, 215, 112892. DOI: 10.1016/j.enconman.2020.112892

Xu, F., Wu, Y., Zhou, Y., Zhou, H., & Liu, X. (2022). Ethical considerations for machine learning in transportation systems. *Transportation Research Part C, Emerging Technologies*, 144, 103796.

Xu, T., Gao, Q., Ge, X., & Lu, J. (2024). The relationship between social media and professional learning from the perspective of pre-service teachers: A survey. *Education and Information Technologies*, 29(2), 2067–2092. DOI: 10.1007/s10639-023-11861-y PMID: 37361740

Yang, C., Xu, W., Nan, Y., Wang, Y., Hu, Y., Gao, C., & Chen, X. (2020). Fabrication and characterization of a high performance polyimide ultrafiltration membrane for dye removal. *Journal of Colloid and Interface Science*, 562, 589–597. DOI: 10.1016/j.jcis.2019.11.075 PMID: 31771878

Yang, X.-S., & Karamanoglu, M. (2013). *Swarm Intelligence and Bio-Inspired Computation. An Overview*. Elsevier.

Yaqoob, A., Aziz, R. M., Verma, N. K., Lalwani, P., Makrariya, A., & Kumar, P. (2023). A review on nature-inspired algorithms for cancer disease prediction and classification. *Mathematics*, 11(5), 1081. DOI: 10.3390/math11051081

Yazdanpanahrostami, A., & Rasouli, K. (2009). Simulation of Tehran Air Pollution Using Artificial Neural Networks. In *World Environmental and Water Resources Congress 2009*. American Society of Civil Engineers. DOI: 10.1061/41036(342)242

Yeomans, H., & Grossmann, I. E. (1999). A systematic modeling framework of superstructure optimization in process synthesis. *Computers &amp. Computers & Chemical Engineering*, 23(6), 709–731. DOI: 10.1016/S0098-1354(99)00003-4

Yeung, M., Sala, E., Schönlieb, C.-B., & Rundo, L. (2021). Focus U-Net. A novel dual attention-gated CNN for polyp segmentation during colonoscopy. *Computers in Biology and Medicine*, 137, 104815. DOI: 10.1016/j.compbiomed.2021.104815 PMID: 34507156

Ye, W., Jiang, W., Tong, Z., Yuan, D., & Xiao, J. (2021). Convolutional neural network for pothole detection in asphalt pavement. *Road Materials and Pavement Design*, 22(1), 42–58. DOI: 10.1080/14680629.2019.1615533

Ying, R., Zhu, R., Luo, S., & He, Q. (2018). Graph convolutional neural networks for semi-supervised learning. *Proceedings of the ACM SIGKDD International Conference on Knowledge Discovery and Data Mining, 22*(3), 2041-2050.

Yu, KH., Beam, A.L. & Kohane, I.S. (2018) Artificial intelligence in healthcare. *Nat Biomed Eng* 2, 719–731 . https.//.DOI: 10.1038/s41551-018-0305-z

Yuan, L. (2024). Where does AI-driven Education, in the Chinese Context and Beyond, go next? *International Journal of Artificial Intelligence in Education*, 34(1), 31–41. DOI: 10.1007/s40593-023-00341-6

Yuen, C. L., & Schlote, N. (2024). Learner Experiences of Mobile Apps and Artificial Intelligence to Support Additional Language Learning in Education. *Journal of Educational Technology Systems*, 52(4), 00472395241238693. DOI: 10.1177/00472395241238693

Yu, H. (2024). The application and challenges of ChatGPT in educational transformation: New demands for teachers' roles. *Heliyon*, 10(2), e24289. DOI: 10.1016/j.heliyon.2024.e24289 PMID: 38298626

Zahra, A., Waheed, Z., Fatima, T., & Khong, K. W. (2024). Leveraging Technology for Environmental Awareness: Insights from Experimental Research with Middle School Students in Malaysia. *RMLE Online: Research in Middle Level Education*, 47(4), 1–16. DOI: 10.1080/19404476.2024.2322046

Zammit, A. (2024). Exploring the Influence of Social Emotional Development on Academic Achievement in Vocational Education: Insights from Student Mentor Interventions. *MCAST Journal of Applied Research & Practice*, 8(1), 80–104. DOI: 10.5604/01.3001.0054.5096

Zaqoot, H. A., Hamada, M., & Miqdad, S. (2018). A Comparative Study Of Ann For Predicting Nitrate Concentration In Groundwater Wells In The Southern Area Of Gaza Strip. *Applied Artificial Intelligence*, 32(7–8), 727–744. DOI: 10.1080/08839514.2018.1506970

Zeiler, M. D. (2012). ADADELTA. an adaptive learning rate method. arXiv preprint arXiv.1212.5701.

Zeineddine, H., Braendle, U., & Farah, A. (2021). Enhancing prediction of student success: Automated machine learning approach. *Computers & Electrical Engineering*, 89, 106903. DOI: 10.1016/j.compeleceng.2020.106903

Zhang, C., Meng, Y., & Ma, X. (2024). Artificial intelligence in EFL speaking: Impact on enjoyment, anxiety, and willingness to communicate. *System*, 121, 103259. DOI: 10.1016/j.system.2024.103259

Zhang, J., Li, Y., & Wang, X. (2018). A deep learning approach for real time traffic anomaly detection using unlabeled streaming data. *IEEE Transactions on Intelligent Transportation Systems*, 19(12), 4628–4640.

Zhang, Q., & Zhuo, Q. (2024). Estimating forest aboveground biomass in tropical forests with Landsat time-series data and recurrent neural network. *International Journal of Remote Sensing*, 45(14), 4764–4787. DOI: 10.1080/01431161.2024.2368930

Zhang, Z., Liu, Q., & Wang, Y. (2018). Road extraction by deep residual u-net. *IEEE Geoscience and Remote Sensing Letters*, 15(5), 749–753. DOI: 10.1109/LGRS.2018.2802944

Zhao, P., Yang, Y., Song, Y., & Huang, H. (2019). Spatial-temporal graph convolutional networks for predicting taxi demand. *Proceedings of the 27th ACM SIGKDD International Conference on Knowledge Discovery and Data Mining*, 3235-3244.

Zhou, J., Sun, L., & Guo, Y. (2018). Deep learning based long-term traffic flow forecasting: From spatial- temporal to spatiotemporal-spectral. *IEEE Transactions on Intelligent Transportation Systems*, 19(7), 2057–2069.

Zhou, Z., Siddiquee, M. M. R., Tajbakhsh, N., & Liang, J. (2019). Unet++. Redesigning skip connections to exploit multiscale features in image segmentation. *IEEE Transactions on Medical Imaging*, 39(6), 1856–1867. DOI: 10.1109/TMI.2019.2959609 PMID: 31841402

Zhu, J.-J., Kang, L., & Anderson, P. R. (2018). Predicting influent biochemical oxygen demand: Balancing energy demand and risk management. *Water Research*, 128, 304–313. DOI: 10.1016/j.watres.2017.10.053 PMID: 29107915

Zola, J., & Zola, J. (2024). *Teaching as if Students Matter: A Guide to Creating Classrooms Based on Relationships and Engaged Learning*. State University of New York Press.

About the Contributors

Mzili Toufik holds the position of Assistant Professor in the Department of Computer Science at the Faculty of Science, Chouaib Doukkali University. Recognized as a distinguished researcher, he has made noteworthy contributions to the fields of metaheuristics, optimization, and scheduling problems. His scholarly impact is evident through numerous publications in esteemed Q1 journals. Additionally, Dr. Mzili Toufik has demonstrated his expertise as a peer reviewer for several prestigious journals, showcasing his commitment to advancing the academic discourse in his field.

***

Eka Apriani, M.Pd. is an English Lecturer at English Tadris Study Program of the Faculty of Tarbiyah in Islamic State Institute of Curup. She is a Editor in Chief in Journal of English Franca and International Journal of Education Research and Development. She is also the secretary of Corolla Education Centre Foundation, Indonesia. She Received her Master Degree (Master of Language Education) from Sriwijaya University in 2013 and Doctor of Education in Bengkulu University 2022. She has published some researchs in international journal and national journal. She has Scopus Id and WOS Id. She is a member of profession organisation: ADRI, TEFLIN, PDPI and ELITE. She is treasurer of ADRI in Bengkulu Province. Her research interest is in Writing skill and ICT in learning English. She write some books: The beauty of Rejang Lebong, Semantics Crossword, Metode Pengajaran BIPA, Drama Manuscript: The Marriage of Siti Zubaidah. She has also some Intelektual Property Right: Semantics Crossword, The beauty of Rejang Lebong, The Marriage of Siti Zubaidah, Metode Pengajaran BIPA.

Esra Demir Unal graduated with honors from Atatürk University Faculty of Medicine (Medicine in English) in 2016. Between 2017 and 2021, worked as a

Research Assistant at Ankara Yıldırım Beyazıt University Faculty of Medicine, Department of Neurology, and received her Doctorate Diploma in 2021. She conducted her doctoral thesis with Özcan EREL, who was awarded the title of the most valuable scientist in the world, and in this context, her thesis was supported within the scope of Ankara Yıldırım Beyazıt University Scientific Research Projects. After receiving PhD, she conducted various studies on Movement Disorders, primarily Stroke and cerebrovascular Diseases and Parkinson's Disease, Huntington's Disease, and Amyotrophic Lateral Sclerosis, and I have also applied various invasive treatments to more than 50 Parkinson's Patients so far. Among the international associations, She has been a member of the European Academy of Neurology since 2020, a member of The International Parkinson and Movement Disorder Society (MDS) since 2020, and a member of The Huntington Study Group (HSG) since 2021. Within the scope of ULAKBİM, she is currently a referee in Abant Medical Journal, Medical Faculty Clinics Journal, and Acta Medica Nicomedia journal. She has been working as an editor at one of the international journals, Applied Progress in Gene Research, and MDPI journals since 2023. She is a member of the Turkish Neurology Association Young Neurologists Working Group, Turkish Neurology Association Artificial Intelligence Informatics Society and the Neurodegenerative Diseases Association Neurologists Working Group Membership, which are national associations. Within the scope of her work so far, she has received first, second, and third place awards in both international and national congresses. She currently runs the Movement Disorders Polyclinic in Ankara Yıldırım Beyazıt University Medical Faculty Yenimahalle Training and Research Hospital.

Ahmed A. Elngar is an Associate Professor and Head of the Computer Science Department at the Faculty of Computers and Artificial Intelligence, Beni-Suef University, Egypt. Dr. Elngar is also, an Associate Professor of Computer Science at the College of Computer Information Technology, American University in the Emirates, United Arab Emirates. Also, Dr. AE is Adjunct Professor at School of Technology, Woxsen University, India. Dr. AE is the Founder and Head of the Scientific Innovation Research Group (SIRG). Dr. AE is a Director of the Technological and Informatics Studies Center (TISC), Faculty of Computers and Artificial Intelligence, Beni-Suef University. Dr. AE has more than 150 scientific research papers published in prestigious international journals and over 35 books covering such diverse topics as data mining, intelligent systems, social networks, and smart environment. Dr. AE is a collaborative researcher He is a member of the Egyptian Mathematical Society (EMS) and International Rough Set Society (IRSS). His other research areas include the Internet of Things (IoT), Network Security, Intrusion Detection, Machine Learning, Data Mining, and Artificial Intelligence. Big Data, Authentication, Cryptology, Healthcare Systems, Automation Systems.

He is an Editor and Reviewer of many international journals around the world. Dr. AE won several awards including the Young Researcher in Computer Science Engineering", from Global Outreach Education Summit and Awards 2019, on 31 January 2019 (Thursday) in Delhi, India. Also, he awards the Best Young Researcher Award (Male) (Below 40 years)", Global Education and Corporate Leadership Awards (GECL-2018), Plot No-8, Shivaji Park, Alwar 301001, Rajasthan

Shugufta Fatima is an Assistant Professor in the Department of Computer Science and Engineering at Stanley College of Engineering and Technology for Women, Hyderabad, India. With 7+ years of experience in teaching, she brings a unique perspective to the academic community. She earned her Bachelor's and Master's degree in Cse discipline from Osmania University.Her academic journey has been marked by excellence, demonstrated through her earning Academic Excellence awards throughout her academics.Her dedication to advancing knowledge in her area of interests which includes Machine Learning, Deep Learning and Artificial Intelligence is evident through her research, teaching and is committed to exploring more . Through her research, teaching, and community engagement efforts, she is making valuable contributions to both the academic community and society at large. She has published papers in reputable academic journals which have been recognized for good citations,guided students with projects.She has got a patent to her name. She conducted many national level workshops,organized events,participated in conferences,fdps and workshops on a scale.As an educator, she is dedicated to fostering critical thinking and intellectual curiosity in her students. She believes in creating an inclusive learning environment where students feel empowered to engage with complex ideas and develop their analytical skills which is happening through mentoring sessions. Through innovative teaching methods and real-world examples, she strives to inspire a passion for learning among her students.

Marlia Mohd Hanafiah is a Professor and Head, Centre for Tropical Climate Change System, Institute of Climate Change, The National University of Malaysia, Malaysia. Areas of Research Expertise: Life Cycle Impact Assessment (LCA) and Environmental Footprinting of Green Materials and Energy, Environmental Engineering, Wastewater Treatment & Water Management, Green Technology and Sustainability. She has a total academic teaching experience of 15+ years with more than 170 publications in reputed journals and online book chapter contributions (Indexed By: SCI, SCIE, SSCI, Scopus, DBLP). She received a research grant and consultation (as project leader & team member) of more than RM 7 million.

Misdi is a faculty member in the English Education Department of Universitas Swadaya Gunung Jati, Cirebon, Indonesia. He undertook his doctorate program in

English Education at Universitas Pendidikan Indonesia, Bandung, Indonesia. His research interests are TEFL methodology, academic writing, and extensive reading. His research articles have been published in some reputable journals such as the Qualitative Report, Qualitative Research Journal and Register Journal.

Muliati is a permanent lecturer at the Faculty of Teacher Training and Education, Universitas Bosowa, Indonesia. She obtained a bachelor's degree in English Education from Universitas Muhammadiyah Makassar and a Master's Degree in English Language Studies from Universitas Hasanuddin, Indonesia. In addition, she received Master of Education degree from Flinders University, Adelaide, South Australia. and she got doctoral degree from State University of Makassar (Indonesia).Her current field is teaching EFL. Her research interests are CBLT, ESP, Digital Learning, teaching methodology, and Artificial Intelligence

Ahmed J. Obaid is an Asst. Professor at the Department of Computer Science, Faculty of Computer Science and Mathematics, University of Kufa, Iraq. Dr. Ahmed holds a Bachelor in Computer Science, degree in – Information Systems from College of Computers, University of Anbar, Iraq (2001-2005), and a Master Degree (M. TECH) of Computer Science Engineering (CSE) from School of Information Technology, Jawaharlal Nehru Technological University, Hyderabad, India (2010-2013), and a Doctor of Philosophy (PhD) in Web Mining from College of Information Technology, University of University of Babylon, Iraq (2013-2017). He is a Certified Web Mining Consultant with over 14 years of experience in working as Faculty Member in University of Kufa, Iraq. He has taught courses in Web Designing, Web Scripting, JavaScript, VB.Net, MATLAB Toolbox's, and other courses on PHP, CMC, and DHTML from more than 10 international organizations and institutes from USA, and India. Dr. Ahmed is a member of Statistical and Information Consultation Center (SICC), University of Kufa, Iraq.

Özen Özer earned her Bachelors and Masters degrees in Mathematics from Trakya University in Edirne, Turkey. She also obtained her Ph.D. in Mathematics from Süleyman Demirel University in Isparta, Turkey. Currently, she holds the position of Full Professor Doctor at the Department of Mathematics within the Faculty of Science and Arts at Kırklareli University. Her specialized research areas encompass a wide range of subjects, including the Theory of Real Quadratic Number Fields with practical applications, Diophantine and Pell Equations, Diophantine Sets, Arithmetic Functions, Fixed Point Theory, p-adic Analysis, q-Analysis, Special Integer Sequences, Nonlinear Analysis, C* Algebra, Matrix Theory, Optimization, Approximation Theory, Cryptography, Machine Learning, Artificial Intelligence, Differential Equations, Mathematical Models, Mathematical

Education, Optimization, Fuzzy Set and Fuzzy Spaces, Statistics and more… With an extensive academic portfolio, she has authored or completed over 90 research papers published in prestigious international journals, alongside her participation in numerous national and international scientific projects. Özen ÖZER has also authored books and contributed chapters to international publishing houses on various subjects such as Number Theory, Algebra, Applied Mathematics, Analysis and Artificial Intelligence. She has further demonstrated her expertise as a keynote speaker at various national and international conferences held across different countries, covering a diverse array of topics. Her involvement extends to being a reviewer for more than 80 distinct publishing houses and journals, and she serves as a member of editorial boards for books, papers, and conference proceedings. Özen ÖZER is enthusiastic about engaging in academic and scientific collaborations, demonstrating a willingness to participate in cooperative endeavors across various domains.

Kiran Sree Pokkuluri has received his B.Tech and M.E in Computer Science and Engineering from JNTU and Anna University, respectively. He has obtained his Ph.D. degree in the area of Artificial Intelligence from JNTU-Hyderabad. He has authored Six Textbooks for UG and PG students of engineering in AI and published more than 100+ Research Articles in various International Journals and Conferences. He has filed and published SIX patents in the areas of Deep Learning & AI. His bibliography was listed in Marquis Who's Who in the World, 29th Edition (2012), USA. Prof Kiran is the Recipient of Bharat Excellence Award from Dr. G.V. Krishna Murthy, Former Election Commissioner of India for two times and recipient of Rashtrya Ratan Award. He was the BOS member of CSE&IT in some universities and autonomous colleges. He also worked as Principal of the N.B.K.R.Institute of Science & Technology (Second Oldest Private Engg College), Vidyanagar, for two years. He has got 18+ years of Teaching Experience and working as Professor in the department of CSE at Shri Vishnu Engineering College for Women(A), Bhimavaram. He has delivered 100+ technical talks on Deep Learning and AI

Krishna Kumari R holds a diverse array of qualifications including M.Sc., M.Phil., M.Ed., TNSET, and a Ph.D. With over 5 years of teaching experience, she has contributed significantly to the academic field. Dr. Krishna Kumari has authored a total of 18 publications, all indexed in SCOPUS, with 11 being published post-Ph.D. Additionally, she has authored 3 books/book chapters and holds 1 patent published and 1 granted. She serves as a reviewer in 4 SCOPUS indexed journals and 2 SCOPUS and ESCI indexed journals. Dr. Krishna Kumari has been a resource person for 3 Faculty Development Programs (FDPs), presented at 2 conferences/symposiums, and chaired 2 conferences. She has also attended 3 workshops and participated in 10 FDPs. Furthermore, Dr. Krishna Kumari is an active member of 10 professional memberships, showcasing her commitment to professional development and academic excellence.

Kishor Kumar Reddy C. is currently working as Associate Professor, Dept. of Computer Science and Engineering, Stanley College of Engineering and Technology for Women, Hyderabad, India. He has research and teaching experience of more than 10 years. He has published more than 50 research papers in National and International Conferences, Book Chapters, and Journals indexed by Scopus and others. He is an author for 2 text books and 7 co-edited books. He acted as the special session chair for Springer FICTA 2020, 2022, SCI 2021, INDIA 2022 and IEEE ICCSEA 2020 conferences. He is the corresponding editor of AMSE 2021 conferences, published by IoP Science JPCS. He is the member of ISTE, CSI, IAENG, UACEE, IACSIT

Muthmainnah Yakin is a highly accomplished Assistant Professor at Universitas Al Asyariah Mandar in West Sulawesi, Indonesia. She has extensive experience as a lecturer, global speaker, and international leader. Muthmainnah has held numerous positions at her university, including Chairman of the Indonesian Language Department, Public Relations Officer, Director of Women's Centre Studies, Deputy Director of Quality Assurance Unit, and currently, Deputy Director of the Language and Character Development Institute. Muthmainnah is also an accomplished author, having written 63 national and International books, published by esteemed publishers such as IGI Global, Springer, Emerald, Wiley, Taylor and Francis https://sinta.kemdikbud.go.id/profile/books She serves as an international board member for various scientific innovation research groups, adjunct faculty of MAAUN Nigeria and she is a member of the United Nations Volunteers roster. Muthmainnah has won over 100 international awards, including Outstanding Professor, Outstanding Leadership, Best International Influencer, Best Emerging Professor of the Year, International Award for Working Women, and SDGs Warrior. She is also an honorary doctorate degree holder from Layahe University and the recipient of the Doctor of Excellence award from the Philippines. Moreover, Muthmainnah is a member of the United Nations volunteers' roster and the advisory board of Indonesian education share to care volunteers. Her research interests include developing literary work, instructional teaching material, assurance of learning (Assure), TESOL (Teaching English to Speakers of Other Languages), ICT in education, digital education, and cybergogy. Muthmainnah is an outstanding scholar with a Sinta (Indonesian Ministry of education and culture) score overall of 2052 and h-index 20 and External examiner for PhD students at 3 A+ accredited campuses in India, Malaysia, Cyprus and Australia. She is a strong believer in international collaboration and teamwork, making her an ideal candidate for various academic and leadership roles. With her extensive experience and outstanding track record, Muthmainnah is open to international collaborations and teamwork.

Index

A

B

C

D

E

F

G

T

U

W

www.ingramcontent.com/pod-product-compliance
Lightning Source LLC
LaVergne TN
LVHW060007031026
849161LV00025B/2190